Data-Driven Cyber Physical Systems

Fangyu Li · Xiaolong Wu · Honggui Han

Data-Driven Cyber Physical Systems

清華大學出版社
Tsinghua University Press

Springer

Fangyu Li
School of Information Science
and Technology
Beijing University of Technology
Beijing, China

Xiaolong Wu
School of Information Science
and Technology
Beijing University of Technology
Beijing, China

Honggui Han
School of Information Science
and Technology
Beijing University of Technology
Beijing, China

ISBN 978-981-96-8708-4 ISBN 978-981-96-8709-1 (eBook)
https://doi.org/10.1007/978-981-96-8709-1

Jointly published with Tsinghua University Press
The print edition is not for sale in China (Mainland). Customers from China (Mainland) please order the print book from: Tsinghua University Press.

This Springer imprint is published by the registered company Springer Nature Singapore Pte Ltd.
The registered company address is: 152 Beach Road, #21-01/04 Gateway East, Singapore 189721, Singapore

To the tireless pursuit of knowledge and innovation, and to the boundless possibilities that the future holds.

Preface I

In recent years, we have witnessed the rapid development of cyber physical systems (CPS), which have transformed the way we live, work, and interact with technology. As the world becomes increasingly digitalized and interconnected, CPS devices and systems are becoming an essential part of our daily lives.

The integration of IoT (Internet of Things), big data, and AI has revolutionized the way we live and work, bringing about new opportunities and challenges in CPS design and development as these technologies become increasingly prevalent. The IoT enables a vast network of connected devices to collect real-time data, while big data offers ways to efficiently process and analyze large volumes of information, and AI automates decision-making processes for quick adaptation and response of CPS to environmental changes. As the integration of these technologies continues to evolve, the design and development of CPS must keep pace to fully realize the potential benefits offered by this exciting frontier.

In this book on data-driven CPS, the author provides a comprehensive guide to the principles, design, and implementation of CPS. With a focus on data-driven approaches, this book covers key techniques such as data acquisition, storage, data management, analysis, modeling, simulation, machine learning, AI, network and distributed computing, and cybersecurity. The book also features case studies and real-world examples from various industries, offering readers insights into the practical applications of CPS.

This book, which presents the latest research trends and future directions in CPS design and development, will be an invaluable resource for researchers, practitioners, and students working in the field of CPS, providing a thorough understanding of data-driven CPS and their applications, as well as insights into the latest research trends and future directions.

I highly recommend this book to anyone who is interested in learning about the exciting field of data-driven cyber physical systems.

Beijing, China
November 2023

Honggui Han

Preface II

The field of cyber physical systems (CPS) has been rapidly evolving over the past few years, driven by advancements in computing, networking, and data analytics. As the world becomes more digitized and interconnected, CPS are playing an increasingly critical role in shaping the way we interact with technology and the world around us. The rise of the Internet of Things (IoT) and big data has brought about new opportunities and challenges in CPS design and development.

In complex systems, where interactions and interdependencies among components are intricate and diverse, data is even more crucial as it provides a basis for understanding system behavior, optimizing performance, and managing risk. As we move towards the data era, the need for CPS that can respond to the large amounts of data being generated becomes increasingly paramount. In this vein, I have written a comprehensive guide aimed at providing researchers, practitioners, and students with a thorough understanding of the principles, design, and implementation of data-driven CPS. In essence, this book should be a valuable resource for those interested in the applicability of CPS in various domains, including healthcare, transportation, energy, and manufacturing. The ideas and knowledge presented herein will enable readers to gain the necessary skills needed to design and develop effective data-driven CPS that can process and analyze large volumes of information quickly and efficiently.

The book takes a data-driven approach to CPS design, covering vital techniques such as data acquisition, analysis, and modeling, AI and machine learning, network and distributed computing, and cybersecurity. The content is structured to cater to readers who are interested in developing a deep understanding of the state-of-the-art technologies and methodologies used in the development of data-driven CPS. It combines theoretical concepts with practical examples that highlight the potential of data-driven CPS in solving real-world problems.

Chapter 1 provides an introduction to CPS, their importance, and the challenges associated with their design and implementation. Chapter 2 discusses the basics of data-driven CPS. Chapter 3 covers the data collection, including data acquisition and evaluation for CPS. Chapter 4 talks about data storage and management. Chapter 5 delves into data integration and fusion for the CPS systems. Chapter 6 explores machine learning and AI techniques and their applications in modeling and simulation in CPS applications. Chapter 7 discusses fault detection and predictive maintenance with real-world case studies from various industries, highlighting the practical applications of CPS. Chapter 8 presents cybersecurity and protection issues in CPS. Finally, Chap. 9 concludes the book by summarizing the key points covered in the previous chapters and offering some final thoughts on the future of data-driven CPS.

As the field of cyber physical systems continues to evolve, data-driven approaches have become increasingly important in developing intelligent and responsive systems. With this book, I aim to provide a comprehensive resource for anyone interested in exploring the principles, design, and implementation of data-driven CPS. Whether you are a researcher, practitioner, or student, this book offers essential knowledge and techniques that will enable you to design effective CPS capable of processing and analyzing large volumes of data in real-time.

Throughout the writing process, I have strived to present the material in a clear, accessible manner, aided by examples and illustrations that elucidate key concepts and techniques. My hope is that readers will find this book useful not only as a technical reference but also as a source of inspiration for future research and development in the field.

I would like to extend my heartfelt appreciation to my colleagues, friends, and family who have supported and encouraged me throughout the writing process. Additionally, I would like to express my deepest gratitude to the reviewers who provided invaluable feedback and suggestions that helped to improve the quality and clarity of this book. Finally, I hope that this book will be a valuable resource for anyone interested in exploring the exciting world of data-driven cyber physical systems and the immense possibilities that they offer.

Beijing, China
November 2023

Fangyu Li

Acknowledgements

Writing a book is a significant undertaking that demands dedication, passion, and perseverance. I am eternally grateful for the unwavering support of the many individuals who have contributed to the completion of this project. This book would not have been possible without the contributions of many people.

I would like to thank those who helped plan and contribute to the book's contents and organization: Junzhu Duan, Junnuo Lin, Yiming Liu, Qiyu Zhang, Xiaoye Fan, and Xuqiang Chen.

I am also immensely grateful to the many colleagues, researchers, and professionals who generously shared their knowledge, expertise, and insights with me. Their feedback, guidance, and contributions have helped to shape this book into what it is today. In particular, I would like to extend my thanks to the members of my research group whose fruitful discussions, ideas, and support have been invaluable throughout this journey.

Finally, I would like to acknowledge the publisher for their trust and support, without which this book would not have come to fruition. Their professionalism, guidance, and expertise have been instrumental in bringing this project to a successful conclusion.

In closing, I am humbled by the support and contributions of all those who have made this book possible. It is a testament to the power of collaboration, and I hope that it will prove to be a valuable resource for readers.

Contents

Acronyms

3C	Computation, Communication, and Control
AADL	Architecture Analysis and Design Language
ADS-B	Automatic Dependent Surveillance-Broadcast
AGV	Automated Guided Vehicle
AI	Artificial Intelligence
ANFIS	Adaptive Neuro-Fuzzy Inference System
ANN	Artificial Neural Networks
APTs	Advanced Persistent Threats
AR	Augmented Reality
CBM	Condition-Based Maintenance
CD	Compact Disc
CEP	Complex Event Processing
CIA	Complex Industrial Application
CMF	Corrective Matrix Factorization
CMTF	Coupled Matrix and Tensor Factorization
CNN	Convolutional Neural Network
C-OODA	Cognitive Observe-Orient-Decide-Art
CPCA	Multi-block Consensus Principal Component Analysis
CPS	Cyber Physical Systems
CPST	Cyber-Physical-Social Thinking
CSTR	Continuously Stirred Tank Reactor
CSV	Comma Separated Values
DDCPS	Data-Driven Cyber Physical Systems
DFEO	Distributed Fault Estimation Observer
DFIG	Data Fusion Information Group
DICA	Dynamic Independent Component Analysis
DNN	Deep Neural Network
DPCA	Dynamic Principal Component Analysis
DRL	Deep Reinforcement Learning
DVD	Digital Video Disc
ELM	Extreme Learning Machine

ETL	Extract, Transform, and Load
FDA	Fisher Discriminant Analysis
FPGA	Field Programmable Gate Array
GANs	Generative Adversarial Networks
GDPR	General Data Protection Regulation
GLR	Generalized Likelihood Ratio
GPS	Global Positioning System
GPU	Graphics Processing Unit
HDD	Hard Disk Drive
HDFS	Hadoop Distributed File System
HID	High Intensity Discharge
HIPAA	Health Insurance Portability and Accountability Act
HLIT	High-Level Fusion
HSMs	Hardware Security Modules
ICA	Independent Component Analysis
IDPSs	Intrusion Detection and Prevention Systems
IDS	Intrusion Detection System
IoT	Internet of Things
IP	Ingress Protection
JDL	Joint Director Laboratories
JSON	JavaScript Object Notation
KLD	Kullback-Leibler Divergence
KNN	K-Nearest Neighbors
KPCA	Kernel Principal Component Analysis
LDA	Linear Discriminant Analysis
LiDAR	Light Detection and Ranging
LKPCA	Local Kernel Principal Component Analysis
LLIT	Low-Level Fusion
LSTM	Long Short-Term Memory
MAE	Mean Absolute Error
MAPE	Mean Absolute Percentage Error
MC	Markov Chain
MEC	Mobile Edge Computing
MES	Manufacturing Execution System
MKL	Multi-kernel Learning
ML	Machine Learning
MLP	Multilayer Perceptron
MLPCA	Multi-level Principal Component Analysis
MQTT	Message Queuing Telemetry Transport
MRE	Mean Relative Error
MRI	Magnetic Resonance Imaging
MSPCA	Multiscale Principal Component Analysis
MSQL	My Structured Query Language
MVCC	Multiversion Concurrency Control
MySQL	My Structured Query Language

Oracle	Oak Ridge Automatic Computer and Logical Engine Relays
PABM	Predictive Analytics-Based Maintenance
PCA	Principal Component Analysis
PDDL	Planning Domain Definition Language
PLS	Partial Least Squares
QoS	Quality of Service
RBAC	Role-Based Access Control
RBM	Boltzmann Machines
RDBMS	Relational Database Management System
RF	Random Forest
RFID	Radio Frequency Identification
RL	Reinforcement Learning
RMSE	Mean Squared Error
RNN	Recurrent Neural Network
RT	Regression Tree
RTOS	Real-Time Operating Systems
RTPS	Real Time Publish-Subscribe
SCADA	Supervisory Control and Data Acquisition
SMC	Statistical Model Checking
SOA	Service-Oriented Architecture
SPC	Statistical Process Control
SQL	Structured Query Language
SR	Semantic Reasoning Methods
SSD	Solid State Drive
SSH	Secure Shell
STDF	State Transition Data Fusion
SVD	Singular Value Decomposition
SVM	Support Vector Machine
TBM	Time-Based Maintenance
TLS	Transport Layer Security
TPCR	Total Principle Component Regression
TT	Tensor Train
UAV	Unmanned Aerial Vehicle
UML	Unified Modeling Language
URLLC	Ultra-Reliable Low-Latency Communication
VR	Virtual Reality
WSN	Wireless Sensor Network
XML	Xtensible Markup Language

Chapter 1
Introduction to Data-Driven Cyber Physical Systems

Abstract We first define and explain cyber physical systems (CPS) and their roles of complex systems in real-time optimization and decision-making. This chapter introduces the concept of data-driven cyber physical systems (DDCPS) and their significance in modern industries. Challenges of developing DDCPS are discussed, including data acquisition, analysis, modeling, and cybersecurity. The importance of artificial intelligence and big data techniques is also highlighted. This chapter provides an overview of the book, which covers key techniques such as data acquisition, analysis, and modeling, simulation, machine learning and AI, network and distributed computing, and cybersecurity. Case studies and real-world examples from various industries are featured, offering readers insights into the practical applications of CPS and DDCPS. We summarize by emphasizing the practical significance of DDCPS and providing readers with a framework for understanding the rest of the book's content.

In recent years, there has been a growing interest in cyber physical systems (CPS), which are systems that integrate physical and computational components to provide enhanced functionality and improved performance [1–3]. These systems are designed to interact with the physical world and can be found in a wide range of applications, from manufacturing and transportation to healthcare and energy management [4–12].

While traditional CPS rely on rule-based systems to control their behavior, the increasing availability of data has led to the development of data-driven approaches that leverage machine learning and artificial intelligence techniques [13, 14]. David Donoho once said that the coming century would undoubtedly be the century of data [15]. By using data to drive analysis and insights, data-driven cyber-physical systems (DDCPS) enable real-time optimization and decision-making, leading to significant improvements in productivity and energy management, among others [16].

The goal of this book is to provide a comprehensive guide to DDCPS systems, covering both the principles and practical aspects of their design, implementation, and applications. This introductory chapter provides an overview of CPS and the role of data in improving their performance, as well as an overview of the book's content and structure. First, we will define what is meant by the term "cyber physical

F. Li et al., *Data-Driven Cyber Physical Systems*,
https://doi.org/10.1007/978-981-96-8709-1_1

system" and provide a brief history of its development. We will also discuss the key characteristics of CPS, including their ability to sense, actuate, and communicate with the physical world. Additionally, we will explore the challenges and opportunities that arise in the design and implementation of these systems. Next, we will introduce the concept of DDCPS and how they differ from traditional rule-based systems. We will explain how data can be used to improve the performance of CPS and provide an overview of the machine learning and artificial intelligence techniques that are commonly used in DDCPS. We will also discuss the benefits and limitations of these techniques, as well as the challenges and open research questions in this field. Finally, we will provide an overview of the book's content and structure, highlighting the key topics and themes that will be covered in each chapter. By the end of this chapter, readers will have a solid understanding of what DDCPS are, how they can be used to improve performance, and what topics will be covered in the following chapters.

1.1 What are Cyber Physical Systems?

As an emerging technology, CPS integrate physical systems with computational and networking capabilities. This integration allows for the creation of intelligent systems that can sense, monitor, and control physical processes in real-time. CPS have applications in a wide range of industries, including manufacturing, healthcare, transportation, and energy, and has the potential to revolutionize the way we interact with the physical world.

At its core, CPS are designed to bridge the gap between the physical and digital worlds, using data and information to optimize physical processes and systems. The physical components of CPS can include everything from sensors and actuators to robots and vehicles, while the digital components can include everything from data analytics and machine learning algorithms to cloud computing platforms and IoT (Internet of Things) devices. By combining these components, CPS enable the seamless integration of physical and digital systems, providing new insights and capabilities that were previously impossible. One of the key advantages of CPS is the ability to provide real-time data and insights into physical systems. The data can be used to optimize processes, detect anomalies and errors, and improve overall system performance. For example, in manufacturing, CPS can be used to monitor production lines and detect defects in real-time, enabling faster and more accurate quality control. In healthcare, CPS can be used to monitor patients and detect potential health issues before they become serious, improving patient outcomes and reducing healthcare costs. Therefore, "data-centric" has also become an indispensable trend for CPS principles and applications.

1.2 Data-Driven Approaches for CPS?

Data-driven approaches for CPS are becoming increasingly popular, as more and more sensors and connected devices become available. These systems are characterized by their ability to collect large volumes of data from multiple sources and use the information/feature to improve performance, reliability, and efficiency. Data fusion and utilization become important approaches for DDCPS, which involves the integration of data from multiple sources to produce a unified view of the system. Combining data from sensors, communication networks, and other sources to improve system performance and reliability is involved.

AI (artificial intelligence) for DDCPS involves the use of algorithms to analyze large data sets and identify patterns that can be used to make predictions or improve system performance. Machine learning has been applied to a variety of CPS applications, including predictive maintenance, fault detection, and optimization of energy consumption. Reinforcement learning for DDCPS involves the use of rewards and penalties to train systems to learn optimal behaviors, which has been used in a variety of applications, including autonomous vehicle control and energy management. Deep learning is another powerful data-driven approach that has been used in CPS, which involves the use of neural networks to analyze complex data sets and identify patterns that can be used to make predictions or improve system performance. AI has been used in applications such as image recognition, natural language processing, and speech recognition. Finally, data visualization is an important tool for DDCPS, as accurate visualization allows operators and users to better understand and interpret large data sets. Visualization tools can help identifying trends and anomalies in the data, and providing insights that can be used to improve system operation and performance.

Overall, data-driven approaches are becoming increasingly important in the development of CPS, since they allow for the analysis and interpretation of large volumes of data that would be difficult or impossible to handle using traditional methods. By leveraging these approaches, researchers and users can develop more powerful and efficient CPS that can provide significant benefits in terms of performance, reliability, and efficiency.

1.3 Importance of DDCPS

DDCPS are increasingly important today due to the vast amounts of data generated by modern sensors and communication networks. These systems rely on advanced algorithms and AI techniques to analyze large volumes of data in real-time, making it possible to improve system performance, reliability, and efficiency.

DDCPS have their abilities to detect anomalies and predict future behavior. By analyzing large datasets, these systems can identify patterns that may indicate a fault or failure in a component, allowing operators to proactively take action before a problem occurs. Additionally, data-driven approaches can be used to optimize system performance, improve energy efficiency, reduce downtime, and increase overall productivity. DDCPS can learn from experience. By leveraging AI techniques, these systems can adapt and improve over time, becoming more intelligent and better able to handle complex tasks. This can lead to significant benefits, such as improved safety, reduced costs, and increased flexibility. In addition, DDCPS can help organizations to make more informed decisions. By providing real-time insights into system performance and operational status, these systems can enable organizations to quickly respond to changing conditions and make more informed decisions about resource allocation and system optimization. Finally, DDCPS can have significant environmental benefits. For example, optimizing energy consumption using data-driven approaches can reduce carbon emissions and help meet sustainability goals. Similarly, data-driven approaches can be used to improve transportation systems, reducing traffic congestion and improving overall air quality. As the amount of data generated by modern systems continues to grow, the importance of data-driven approaches and DDCPS will only increase in the years to come.

1.4 Key Challenges in DDCPS

Several key challenges that must be addressed to realize their full potentials exist for DDCPS. In this section, we explore some main challenges faced by users and operators of these systems.

One of the foremost and formidable challenges that loom large in the realm of DDCPS is the herculean task of effectively handling the prodigious volumes of data that these systems generate in real-time. This challenge acquires paramount significance given that the cornerstone of DDCPS lies in its ability to harness data for real-time decision-making, optimization, and responsiveness. Navigating this data deluge demands a multifaceted strategy, marked by the convergence of advanced algorithms, high-performance computing systems, and judicious data management techniques, all orchestrated in harmonious concert. At the heart of this challenge lies the exigent need for advanced algorithms capable of processing data streams at astonishing speeds while maintaining the integrity of information. The advent of AI techniques, and real-time analytics has imbued DDCPS with the requisite analytical prowess to discern patterns, detect anomalies, and make informed decisions within fractions of a second. These algorithms are the backbone of DDCPS, leveraging their computational finesse to derive actionable insights from raw data streams, thus transforming data into a strategic asset rather than an overwhelming liability.

However, the computational demands imposed by real-time data processing necessitate more than just algorithms; they necessitate high-performance computing systems. These systems, fortified with cutting-edge processors, memory architectures, and parallel computing capabilities, become the workhorses that tirelessly crunch through the torrents of data. Whether it's optimizing traffic flow in smart cities, ensuring the precise delivery of medical treatments in healthcare settings, or fine-tuning manufacturing processes in Industry 4.0, high-performance computing systems are the linchpin upon which the real-time efficacy of DDCPS hinges.

Furthermore, efficient data storage and management systems assume pivotal roles in ensuring that the vast reservoirs of data remain accessible, retrievable, and responsive. Fast and scalable storage solutions are indispensable to accommodate the continuous influx of data streams. Simultaneously, data management systems orchestrate the organization, indexing, and retrieval of data with precision, enabling DDCPS to access the right information at the right time without succumbing to the pitfalls of data sprawl and latency. In essence, the challenge of handling large volumes of data in real-time is emblematic of the dynamic landscape of DDCPS, where the orchestration of advanced algorithms and high-performance computing systems converges to transform raw data into actionable insights. This pivotal challenge underscores the essence of DDCPS as a discipline where the capacity to navigate the data deluge with finesse and alacrity ultimately delineates the boundaries of operational excellence and system performance. The challenge of data integration in DDCPS underscores the multidimensional nature of this domain, which delves into the intricacies of heterogeneous data sources and their convergence into a coherent and actionable model of the system. Through the meticulous orchestration of data fusion techniques, DDCPS seeks not only to harmonize diverse data streams but also to enrich decision-making with insights that are truly representative of the complex interplay within these data-driven cyber-physical ecosystems.

Beyond the management of voluminous real-time data, DDCPS grapple with another formidable challenge—the seamless integration of data hailing from an array of diverse sources. These sources encompass an intricate web of sensors, communication networks, and other data origins, each employing distinct formats, protocols, and semantic nuances. The essence of this challenge lies in harmonizing this cacophony of disparate data streams into a coherent and comprehensive model of the system, ensuring that no valuable insights are lost in the process. The mosaic of data sources in DDCPS often presents itself as a mosaic of heterogeneity. Sensors, for instance, may vary in terms of their data types, sampling frequencies, precision levels, and communication protocols. Communication networks might deploy different transmission speeds, latency characteristics, and routing algorithms, further adding to the intricacy of data integration. These disparities, although reflective of the diverse nature of data sources, also pose significant hurdles when aiming to fuse this multifarious data into a unified and meaningful representation.

To surmount these challenges, DDCPS endeavor to develop robust data fusion techniques that serve as the linchpin of effective integration. These techniques act as sophisticated translators and assimilators, adept at deciphering the idiosyncrasies of each data source. They extract salient information from the plethora of inputs, whether it's extracting spatial-temporal patterns from sensor data or reconciling disparate data formats and units of measurement. Data fusion algorithms are meticulously designed to combine these disparate data sources into a holistic model that encapsulates the system's behavior, thus revealing a more profound understanding of its dynamics. Furthermore, data fusion is not confined to merely aggregating data; it extends its reach to the realm of uncertainty management. DDCPS grapple with data that is often imbued with noise, incompleteness, or even conflicting information. Robust data fusion techniques must possess the sophistication to weigh the reliability of different data sources, assign confidence levels and appropriately handle outliers or discrepancies.

In the intricate terrain of DDCPS, an intimately related challenge is to safeguard data quality and integrity. The very foundation upon which data-driven insights are built, data quality is an indispensable cornerstone that ensures the reliability, accuracy, and consistency of the data harnessed by DDCPS. It is an intricate tapestry woven from diverse threads, each carrying its own unique attributes and quirks, thereby necessitating a vigilant approach to data cleaning, preprocessing, and quality control. The challenge arises from the inherent heterogeneity of data sources, each of which may introduce its own idiosyncratic imperfections. Incomplete data, where crucial fields are missing or only partially recorded, can impede the comprehensive analysis of the system. Inconsistencies may emerge when different data sources provide conflicting or contradictory information, leading to ambiguities that hinder the generation of reliable insights. Accuracy concerns stem from the potential for erroneous data entry, sensor malfunctions, or data transmission errors, all of which may taint the veracity of the information. Indeed, in the dynamic landscape of DDCPS, the ever-escalating volume of data collection and analysis unfurls a parallel challenge that stands as paramount—the imperative to fortify security and preserve privacy. As the data reservoirs swell and the networked ecosystem burgeons, the specter of security breaches and privacy infringements looms ever larger. DDCPS operates in a realm where data is not merely a resource but a potent asset, rendering it imperative to safeguard its sanctity through multifaceted security and privacy measures.

To confront this challenge head-on, DDCPS must meticulously orchestrate a symphony of data cleaning and preprocessing techniques. Data cleaning involves the identification and rectification of outliers, missing values, and anomalies that mar the dataset. Techniques range from statistical approaches to machine learning-driven methods, each tailored to the specific characteristics of the data in question. Data preprocessing encompasses a broader spectrum of activities, encompassing data transformation, normalization, and feature engineering, all aimed at preparing the data for downstream analysis and modeling.

Integral to the pursuit of data quality is the imposition of stringent quality control measures. These measures serve as sentinels, guarding the integrity of the data as it traverses the various stages of its lifecycle. They encompass comprehensive data validation processes, data validation checks, and data auditing mechanisms that meticulously scrutinize incoming data streams for inaccuracies, inconsistencies, or discrepancies. By fostering a culture of data governance and best practices, DDCPS ensures that data remains not just a raw resource but a refined and reliable asset that underpins decision-making, insights generation, and system optimization.

Security, in the context of DDCPS, encompasses the protective envelope that shrouds data from unauthorized access, tampering, or malevolent intent. Encryption, as a vanguard technology, plays a pivotal role in rendering data indecipherable to prying eyes. Advanced encryption algorithms ensure that data, whether in transit or at rest, remains cloaked in a cryptographic veil, accessible only to those equipped with the requisite keys. Furthermore, robust access controls form an impregnable barrier, dictating who can access data, when, and under what circumstances. These access controls are multifaceted, encompassing authentication mechanisms, authorization protocols, and role-based permissions that delineate the boundaries of data access.

Privacy, on the other hand, is the custodian of individual liberties and data sovereignty. In a world where personal and sensitive information intertwine with the data fabric, privacy protection is an ethical and legal imperative. DDCPS navigates these turbulent waters through stringent measures that anonymize, pseudonymize, or obfuscate personally identifiable information (PII). Privacy by design principles ensure that the collection and handling of data are rooted in privacy-conscious strategies. Consent mechanisms empower individuals to determine the fate of their data, whether it is anonymized, retained, or deleted, reinforcing the ethos of informed consent and data ownership.

Monitoring systems constitute the vigilant guardians of DDCPS security and privacy. These systems remain perpetually vigilant, surveilling the network and data transactions for anomalies, intrusion attempts, or deviations from established security and privacy policies. Real-time alerts and automated responses ensure that any breach or irregularity is addressed with expediency, thus minimizing the window of vulnerability.

In essence, the challenge of preserving data quality and integrity is emblematic of DDCPS's commitment to ensuring that the insights derived from data are not merely artifacts but genuine reflections of the system's behavior. It underscores the rigorous processes, methodologies, and quality assurance protocols that define the domain's commitment to reliability and accuracy in an era where data reigns supreme. The challenge of ensuring security and privacy in DDCPS underscores the domain's commitment to preserving the integrity and trustworthiness of data-driven insights. In an age where data is both a catalyst for innovation and a potential liability, the multifaceted security and privacy measures delineate the boundaries within which DDCPS operate responsibly and ethically. These measures forge a future where data-driven advancements coexist harmoniously with data security and privacy, ushering in an era of trust and reliability.

Amid the complex landscape of DDCPS, another challenge that commands unwavering attention revolves around the integration of human operators into the very heart of these systems. While machine learning and data-driven algorithms have emerged as formidable tools for insights generation and automation, the indispensability of human intelligence, intuition, and oversight remains resolute. Bridging this human-machine chasm requires the careful calibration of user-friendly interfaces, training programs, and support mechanisms that empower human operators to interact effectively with these data-driven systems and harness their capabilities to their fullest extent.

User-friendly interfaces: At the nucleus of this challenge lies the development of user-friendly interfaces that serve as the conduits through which human operators commune with the data-driven system. These interfaces must be intuitive, ergonomic, and tailored to the specific needs and competencies of the operators. Considerations span from the design of dashboard layouts and visualization schemes to the provision of interactive tools that enable operators to query, explore, and interact with data seamlessly. A paramount goal is to demystify the complexities of DDCPS, fostering an environment where operators can translate data-driven insights into actionable decisions with ease.

Training and skill enhancement: The empowerment of human operators hinges on robust training and skill enhancement programs that nurture a deep understanding of DDCPS principles, tools, and workflows. Adequate training ensures that operators are not only proficient in operating the system but also in interpreting and critically evaluating the insights generated. Training programs encompass a spectrum of pedagogical approaches, from hands-on simulations to scenario-based exercises that mirror real-world challenges. Ongoing skill enhancement is equally vital, given the evolving nature of DDCPS technologies. Continuous learning ensures that operators remain adept at harnessing the latest advancements and optimizing system performance.

Support ecosystem: DDCPS embark on a journey where human-machine collaboration becomes increasingly intricate. To navigate this journey successfully, a robust support ecosystem is indispensable. It includes access to help desks, expert guidance, and troubleshooting resources that can address queries, challenges, and uncertainties in real-time. The support ecosystem serves as a lifeline, ensuring that operators can navigate the system's intricacies with confidence and that any impediments or bottlenecks are swiftly resolved.

The challenge of integrating human operators into DDCPS underscores the domain's commitment to a harmonious synergy between human intelligence and data-driven automation. It is a recognition that DDCPS are not just about algorithms and sensors but also about the individuals who wield these technologies to drive meaningful outcomes. By cultivating user-friendly interfaces, investing in comprehensive training, and fortifying the support infrastructure, DDCPS foster a future where human-machine collaboration transcends the boundaries of potential, enabling systems to operate with unprecedented proficiency and efficacy.

In the intricate tapestry of DDCPS, a multitude of challenges unfurl, each presenting its unique complexity and significance. To harness the full potential of DDCPS, these challenges must be confronted head-on, recognizing that their resolution paves the path toward more profound insights, enhanced performance, and heightened efficiency. These challenges are not mere hurdles; they are the crucibles within which innovation is forged, shaping the future landscape of DDCPS. The challenges in DDCPS are shown in Fig. 1.1.

Handling large data volumes in real-time: The challenge of managing large volumes of data in real-time encapsulates the essence of DDCPS. As these systems amass torrents of data with every passing moment, the imperative lies in the development of advanced algorithms and high-performance computing systems capable of processing data streams at lightning speeds. Scalable storage solutions and efficient data management orchestrate the stage for data to be not just processed but transformed into actionable insights, steering DDCPS toward optimal decision-making and real-time responsiveness.

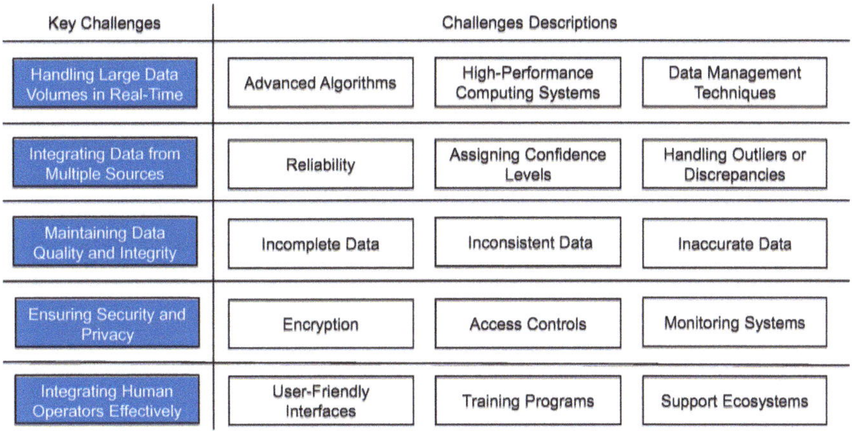

Fig. 1.1 The challenges in DDCPS. There are five core challenges in DDCPS, namely handling large data volumes in real-time, integrating data from multiple sources, maintaining data quality and intergrity, ensuring security and privacy, integrating human operators effectively

Integrating data from multiple sources: The tapestry of data in DDCPS is woven from diverse threads, each originating from a different source with its distinct format, protocol, and semantics. The challenge is to harmonize this symphony of heterogeneity, developing robust data fusion techniques that discern relevant information from each source and amalgamate it into a unified model of the system. Through data fusion, DDCPS gain the ability to extract meaningful insights from a cacophony of inputs, enriching its understanding of system dynamics.

Maintaining data quality and integrity: Data quality and integrity stand as the sentinels that guard the sanctity of insights derived from DDCPS. The challenge arises from the potential for incomplete, inconsistent, or inaccurate data that can hinder the generation of meaningful insights. Data cleaning and preprocessing techniques, coupled with stringent quality control measures, are indispensable in the quest to ensure that data remains accurate and reliable. The integrity of insights is a reflection of the integrity of the data that underpins them.

Ensuring security and privacy: The evolving landscape of DDCPS teems with data that is both an asset and a target. Security breaches and privacy violations are pervasive threats. To counteract these risks, DDCPS deploy an arsenal of security measures encompassing encryption, access controls, monitoring systems, and privacy-preserving techniques. These measures ensure that data remains impervious to malevolent intent while preserving individual liberties and data sovereignty.

Integrating human operators effectively: DDCPS acknowledge that human operators remain pivotal components of many systems. To harness the symbiotic potential of human-machine collaboration, user-friendly interfaces, comprehensive training programs, and robust support ecosystems are indispensable. By empowering operators with the tools, knowledge, and resources they need, DDCPS transcends the confines of automation to create systems where human expertise augments data-driven intelligence.

In summary, DDCPS challenges are not impediments but crucibles for innovation and progress. They are the facets that define the contours of a future where systems operate with unprecedented acumen, responsiveness, and reliability. By addressing these challenges with diligence and ingenuity, DDCPS pioneers chart a course toward systems that not only fulfill their potential but also redefine the boundaries of what is possible in the data-driven landscape.

1.5 Applications of DDCPS

DDCPS have a wide range of applications across many different industries and domains. In this section, we explore some of the key applications of these systems and the benefits they provide. These systems enable organizations to obtain real-time insights into critical processes, optimize performance, improve efficiency and profitability, and achieve sustainability goals. As the importance of data-driven approaches continues to grow, we can expect to see even greater adoption of these systems across a wide range of industries and applications.

One important application of DDCPS is in the field of manufacturing. By integrating data from sensors, communication networks, and other sources, manufacturers can obtain real-time insights into the performance of their equipment and processes. This enables them to identify and address problems quickly, optimize production processes, and reduce downtime, leading to improvements in efficiency, productivity, and profitability.

Another important application of DDCPS is in transportation and logistics. By leveraging data from GPS systems, traffic sensors, and other sources, these systems can improve the efficiency and safety of transportation networks, reducing congestion, saving time, and reducing fuel consumption. Additionally, by using real-time data from logistics networks, companies can optimize delivery routes, reduce shipping times, and improve product quality, leading to significant cost savings and improved customer satisfaction.

In the field of healthcare, DDCPS are also finding important applications. By analyzing data from medical sensors, electronic health records, and other sources, these systems can help clinicians to diagnose and treat patients more effectively, reduce errors, and improve patient outcomes. Additionally, by using predictive analytics, these systems can identify patients who are at high risk for certain conditions, allowing healthcare providers to intervene early and prevent serious health problems.

Another important application of DDCPS is in smart cities. By integrating data from sensors, communication networks, and other sources, these systems can provide real-time insights into the performance of critical infrastructure, such as water and power systems, transportation networks, and public safety systems. This allows city managers to optimize resource allocation, reduce costs, and improve quality of life for residents.

Finally, DDCPS have important applications in environmental monitoring and sustainability. By analyzing data from sensors and other sources, these systems can provide real-time insights into environmental conditions, allowing regulators and industry leaders to take action to reduce pollution and mitigate the effects of climate change. Additionally, by optimizing energy consumption and other resources, these systems can help organizations achieve their sustainability goals and reduce their environmental impact.

1.6 Evolution of Data-Driven Approaches in Cyber Physical Systems

The evolutionary trajectory of data-driven approaches in CPS spans a dynamic tapestry that has unfolded over the years, traversing distinct stages that mirror the growth of technology and the imperatives of complex systems. In the embryonic phases, CPS design was predominantly anchored in model-based techniques, where intricate mathematical models sought to encapsulate the behavior of these systems. These models, though proficient in capturing system dynamics, grappled with the sheer complexity and variability intrinsic to CPS, prompting the quest for innovative approaches that could surmount these limitations.

The limitations of early model-based techniques ignited a paradigm shift that catalyzed the ascent of data-driven approaches in the CPS landscape. Recognizing the innate capacity of data to unveil hidden insights, these approaches harnessed the power of real-world data streams to engender a more nuanced understanding of CPS behavior. Data-driven methodologies, fueled by the burgeoning troves of sensor data, commenced a transformative journey toward adaptive systems that could dynamically respond to real-world complexities.

The subsequent chapters in the evolution of data-driven CPS witnessed a convergence of disciplines, as machine learning and AI techniques came to the forefront. These methodologies endowed CPS with the ability to learn from data, paving the way for systems that could seamlessly adapt and optimize their performance. This metamorphosis ushered in a new era of responsiveness, where CPS could glean insights from data patterns, anticipate anomalies, and recalibrate responses in real time. The evolutionary trajectory of data-driven CPS remains a testament to the field's unceasing quest for innovations that resonate with the inherent intricacies of CPS, ultimately culminating in systems that stand at the confluence of data, intelligence, and adaptability. The development stages of DDCPS are shown in Fig. 1.2.

Amid the nascent stages of DDCPS, the thrust was on harnessing the prowess of statistical methods to delve into the rich tapestry of data emanating from the physical fabric of these systems. The overarching objective was to distill meaningful insights

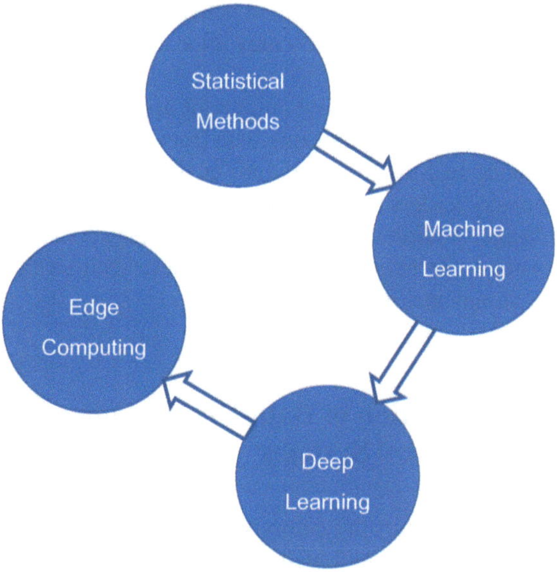

Fig. 1.2 The development stages of DDCPS. The development of DDCPS went through four stages: based on statistical methods, based on machine learning methods, based on deep learning methods and based on edge computing

from the intricate data streams, unveiling patterns and trends that could be harnessed to orchestrate profound enhancements in system performance. Statistical methods emerged as the guiding compass, navigating through the data deluge, as they sought to decipher the hidden language of these intricate systems.

At the core of this statistical voyage lay the art of identifying patterns, an endeavor that bore testament to the transformative power of data-driven insights. By meticulously analyzing time-series data and orchestrating regression analyses, the burgeoning arsenal of statistical tools unveiled the intricate correlations woven within the system. This illuminating correlation was not confined to mere statistical relationships; it bore the promise of unraveling the intricate dance between input variables and their cascading impact on the output of these complex systems.

The marriage of time-series analysis and regression techniques ushered in a new era of comprehension, transcending the mere interpretation of data and paving the path toward data-informed decision-making. The historical context embedded within time-series data illuminated temporal trends, unveiling the story of how system dynamics evolved over time. On the other hand, regression analysis untangled the multivariate interplay of variables, exposing how changes in one parameter could reverberate across the system—insights that proved invaluable in optimizing performance, enhancing efficiency, and charting a course toward better-informed system control strategies. In the intricate interplay of statistical methods, DDCPS found a fertile ground for its nascent exploration, setting the stage for the unfolding tapestry of data-driven intelligence.

As the data-driven landscape of CPS evolved, a resounding imperative emerged to grapple with the exponential surge in data volume and complexity. The sheer deluge of information necessitated the advent of more intricate methodologies capable of comprehending, deciphering, and harnessing the multifaceted data streams. In response, the spotlight turned to machine learning techniques, a transformative paradigm that harnessed the power of algorithms to glean insights, learn patterns, and unravel intricate relationships within the data.

The introduction of machine learning marked a pivotal inflection point, where CPS systems transitioned from passive data recipients to active learners. Supervised learning emerged as a foundational pillar, guiding the system to generalize from labeled examples and make informed predictions. Unsupervised learning, on the other hand, delved into the unknown, uncovering hidden patterns and clusters within unlabeled data streams. Reinforcement learning ushered in a new era, empowering CPS to make sequential decisions through interaction with its environment, amplifying the system's adaptability and intelligence.

However, the evolutionary journey of data-driven CPS did not rest with machine learning alone. Recent years have witnessed the emergence of deep learning as a formidable contender, revolutionizing the handling of complex CPS data. Deep learning, epitomized by neural networks with multiple layers, transcends the surface to extract intricate features buried within the data's labyrinth. These features are the building blocks of predictive models and decision-making frameworks, casting a spotlight on the system's behavior and dynamics. This newfound capability has ignited a renaissance across a myriad of CPS domains, from autonomous driving to industrial automation, as well as healthcare, where deep learning has catalyzed breakthroughs in diagnostics, treatment optimization, and patient care.

In a convergence of visionary insight and technological innovation, data-driven CPS systems have embraced the orchestration of machine learning and deep learning techniques. These methodologies collectively surmount the complexities of data, shaping systems that are not only perceptive but also prescient. As the synergy of data and intelligence continues to unfold, the trajectory of CPS evolution moves steadfastly toward a future where systems navigate the realms of possibility with unprecedented acumen and adaptability.

Another pivotal stride in the evolutionary narrative of DDCPS is the burgeoning embrace of edge computing, a paradigm shift that redefines the locus of data processing. Edge computing, in essence, entails the localized processing of data at the network's periphery, in close proximity to the very source from which it emanates. This transformative approach stands as a testament to the industry's determination to address the latency predicament that often mars real-time data processing, thereby catalyzing a seismic enhancement in system performance.

The allure of edge computing rests in its ability to minimize the chasm between data generation and processing. By virtue of processing data at the edge of the network, in immediate proximity to the sensors and devices that generate it, the latency hurdle is significantly mitigated. This orchestration ushers in a paradigm of enhanced real-time performance, enabling systems to respond with unprecedented immediacy to evolving scenarios. For CPS that operate in dynamic environments, such as autonomous vehicles navigating bustling urban landscapes or industrial systems responding to shifting demands, this is a game-changer. Edge computing addresses the temporal chasm that can hinder instantaneous decision-making, imbuing CPS with newfound agility.

The synergy between edge computing and DDCPS is particularly pronounced in domains that demand real-time decision-making prowess. In CPS applications where split-second responses translate into tangible outcomes—be it optimizing manufacturing processes, navigating complex traffic systems, or orchestrating healthcare interventions—the strategic deployment of edge computing stands as a linchpin. By seamlessly fusing real-time data processing with DDCPS's data-driven insights, these systems transform into nimble entities, deftly navigating the intricacies of their operational landscape.

In the symphony of DDCPS evolution, the crescendo of progress resonates through every chord. From the nascent stages of statistical methods to the era of machine learning, deep learning, and the vanguard of edge computing, each stride shapes CPS into adaptive, intelligent entities. This intricate interplay fosters not only a deeper comprehension of system dynamics but also a tangible amplification in decision-making precision. As data-driven paradigms continue to mature and technology evolves, the promise of CPS blossoms into a reality where systems respond, adapt, and flourish with unparalleled prowess.

1.7 How Can Data be Used to Improve Cyber Physical Systems?

The integration of data with CPS has ushered in a new era of technological innovation, offering boundless opportunities for the development of intelligent systems that can seamlessly learn, adapt, and respond to dynamic environments. CPS, at their core, encompass a synergy of physical components—including sensors, actuators, and controllers—all interconnected within a network and managed by sophisticated software. This integration of physical and digital elements is at the heart of the CPS paradigm, and data plays a pivotal role in unlocking its true potential. The symbiotic fusion of data and advanced AI techniques drives a revolution in CPS design. The data-driven CPS paradigm equips systems with cognitive abilities, enabling them to decipher the intricate language of data and adapt their behavior accordingly. This approach is poised to redefine conventional boundaries across industries, empowering CPS to unlock new frontiers of optimization, fault detection, and predictive insights.

With data serving as the lifeblood of these systems, CPS can transcend their static, rule-based nature and evolve into agile entities capable of autonomous optimization and real-time decision-making. The sensors and actuators embedded within the physical infrastructure continually generate streams of data, which, when effectively harnessed, can provide a holistic view of the system's behavior, performance, and environment. This wealth of information empowers CPS to adapt on the fly, making informed decisions that enhance efficiency, productivity, and safety.

Harnessing the power of data-driven analysis and insights, CPS can be fine-tuned to achieve levels of performance that were previously unimaginable. By continuously gathering, processing, and interpreting data from various sources, CPS can identify patterns, trends, and anomalies. These insights can then be leveraged to optimize system parameters, reduce energy consumption, minimize waste, and mitigate potential faults. This self-optimization is particularly valuable in dynamic environments where conditions fluctuate rapidly, allowing CPS to adapt and respond without human intervention.

The integration of data transforms CPS from passive observers to active participants in their operational contexts. The real-time analysis of data enables these systems to be proactive rather than reactive, enhancing their ability to anticipate changes and make informed decisions. This not only improves their overall performance but also augments their resilience and adaptability in the face of uncertainties. As the era of data-driven CPS unfolds, the integration of data becomes the bridge between the physical and digital realms, enabling a new generation of intelligent systems that redefine the boundaries of possibility.

Data-driven approaches to CPS design constitute a groundbreaking paradigm shift, employing the prowess of AI techniques to unravel insights from the deluge of data emanating from the system's physical components. This transformative approach, hinged on data-driven intelligence, unlocks a new dimension of system optimization, adaptability, and predictive prowess that was hitherto unattainable through conventional means. The marriage of data and advanced algorithms propels CPS into an era of enhanced performance and efficiency.

By delving into vast datasets, DDCPS can unearth hidden patterns and intricate relationships within the data, transforming raw information into actionable knowledge. These insights become the guiding lights for optimizing system performance, identifying deviations from normal behavior, and envisioning future trajectories. The AI-infused techniques decode the underlying dynamics of the system, enabling real-time adjustments to system parameters, resource allocations, and control strategies. In this way, data-driven CPS transcends static rule-based systems, dynamically adapting to evolving circumstances.

Illustratively, consider the realm of manufacturing, where the integration of data-driven CPS has led to transformative outcomes. Within a manufacturing plant, the data-driven approach facilitates real-time optimization of production processes, leading to enhanced throughput, minimized waste, and reductions in downtime. By analyzing data from sensors and other sources, CPS can autonomously adjust manufacturing parameters, optimize machine performance, and even anticipate equipment maintenance needs. This dynamic response to real-time data translates into tangible benefits, such as increased productivity and resource efficiency, which are critical in modern, competitive manufacturing environments.

An overarching advantage of DDCPS lies in their unparalleled capacity to drive real-time decision-making, ushering in a new era of agility and precision across a spectrum of applications. The integration of data-derived insights into the operational fabric of CPS empowers swift and well-informed decision-making, unraveling a tapestry of benefits that span industries and domains. By leveraging real-time data streams, DDCPS transcends the traditional boundaries of response times, leading to an era of heightened efficiency and productivity.

In healthcare, the potency of DDCPS becomes particularly apparent. The ability to monitor patients' vital signs in real time constitutes a paradigm shift in patient care. As sensors continuously capture crucial health metrics, DDCPS can swiftly detect deviations from normal ranges and trigger instant alerts for medical staff. This proactive approach enables early intervention and personalized care, mitigating potential health crises before they escalate. The integration of DDCPS thus culminates in safer, more efficient healthcare practices that prioritize patient well-being.

Moreover, the transformative impact of DDCPS extends its reach to transportation systems, redefining the way we navigate bustling urban landscapes. In the realm of traffic management, DDCPS emerges as a powerful ally in the quest for seamless mobility. By ingesting real-time data on traffic patterns, road conditions, and vehicle movements, DDCPS can orchestrate real-time optimizations that ease congestion, enhance traffic flow, and alleviate the often-frustrating gridlock of urban environments. This not only improves the efficiency of transportation systems but also translates into tangible benefits for citizens, reducing travel times and minimizing environmental footprints.

In the grand tapestry of DDCPS, the thread of real-time decision-making weaves a narrative of transformative potential. From healthcare to transportation, the power of data-driven insights shapes industries in ways that were previously unimaginable. By enabling swift and precise responses to dynamic situations, DDCPS promise to redefine efficiency, safety, and quality of life in a world increasingly reliant on interconnected systems.

1.8 Overview of the Book

Our book stands as an invaluable beacon, illuminating the intricate landscape of CPS principles, design methodologies, and implementation strategies with a special focus on the transformative realm of data-driven approaches. Within its pages, readers are ushered into a multidimensional journey that unveils the convergence of physical and digital realms, igniting a profound exploration into the fusion of data and intelligent systems.

Structured across nine meticulously curated chapters, each exuding a wealth of insights, the book crystallizes the quintessence of data-driven CPS systems. Its structure serves as a roadmap that guides readers through the multifaceted terrain, nurturing a deep understanding of core concepts and fostering a practical proficiency that extends to real-world applications. With each chapter uniquely positioned to unravel intricate facets of the field, the book attunes readers to the nuanced interplay between theoretical underpinnings and pragmatic implementations.

As the canvas of each chapter unfurls, readers are invited to embark on a voyage through the layers of DDCPS, from foundational principles to cutting-edge advancements. This comprehensive exploration encapsulates not only the intricacies of machine learning and AI techniques but also delves into critical elements such as data acquisition, analysis, and cybersecurity–elements that serve as the very lifeblood of a resilient CPS ecosystem.

Imbued with a dynamic interplay of theoretical insights and real-world exemplars, the book paints a vivid tapestry of possibilities that emerge at the intersection of data-driven intelligence and CPS. The journey navigates through real-world case studies, casting light upon diverse industries where DDCPS systems wield transformative power. From manufacturing to healthcare, transportation to energy, the book's rich tapestry of examples unravels how data-driven approaches enhance efficiency, reduce costs, and elevate performance across an expansive spectrum of applications.

In essence, DDCPS transcend the confines of a mere textbook, metamorphosing into a guiding star that empowers researchers, practitioners, and aspiring minds alike. The collective wisdom woven within its chapters epitomizes the relentless pursuit of innovation, embodying the essence of a paradigm that drives a future where data and intelligence converge to shape the CPS destiny.

Competing Interests

1. The research is supported by the National Key Research and Development Project under Grant 2023YFB3307300, the National Natural Science Foundation of China under Grants 62373014, 62422301, 62125301, and 92467205.
2. The authors have no conflicts of interest to declare that are relevant to the content of this chapter.

References

1. R. Rajkumar, I. Lee, L. Sha, and J. Stankovic, "Cyber-physical systems: the next computing revolution," in *Proceedings of the 47th design automation conference*, pp. 731–736, 2010.
2. R. Alur, *Principles of cyber-physical systems*. MIT press, 2015.
3. T. Sanislav and L. Miclea, "Cyber-physical systems-concept, challenges and research areas," *Journal of Control Engineering and Applied Informatics*, vol. 14, no. 2, pp. 28–33, 2012.
4. G. Xiong, F. Zhu, X. Liu, X. Dong, W. Huang, S. Chen, and K. Zhao, "Cyber-physical-social system in intelligent transportation," *IEEE/CAA Journal of Automatica Sinica*, vol. 2, no. 3, pp. 320–333, 2015.
5. Y. Zhang, M. Qiu, C.-W. Tsai, M. M. Hassan, and A. Alamri, "Health-cps: Healthcare cyber-physical system assisted by cloud and big data," *IEEE Systems Journal*, vol. 11, no. 1, pp. 88–95, 2015.
6. L. Monostori, B. Kádár, T. Bauernhansl, S. Kondoh, S. Kumara, G. Reinhart, O. Sauer, G. Schuh, W. Sihn, and K. Ueda, "Cyber-physical systems in manufacturing," *Cirp Annals*, vol. 65, no. 2, pp. 621–641, 2016.
7. A. W. Colombo, S. Karnouskos, O. Kaynak, Y. Shi, and S. Yin, "Industrial cyberphysical systems: A backbone of the fourth industrial revolution," *IEEE Industrial Electronics Magazine*, vol. 11, no. 1, pp. 6–16, 2017.

8. V. Gunes, S. Peter, T. Givargis, and F. Vahid, "A survey on concepts, applications, and challenges in cyber-physical systems.," *KSII Trans. Internet Inf. Syst.*, vol. 8, no. 12, pp. 4242–4268, 2014.
9. Y. Lu, "Cyber physical system (cps)-based industry 4.0: A survey," *Journal of Industrial Integration and Management*, vol. 2, no. 03, p. 1750014, 2017.
10. S. K. Khaitan and J. D. McCalley, "Design techniques and applications of cyberphysical systems: A survey," *IEEE systems journal*, vol. 9, no. 2, pp. 350–365, 2014.
11. L. Wang, M. Törngren, and M. Onori, "Current status and advancement of cyber-physical systems in manufacturing," *Journal of manufacturing systems*, vol. 37, pp. 517–527, 2015.
12. O. Inderwildi, C. Zhang, X. Wang, and M. Kraft, "The impact of intelligent cyber-physical systems on the decarbonization of energy," *Energy & Environmental Science*, vol. 13, no. 3, pp. 744–771, 2020.
13. R. Mitchell and R. Chen, "Behavior-rule based intrusion detection systems for safety critical smart grid applications," *IEEE Transactions on Smart Grid*, vol. 4, no. 3, pp. 1254–1263, 2013.
14. A. Villalonga, G. Beruvides, F. Castano, and R. E. Haber, "Cloud-based industrial cyber–physical system for data-driven reasoning: A review and use case on an industry 4.0 pilot line," *IEEE Transactions on Industrial Informatics*, vol. 16, no. 9, pp. 5975–5984, 2020.
15. D. L. Donoho *et al.*, "High-dimensional data analysis: The curses and blessings of dimensionality," *AMS math challenges lecture*, vol. 1, no. 2000, p. 32, 2000.
16. M. Schmidt and C. Åhlund, "Smart buildings as cyber-physical systems: Data-driven predictive control strategies for energy efficiency," *Renewable and Sustainable Energy Reviews*, vol. 90, pp. 742–756, 2018.

Chapter 2
Fundamentals of Data-Driven Cyber Physical Systems

Abstract Data-driven cyber physical systems (DDCPS) are cutting-edge digital systems that revolutionize the way we interact with and integrate cyber space and the physical world. DDCPS consist of a sophisticated blend of computation, communication, control, and physical elements, all seamlessly woven together in a highly integrated framework. In this section, we begin with giving separate definitions of CPS and DDCPS from different perspectives. We then introduce the characteristics of DDCPS which include networked communication, scalability, heterogeneity, interdisciplinary, real-time processing and real-time decision-making. Next, components of DDCPS containing sensing parts, computational parts, communication parts and control parts are described. From theory to reality, we list some of the applications of DDCPS in different fields. Finally, the challenges that DDCPS will face in the future in data storage, integration, communication, cyber security and system stability are discussed.

This chapter commences by establishing distinct definitions for both CPS and DDCPS, each examined from diverse vantage points. The subsequent focus turns to unveiling the defining characteristics of DDCPS. To provide a holistic view, the chapter delves into the intricate components that comprise DDCPS, detailing the integral roles of sensing, computation, communication, and control within DDCPS. Bridging the theoretical to the practical, we showcase a spectrum of real-world applications across various fields where CPS have been harnessed to remarkable effect. The chapter concludes by illuminating the hurdles that DDCPS are poised to confront in the future.

F. Li et al., *Data-Driven Cyber Physical Systems*,
https://doi.org/10.1007/978-981-96-8709-1_2

2.1 Definitions

2.1.1 Definitions of CPS

CPS represent an innovative class of systems that seamlessly integrate computational capabilities (including data analysis and processing, decision-making and control, networking and communication, software systems, and applications) with physical functionalities (such as sensing and actuation, physical processes, and dynamics) as Fig. 2.1. This integration enables a wide range of interactions with the real world [1–5]. This interaction can take various forms, such as direct physical interaction with smart objects, intuitive gesture-based interfaces, voice commands, touchscreens, augmented reality, or even brain-computer interfaces [6–10]. CPS utilize computational capabilities to analyze and process large volumes of data collected from sensors and other sources. Computational techniques such as data mining, machine learning, statistical analysis, and pattern recognition are employed to extract meaningful insights from raw data. CPS incorporate computational algorithms and models to make intelligent decisions and control physical processes. The algorithms employed in CPS can range from simple rule-based control to more complex optimization, prediction, and adaptive control techniques.

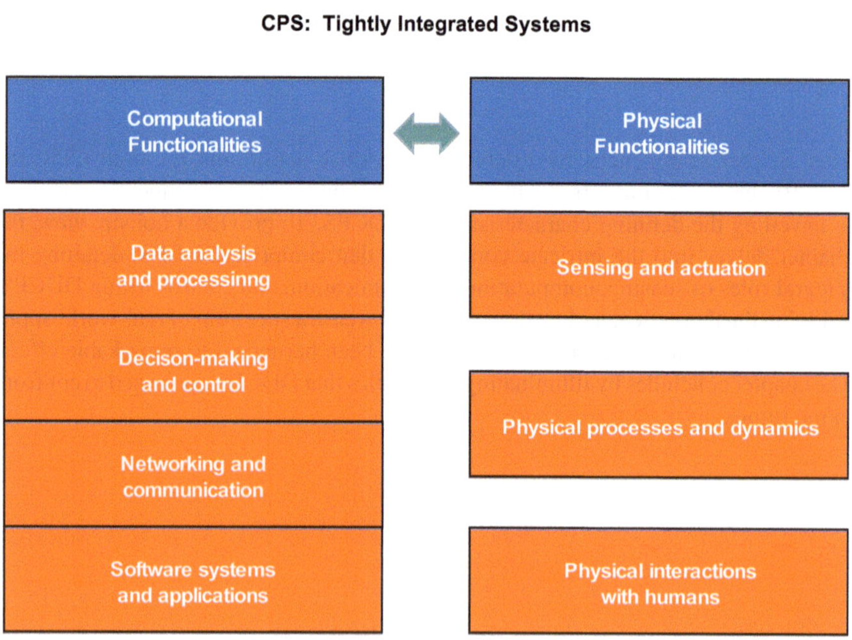

Fig. 2.1 CPS represent tightly integrated systems that combine computational and physical functionalities

CPS utilize computational capabilities to analyze and process large volumes of data collected from sensors and other sources. Computational techniques such as data mining, machine learning, statistical analysis, and pattern recognition are employed to extract meaningful insights from raw data. CPS incorporate computational algorithms and models to make intelligent decisions and control physical processes. The algorithms employed in CPS can range from simple rule-based control to more complex optimization, prediction, and adaptive control techniques.

On the one hand, CPS heavily rely on software systems and applications for data management, computation execution, and user interface provision. The orchestration and supervision of intricate computer systems entail the development of customized software architectures, algorithms, and user interfaces tailored to the specific CPS application. On the other hand, the physical aspect of CPS involves the utilization of sensors and actuators that interact with the physical surroundings. Sensors capture data from the physical environment, encompassing parameters like temperature, pressure, motion, and chemical properties. In contrast, actuators facilitate physical actions and manipulate the environment based on control signals generated by the computational components. This dual interplay of software and physical components characterizes the essence of CPS.

CPS interact with and control physical processes in various domains, including manufacturing, transportation, healthcare, energy, and infrastructure [7–10]. Understanding the dynamics of these physical systems and their behavior is crucial for effective control and optimization. CPS facilitate physical interactions with real world through interfaces and devices. These can include touchscreens, gesture recognition, voice interfaces, wearable devices, haptic feedback systems, and augmented reality interfaces. Physical interactions with real world enable intuitive and seamless engagement with CPS. Networking and communication protocols are integral to CPS, enabling seamless connectivity and information exchange between different components of the system. The interoperability, in turn, facilitates real-time collaboration and coordination.

CPS facilitate collaboration and cooperation, allowing for shared control, coordination, and communication. While CPS have been defined by the academy from different perspectives, it is generally agreed that CPS are an interdisciplinary field that involves the integration of physical and computational components. As CPS continue to evolve, the scientific community is striving to find common ground in defining this field and exploring its full potential.

We will introduce the definitions of CPS from four aspects: development history, different perspectives, components of CPS, comparisons with other systems as Fig. 2.2.

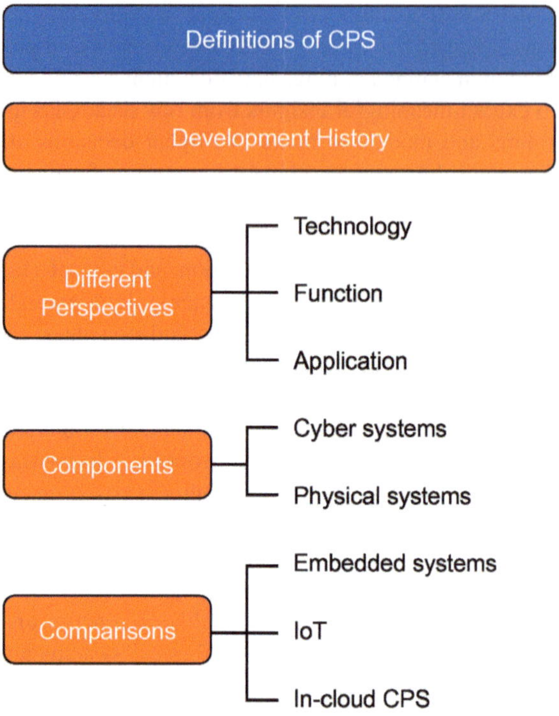

Fig. 2.2 An overview of definitions of CPS. This chapter starts from the development history of the definitions of CPS. Next, CPS from different perspectives are introduced: technology, function and application. This chapter also introduces the components of CPS briefly. Finally, CPS are compared with other systems: embedded systems, IoT, In-cloud CPS

2.1.1.1 Development History

CPS is a general term for the next generation of engineered systems that tightly integrate computing, communication, and control technologies in order to achieve stability, performance, reliability, robustness, and efficiency when dealing with physical systems across a wide range of application domains, including transportation, energy, medicine, and defence [11]. The development history of the definitions of CPS can be seen as Fig. 2.3.

CPS have been in use since the 1970s when microprocessors first appeared [12]. The concept of CPS emerged in the late 1990s and early 2000s [13]. During this time, there was a growing integration of sensors, actuators, and computer systems into physical systems, creating a tight coupling between physical processes and computational systems. The design of these systems aimed at real-time monitoring, feedback, and control. In 2008, the National Science Foundation (NSF) in the United States established the CPS research program. The NSF conducted systematic research on CPS and defined it as "intelligent, distributed systems of sensors and actuators, designed to monitor and control physical processes" [14]. The following is how the NSF defines and explains cyber physical systems: "The term cyber-physical systems refers to the tight conjoining of and coordination between computational and physical resources". This development and evolution of CPS have been influenced by

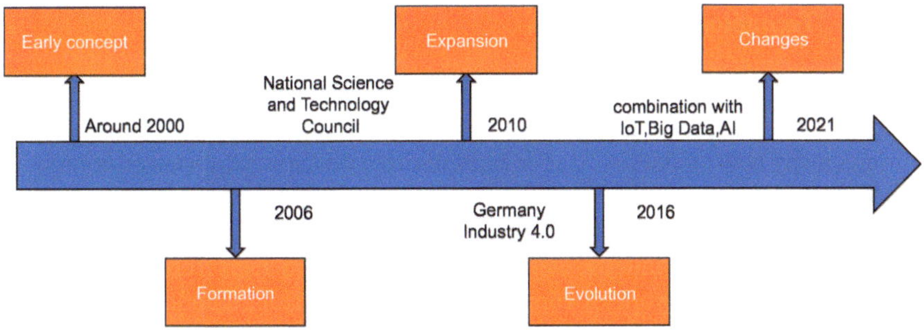

Fig. 2.3 The notion of CPS surfaced in the late 1990s and early 2000s. In 2008, the National Science Foundation proposed a definition of CPS. In 2010, the National Science and Technology Council expanded the scope of CPS. In 2016, Germany formulated its Industry 4.0 strategy, emphasizing the importance of CPS. Until 2021, the definitions of CPS still changes because of the combination with IoT, Big Data, AI

technological advancements and the increasing demand for smart and interconnected systems. Over time, the definition of CPS expanded to encompass a broader range of domains. In 2010, the National Science and Technology Council (NSTC) published the "Cyber Physical Systems Research and Development Roadmap". This roadmap expanded the scope of CPS, applying it to areas such as transportation systems, smart grids, healthcare, and smart manufacturing. In 2016, Germany formulated its Industry 4.0 strategy, emphasizing the importance of CPS. Germany redefined CPS as the "integration of physical and virtual worlds" and combined it with technologies such as the Internet of Things (IoT), big data, and artificial intelligence to achieve intelligent, adaptive, and autonomous systems.

The definition of CPS still varies across different domains and organizations. With technological advancements and deeper research, the definition of CPS continues to evolve. For instance, some scholars define CPS as the integration of physical systems with computing and communication technologies to achieve high automation, real-time decision-making, and autonomy [15–17]. Others may consider CPS as systems where multiple physical and virtual entities interact with each other [18–21]. CPS have evolved from the integration of sensors and actuators with computational systems to a broader concept that involves the integration of physical and virtual worlds, enabling intelligent and adaptive systems. Over time, the integration of sensing, computing, and communication technologies has enabled CPS to connect the physical and digital worlds, creating new opportunities for innovation and improving efficiency in various industries. The tight integration of physical and computational components that work together to achieve a common goal is a defining feature of CPS.

2.1.1.2 Technology Perspectives

From technological perspectives, CPS represent a new generation of systems with integrated computational and physical capabilities that can interact with real world through many new modalities. At the heart of these systems are advanced computing applications that monitor and control the physical components of a system, while providing feedback to the computational elements. The intricate interplay between the physical and computational elements is what sets CPS apart from other systems. CPS are not just a network of computational and physical components, but a network that is capable of monitoring and controlling physical processes. Supervising is achieved through the use of embedded computing devices such as sensors and controllers. These devices are often connected through networks, including the Internet and wireless networks, enabling real-time data collection and control of the physical processes. The integration of physical processes and algorithms in CPS involves a high degree of complexity. The algorithms running on these systems are often responsible for critical tasks that require high reliability, efficiency, and safety. Furthermore, the software must be able to interact with the physical processes in a way sensitive to the timing and dynamics of these processes. In essence, from technological perspectives, CPS are sophisticated systems that tightly integrate computational and physical components, enabling them to monitor and control physical processes with a high degree of precision and reliability. This integration is facilitated by advanced algorithms running on embedded computing devices, which are often networked to enable real-time data collection and control.

2.1.1.3 Function Perspectives

From functional standpoints, CPS serve as a bridge between the digital and physical worlds, enabling a level of interaction and control that was previously unattainable. CPS are designed to manage and manipulate the physical world in real-time, offering a high degree of precision and reliability. One of the defining characteristics of CPS is their ability to provide real-time, reliable, and effective feedback which is not just a simple one-way flow of information. Instead, CPS are capable of creating a continuous loop of data collection, analysis, and action. CPS can gather data from the physical world, process it in real-time, make decisions based on this data, and then act upon these decisions to influence the physical world. The feedback loop allows CPS to adapt and respond to changes in the physical environment quickly and accurately. Moreover, the ability of CPS to respond to changes in the physical world is not a passive process. Instead, the adaptability is an active, dynamic interaction that involves constant monitoring, decision-making, and action. When changes occur in the physical environment, CPS can detect these changes, analyze their implications, decide on the best course of action, and then execute this action to achieve the desired outcome. The dynamic interaction allows CPS to maintain control over the physical world, even in the face of unexpected changes and disturbances. CPS are not just systems that integrate the physical and digital worlds. They are also systems that can actively control and manipulate the physical world through a continuous loop of data collection, analysis, decision-making, and action. The ability to interact with the physical world in real-time, and to adapt and respond to changes, sets CPS apart from traditional computing systems and opens up a wide range of new possibilities and applications.

2.1.1.4 Application Perspectives

From application standpoints, CPS have found their ways into a broad spectrum of industries, revolutionizing traditional processes and systems with their unique capabilities. They are no longer confined to research labs but are actively shaping our world in numerous sectors, including smart factory, smart grids, agriculture, healthcare, intelligent transportation and smart manufacturing. In the realm of smart grids, CPS play a pivotal role in managing and optimizing energy distribution, ensuring a more efficient and sustainable use of resources. CPS enable real-time monitoring and control of energy flows, thus facilitating the integration of renewable energy sources, and enhancing the reliability and resilience of the power grid. In the field of autonomous driving, CPS are the backbone of the technology that allows vehicles to navigate without human intervention. They integrate and process data from a variety of sensors and systems in real-time, making decisions and controlling the vehicle's movements to ensure safe and efficient transportation. Within the healthcare domain, CPS are transforming patient care and medical procedures. They are used in telemedicine, remote patient monitoring, robotic surgery, and personalized medicine, providing more accurate diagnoses, more effective treatments, and improved patient outcomes. In the sphere of smart manufacturing, CPS are key to Industry 4.0, enabling the automation and digitization of production processes. They allow for real-time monitoring and control of production lines, predictive maintenance of machinery, and flexible and customized production, leading to increased efficiency and productivity. In all these applications, CPS provide services that are not only efficient but also reliable and secure. They are designed to operate in a robust manner, even in the face of uncertainties and disturbances in the physical environment. They also incorporate advanced security measures to protect against cyber threats, ensuring the integrity and confidentiality of data. In summary, from an application perspective, CPS are transforming various sectors by enabling real-time, reliable, and secure interaction and control of physical processes. Their wide-ranging applications in smart grids, autonomous driving, healthcare, and smart manufacturing are testament to their potential to reshape our world and pave the way for future technological advancements.

2.1.1.5 Cyber Systems and Physical Systems

CPS refers to an advanced set of interconnected systems that include cyber and physical components, controlled by customized semantic regulations [22]. The cyber

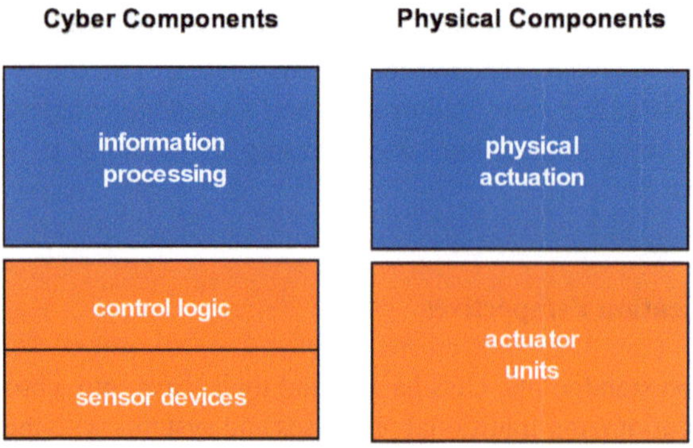

Fig. 2.4 CPS is an advanced set of interconnected systems with cyber and physical components that are governed by customised semantic regulations. The cyber segment includes control logic and sensor devices, whereas the physical segment includes actuator units

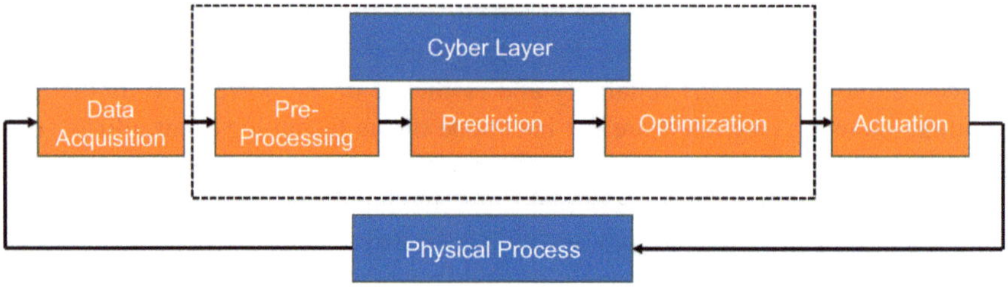

Fig. 2.5 A control loop of CPS. CPS acquire data from physical process and pre-processing of the data in cyber layer. Prediction and optimization are also in cyber layer. Then, cyber layer gives command to actuate physical process

segment comprises control logic and sensor devices, while the physical part encompasses actuator units as Fig. 2.4. CPS enables the collaboration and coordination of disparate systems to achieve complex tasks. Cyber and physical systems share common components of information processing and physical equipment as Fig. 2.5. In the case of cyber systems, the algorithms responsible for processing information holds primary significance, while the hardware plays a supportive role. Conversely, in industrial control systems, the information processing system serves as a support system for the equipment being managed.

Physical systems are collections of actuator units, whereas cyber systems are collections of control logic and sensor units [23]. Physical and engineered systems that have a computational and communication core that monitors, coordinates, controls, and integrates their operations are referred to as CPS. From the nanoworld to large-scale wide-area systems of systems, this tight relationship between the cyber and physical will be manifested [24].

2.1.1.6 Compared to Embedded Systems

CPS will change how we engage with the physical world around us, much as the internet changed how people interact with one another. Systems that combine compute with physical operations are known as CPS. Physical processes are monitored and controlled by embedded computers and networks, typically through feedback loops where computations are affected by physical processes and vice versa. In the physical world, time moves unstoppable forward and concurrency is inherent. These two characteristics are not present in the computing and networking abstractions used today.

CPS are founded on the idea that self-learning and self-driving machines can be built by combining computing (sensing, analysing, forecasting, comprehending), communication (interaction, intervention, interface monitoring), and control (inter-operate, develop, evidence-based certification). A cyber-physical system seamlessly integrates traditional physical elements with digital counterparts. This system involves the integration of biological items and natural systems that constitute the physical component, along with technologically created systems for transportation and energy production. CPS operate in real-time, actively processing data and engaging in communication with the surrounding environment. The computational components of CPS encompass data handling, communication, and control mechanisms that heavily rely on computational methods. Software and digital systems utilize ADCs, DACs, and digital communication networks to establish connections with hardware components, facilitating the necessary calculations. The artificial structures within CPS can be either discrete-time or event-driven systems. The computational part of CPS is intertwined with the physical part and dispersed across the system. The combination of continuous and discrete dynamics makes CPS extremely complicated. Physical parts, digital parts, and the infrastructure required to link them all together constitute CPS. Systems that exist in the physical world are included in the category of "physical components". The cyber parts of a system are the computer programmes that make it work, like a communication or control algorithm. Interfaces, converters, signal conditioners, and networks are all examples of the kinds of subsystems utilised to link them together.

2.1.1.7 Compared to IoT

CPS and IoT are two important technological concepts that often get confused due to their overlapping characteristics. However, they are distinct in their core features, purposes, and applications. On the one hand, CPS are integrations of computation, networking, and physical processes. Embedded computers and networks monitor and control the physical processes, with feedback loops where physical processes affect computations and vice versa. CPS involve a deeper level of interaction with and control over physical processes compared to general IoT applications. The aim is to improve efficiency, capability, and adaptability beyond what can be achieved by either physical or cyber components alone. On the other hand, IoT is a network of physical objects—"things"—that have sensors, software, and other technologies embedded in them for the purpose of connecting and exchanging data with other devices and systems over the internet. These "things" include everyday household items like refrigerators, thermostats, and watches, which can be controlled remotely via network connections.

Firstly, in CPS, the integration of computation, networking, and physical processes is much deeper. It's not just about connecting devices and enabling data exchange, but also about real-time monitoring, control, and optimization of physical processes. In IoT, the main focus is on connectivity and data exchange between devices. Secondly, on the one hand, CPS primarily target mission-critical applications where the coordination between the cyber and physical components is crucial, such as autonomous vehicles, industrial control systems, health monitoring systems, and smart grids. IoT, on the other hand, is often more consumer-oriented, enabling smart homes, wearable devices, and similar conveniences. Thirdly, CPS typically involve more complex systems, often with stringent requirements for real-time operation, safety, reliability, and resiliency. In contrast, IoT applications may not have such high requirements, as they often focus on improving everyday life rather than controlling critical physical processes. In summary, while CPS and IoT both involve the integration of physical devices and cyber technologies, they differ in their levels of integration, their focus and applications, and the complexity of the systems they involve. CPS involves deeper integration, more mission-critical applications, and more complex system requirements, while IoT is more about connectivity and data exchange between devices, often with a focus on consumer applications.

CPS are designed to tightly integrate and coordinate physical processes with computational elements. This means CPS are more closely associated with the real world in terms of monitoring, controlling, and responding to physical events and phenomena. In CPS, computer-based algorithms monitor physical processes (like temperature, pressure, or speed) and adjust these processes in real-time based on the data collected. This could be anything from controlling the temperature in a building based on occupancy and weather data, to an autonomous vehicle adjusting its speed and direction based on road conditions and surrounding objects. In contrast, while IoT also connects the physical world to the digital world, the emphasis is typically more on data collection, remote control, and automation of everyday devices rather than tight real-time integration and control of physical processes. So, while both CPS and IoT bridge the gap between the physical and digital world, CPS often do so in a more tightly coupled, real-time, and control-oriented manner, making it more focused on the connection with the real world.

2.1.1.8 Compared to In-Cloud CPS

Consolidating and sharing physical services amongst applications, auto-scaling computing and communication, and developing and sustaining numerous in-cloud CPS dynamically at the same time are all new perspectives made possible by cloud computing [25].

In order to provide smarter services and enable better control decisions to respond to physical phenomena, a CPS application will acquire global visibility to all CPS on devices deployed across a wide area (for example, a region or a city). Additionally, CPS apps can exchange services with one another to enhance both functional and non-functional characteristics of systems, such as security, intelligence, and speed of service supply. In addition to this, heterogeneous services provided by several CPS can be combined to create new CPS ecosystems while collaborating with current applications.

Instead of the conventional idea of platform- or software-as-a-service, cloud computing becomes a metaphor for the acquisition of information by CPS as a service. By consolidating and sharing physical hardware among various applications, sensing and actuating resources on-demand, customizability through modular composition, rapid development and scalability, resiliency, and performance based on user needs, cloud computing enables the consideration of new features in the design and operation of CPS. The in-cloud CPS continues to develop in the following dimensions: collaborative sensing and actuating through distributed sensing and control; seamless and cross-layer integration of heterogeneous CPS services; and virtual representation of physical system components and services (such as sensing and actuation services). Multiple end-user applications can process intricate distributed coordination models simultaneously with in-cloud CPS development. In this setting, distributed control is especially interesting since it may provide fresh difficulties including dependencies between autonomous control entities and distributed algorithms for coordination.

The term "virtualization of CPS services" describes how CPS are represented virtually in terms of their various parts and functions, such as their sensors, actuators, controls, and processes. It also promises processing of data gathered from a potentially large number of physical devices in a scalable, reliable, and on-demand way. Access to those services and components through a pertinent service virtualization method is also promised.

2.1.2 Definitions of DDCPS

DDCPS are an evolution of traditional CPS. While CPS refer to systems that integrate physical processes with computational and networking capabilities, DDCPS extend this concept by incorporating data-driven approaches into the system design and operation. Specifically speaking, In CPS, the design process involves modeling the physical components and their interactions with each other and the environment. In DDCPS, data-driven methods introduce a more comprehensive and accurate understanding of these interactions which is achieved by collecting and analyzing real-world data from sensors, devices, and other sources. Instead of relying solely on theoretical models, DDCPS leverage real-world data to create more accurate models of the physical processes. Data-driven approaches allow for more informed design decisions. For instance, historical data can be used to optimize the placement of sensors, actuators, and other components, leading to improved system performance and resource efficiency. Real-world systems often face variability due to changing conditions. Data-driven methods enable the system to adapt and reconfigure itself based on incoming data. Once DDCPS are deployed, data-driven methodologies continue to play a crucial role in their operation and maintenance. DDCPS continuously collect and process real-time data that provide insights into the current state of the system from sensors and other sources. By analyzing historical and real-time data, DDCPS can predict potential issues or anomalies before they escalate. DDCPS can dynamically adjust their control strategies based on incoming data. In essence, DDCPS combine the physical world with data-driven intelligence to enhance system design and operation. By utilizing real-world data, it enables more accurate modeling, real-time optimization, predictive analysis, and adaptive decision-making, ultimately leading to more efficient, flexible, and resilient CPS.

In DDCPS, extensive data collected from the physical world is processed and analyzed to inform and improve the decision-making processes within the system. This often involves techniques from machine learning and artificial intelligence to extract insights and patterns from the data. These insights are then used to optimize the operation of the system, predict future behavior, and make the system more robust and resilient to changes in the environment. For instance, in an autonomous vehicle (which can be considered as DDCPS), real-time data from various sensors (e.g., lidar, radar, cameras) is constantly analyzed to understand the surrounding environment, predict the actions of other road users, and make decisions about how to control the vehicle. This is a data-driven process, as the decisions are based on the analysis of the sensor data rather than pre-determined rules or models. By bringing data-driven approaches into the realm of CPS, DDCPS have the potential to significantly improve the efficiency, reliability, and adaptability of these systems. However, they also present new challenges in terms of data management, real-time data processing, and the integration of machine learning and AI techniques into the system design and operation.

2.1.2.1 Physical World Perspectives

The concept of DDCPS revolves around the integration of computations and communication with physical processes, empowering physical systems with enhanced capabilities. These systems encompass embedded computer systems that form an essential component of the physical infrastructure, necessitating their rapid and accurate responses to intricate sequences of external events. These events originate from the "physical world" itself, serving as the driving force behind system behavior. As the environment evolves, the system must adapt accordingly, highlighting the criticality of accurately describing and analyzing the behavior of the physical world during the specification of embedded systems.

Insufficient scrutiny of the external environment during system development can result in systems that fall short of meeting the fundamental expectations of both users and developers. It is important to emphasize that computers, until now, lack inherent intelligence; their operations are limited to executing pre-defined instructions. Consequently, when confronted with an unfamiliar environment, computers encounter difficulties. Notably, the existing methodologies often conflate the analysis of the physical world's behavior with the internal behavior of the system, leading to potential misunderstandings and shortcomings in system design.

2.1.2.2 Function Perspectives

It is important to recognize that the fundamental aim of constructing DDCPS is to actualize the desired functionality as demanded by the user. Thus, we advocate adopting a function-centric perspective when defining the system's purpose. In general, functional requirements serve as a means to describe the algorithm's input-output behavior. Functional analysis encompasses a comprehensive multilevel hierarchical specification, which consistently results in a well-structured hierarchical design.

To facilitate a clear understanding and effective communication of the system's intended functionality, it is customary to generate functional diagrams. These visual representations provide a pictorial depiction of the desired functions that will be integrated into the candidate system. By aligning the algorithm requirements with these functional diagrams, a common practice in the field, we can effectively capture the essence of the system's purpose and ensure its successful implementation.

By emphasizing the significance of a function-oriented approach, we underscore the need to accurately define and align the desired functionality with the software requirements. This approach not only enables a thorough understanding of the system's objectives but also facilitates the creation of detailed functional specifications and a well-organized hierarchical design structure.

2.1.2.3 Data Perspectives

A data view captures the static relation between the object and data. Data view emphasizes the data structure of the system using objects, attributes, operations, and relationships. Data processing can be generally divided into four phases: data generation, data acquisition, data storage, and data analysis. Data generation is the first step of big data. As the second phase of the big data system, big data acquisition includes data collection, data transmission, and data pre-processing. Data collection is to utilize special data collection techniques to acquire raw data from a specific data generation environment. Upon completion of raw data collection, data will be transferred to a data storage infrastructure for processing and analysis. Big data storage refers to the storage and management of large-scale data sets while achieving reliability and availability of data accessing. The storage infrastructure needs to provide information storage service with reliable storage space; it must provide a powerful access interface for query and analysis of a large amount of data. The analysis of big data mainly involves analytical methods for traditional data and big data, analytical architecture for big data, and software used for mining and analysis of big data. Data analysis is the final and the most important phase in the value chain of big data, with the purpose of extracting useful values, providing suggestions or decisions.

2.1.2.4 Hardware Perspectives

The hardware perspective in defining DDCPS revolves around ensuring compatibility and optimization within a specific target hardware environment. When developing such systems, it is crucial for developers to make informed decisions regarding the hardware components to be used and how they should be employed to implement the given specifications. In light of this, we propose a hardware-oriented view that aids in enhancing the implementation process.

Existing methodologies often exhibit gaps between the specification, software design, and implementation stages, lacking in-depth details specifically addressing the implementation phase. However, in the context of multi-core trustworthy aerospace CPS, this implementation stage holds immense importance as it involves mapping the application onto a parallel configuration.

The adoption of a parallel configuration has become increasingly critical for numerous time-critical applications. This configuration primarily entails the allocation of tasks across multiple processors, aiming to enhance the system's response time. By leveraging the potential of parallel computing, the system can effectively distribute and execute tasks concurrently, resulting in improved performance and real-time responsiveness.

Emphasizing the hardware view allows for meticulous consideration of the target hardware environment, enabling developers to make informed decisions regarding hardware selection and utilization. Bridging the gaps between specification, software design, and implementation, this approach addresses the crucial stage of mapping the application onto a parallel configuration, thereby maximizing the potential of multi-core trustworthy aerospace CPS in time-sensitive scenarios.

2.2 Characteristics of DDCPS

We anticipate that DDCPS of the future will be significantly more advanced than those of the present in terms of their safety, interdisciplinary nature, real-time processing, data-driven decision-making, networked communication and so on.

DDCPS are complex systems where the characteristics including networked communication, scalability, heterogeneity, interdisciplinary, real-time processing and real-time decision-making are deeply interconnected as Fig. 2.6. Firstly, networked communication forms the backbone of these systems, enabling physical devices to connect and exchange data. The ability to communicate across the network directly influences the scalability of the system. The system can only expand by adding

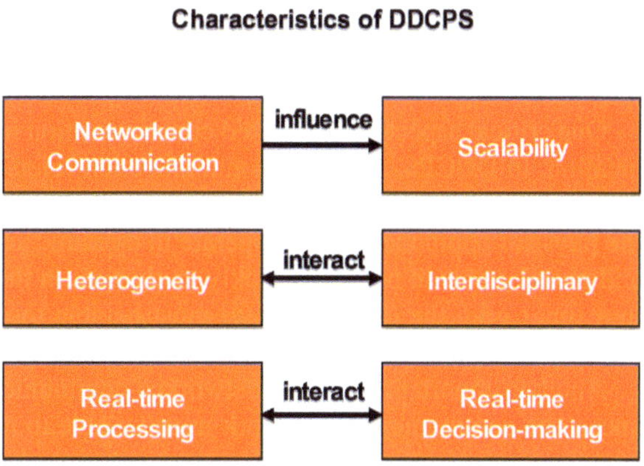

Fig. 2.6 Characteristics of DDCPS. Networked communication can influence scalability of DDCPS, heterogeneity interact with interdisciplinary so as real-time processing and real-time decision making

more devices if effective communication between devices is possible. Next, we consider heterogeneity. There may be a variety of devices in DDCPS, each with different functions, operating systems, and communication protocols. This heterogeneity adds complexity to the system but also makes them more powerful and flexible. Each device can provide unique value to the system based on its characteristics and capabilities. Then there's the interdisciplinary nature of DDCPS. It's a cross-disciplinary field involving computer science, electrical engineering, control theory, network science, and more. This interdisciplinary nature allows the system to be optimized and improved from multiple angles and levels. Finally, we have real-time processing and real-time decision-making. In DDCPS, data needs to be processed and analyzed in a short time for the system to make quick decisions. This real-time aspect is a crucial characteristic of CPS, enabling the systems to react swiftly to environmental changes, thereby improving efficiency and performance. In summary, these characteristics are interdependent and influence each other. They collectively contribute to the complexity and uniqueness of DDCPS.

2.2.1 Networked Communication

Networked communication is a critical feature of DDCPS that enables seamless communication between different components of the system. In DDCPS, the communication network is responsible for transmitting data from the physical components to the computational components, and vice versa. The data is then analyzed and processed by the computational components to make decisions and control the physical components.

Networked communication in DDCPS can take many forms, including wired or wireless networks, local area networks (LAN), wide area networks (WAN), and the Internet [26, 27]. Communication in DDCPS is typically bi-directional, meaning that data is transmitted from the physical components to the computational components, and then control signals are transmitted back to the physical components to adjust their behavior.

One critical aspect of networked communication in DDCPS is real-time communication. In DDCPS, real-time communication refers to the ability of the system to transmit data and control signals with minimal latency. This is essential in DDCPS, as the physical components often operate in dynamic and unpredictable environments and must be able to respond quickly to changes in their environment. For example, in a smart grid, the control system must be able to respond quickly to changes in electricity demand to ensure reliable and efficient operation.

Another important aspect of networked communication in DDCPS is fault tolerance. The communication network must be able to tolerate faults such as communication errors, hardware failures, or network congestion without compromising the overall performance of the system. This is critical in DDCPS, as any disruption in communication can lead to system failure and potentially hazardous conditions. For instance, in an automated manufacturing plant, a communication failure between the control system and a robotic arm could result in a safety hazard or product defects.

Networked control systems (NCS) are considered to be a prominent subcategory of DDCPS, in which networks provide a means for communication among sensors, actuators, and controllers [28, 29]. As networking technology continues to increase in bandwidth capability and decrease in price, NCS will become more pervasive and have the potential to radically change the way the physical systems interact with each other and their environment [30]. Communication networks have been used to exchange information among sensors, controllers, and actuators in digital control systems since the 1970s, originally motivated by the need of reducing costs for cabling, modularizing, and integrating system flexibly in the automobile industry. Today, many other applications of NCS can be easily found in diverse industries.

In summary, networked communication is a vital feature of DDCPS that enables the system to function as a cohesive and integrated entity. It provides real-time data transmission, facilitates control signal feedback, and ensures fault tolerance, making DDCPS reliable and safe [3].

2.2.2 *Scalability*

The scalability of DDCPS refers to the system ability to effectively handle increasing amounts of data, complexity, and resource demands while maintaining or improving the system performance, efficiency, and functionality [9, 31–35]. In other words, scalable DDCPS can accommodate growth without significant degradation in their capabilities or responsiveness. As the scale of DDCPS expands, the number of interconnected devices also increases. This escalation demands efficient communication protocols capable of managing data traffic, processing capabilities, and network bandwidth. Without effective scalability solutions, the performance of the system can degrade significantly, impacting the overall system efficiency.

The scalability of DDCPS is a crucial aspect that determines the system's ability to handle increasing volumes of data, adapt to growing demands, and maintain performance as the system expands. Achieving scalability in data-driven DDCPS involves several key considerations.

Firstly, storage scalability is essential to handle the growing amounts of data generated by sensors, devices, and other sources within the DDCPS. This requires the implementation of scalable storage solutions such as distributed databases, data lakes, or cloud-based storage systems. By leveraging these technologies, DDCPS can efficiently store and retrieve vast amounts of data, ensuring that the system can handle the increasing data requirements without compromising performance.

Secondly, processing scalability is vital to handle the computational demands of analyzing and deriving insights from large volumes of data. DDCPS often employ complex analytics algorithms and machine learning techniques to extract valuable information from the collected data. To achieve processing scalability, DDCPS can leverage distributed computing frameworks such as Apache Hadoop or Apache Spark. These frameworks enable parallel processing of data across multiple computing nodes, allowing the system to scale horizontally and handle the computational load efficiently.

Thirdly, communication scalability is crucial in DDCPS, as the system relies on seamless data exchange and communication between various components. As the system grows, the number of data sources, sensors, actuators, and control devices increases, necessitating a scalable communication infrastructure. This can be achieved through the use of scalable networking protocols, such as Message Queue Telemetry Transport (MQTT) or Advanced Message Queuing Protocol (AMQP), which enable efficient and reliable data transmission between distributed components [36, 37].

Achieving scalability in DDCPS often requires careful architectural design, efficient algorithms, distributed computing, cloud computing, and possibly leveraging technologies like edge computing to offload processing closer to data sources. Ensuring scalability is essential for DDCPS to remain effective and viable as it grows to accommodate larger datasets and more complex tasks.

Overall, achieving scalability in DDCPS is a multifaceted challenge that involves addressing storage, processing, communication, and analytics aspects. By employing scalable storage systems, distributed computing frameworks, scalable communication protocols, and flexible data processing pipelines, DDCPS can effectively handle increasing data volumes, computational demands, and evolving system requirements, enabling the system to grow and adapt to changing circumstances while maintaining optimal performance.

2.2.3 Heterogeneity

The heterogeneity characteristic of DDCPS refers to the diverse and varied nature of these components, data sources, and technologies involved within such systems.

In more specific terms, heterogeneity in DDCPS can be defined as follows: diversity of components, variety of data sources, heterogeneous data types, technological heterogeneity, interdisciplinary nature. DDCPS often comprises a wide range of components, including digital computing elements (such as processors, controllers, and software) and physical components (such as sensors, actuators, and physical infrastructure). These components may come from different manufacturers, utilize various communication protocols, and have distinct computational capabilities. DDCPS rely on data from a multitude of sources, including digital sensors, real-time data streams, historical databases, and external data feeds. The data from variety sources can vary in terms of format, frequency, and accuracy. The data processed and generated in DDCPS can vary widely, including structured and unstructured data, time-series data, images, and even video streams. Handling these different data types is a key aspect of managing heterogeneity. DDCPS may incorporate different technologies and platforms, such as IoT (Internet of Things) devices, cloud computing, edge computing, and various communication protocols (e.g., Wi-Fi, Bluetooth, radio-frequency identification (RFID)). These technologies often have distinct architectures and capabilities. DDCPS bridge the gap between the digital and physical worlds, making it inherently interdisciplinary. It combines principles from computer science, data analytics, control theory, communication systems, and physical sciences (e.g., physics, biology) to design and operate integrated systems.

The heterogeneity characteristic of DDCPS underscores the diversity and complexity inherent in these systems, where digital and physical components, diverse data sources, and various technologies must work together cohesively. Managing and harnessing this heterogeneity is a fundamental aspect of designing, deploying, and maintaining effective DDCPS solutions.

The continued expansion and development of DDCPS, characterized by their heterogeneity, has led to an increase in problems stemming from the interaction of various domains. These problems have caused a loss of compositionality, which has hampered the process of system integration within DDCPS [38]. Compositionality here implies the ability to seamlessly integrate different heterogeneous components into the system. Due to the heterogeneity of DDCPS, the traditional ad hoc approach to composition, where individual components are built and integrated, becomes more complex and challenging. Processes within DDCPS often run in an ordered manner and are represented as difference equations, reflecting the data-driven and computational aspects. However, the heterogeneity of DDCPS extends beyond just computational components; it also encompasses physical components, such as sensors and actuators. This diverse set of components and their interactions introduces complexity that needs to be addressed [39]. Abstractions are essential in handling the heterogeneity of DDCPS and supporting necessary analysis and synthesis tasks. These abstractions must capture the dynamic behaviors of both digital and physical components accurately. The range of necessary physical behaviors, such as power consumption and size, which are inherent to the physical domain, also adds to the complexity of these abstractions. Furthermore, modeling interactions between components within DDCPS necessitates the development of specialized abstractions. The abstractions used for modeling interactions in digital communication systems are fundamentally different from those used to model interactions between physical objects, highlighting the challenge posed by the heterogeneity of DDCPS [40].

2.2.4 Interdisciplinary

The interdisciplinary nature of DDCPS refers to the fact that DDCPS research and development involves multiple academic disciplines and fields of study, including computer science, electrical engineering, mechanical engineering, control theory, optimization, communication systems, and more [41–46]. DDCPS are interdisciplinary in nature, combining elements of computer science, electrical engineering, and mechanical engineering. The interdisciplinary nature of DDCPS research and development requires collaboration and communication among researchers and practitioners from different disciplines. It also presents challenges in developing shared terminology, methods, and tools to facilitate effective collaboration. However, the benefits of interdisciplinary research and development are significant, leading to more innovative and effective solutions to complex problems.

DDCPS are complex systems that involve the integration of physical systems with computational and communication systems. This integration requires expertise from multiple disciplines to develop efficient and effective solutions. For example, developing an autonomous vehicle involves expertise in mechanical engineering to design the physical components of the vehicle, as well as computer science and control theory to develop the algorithms and control systems that govern the vehicle's behavior. Furthermore, DDCPS often require new approaches to problem-solving that transcend traditional disciplinary boundaries. For example, designing a data-driven cyber-physical system to optimize energy efficiency in a smart home requires a multidisciplinary approach that combines knowledge of building design, energy systems, communication systems, and data analytics.

2.2.5 Real-Time Processing

DDCPS require real-time processing of data to respond to changes in the physical environment. Real-time processing is a critical characteristic of DDCPS that enables them to respond quickly and effectively to changes in the physical environment. Real-time processing involves the ability to collect, process, and analyze data in near real-time, typically within milliseconds to seconds.

DDCPS rely on sensors and other physical data sources to continuously monitor the physical environment. For example, in a smart building data-driven cyber-physical system, sensors may monitor temperature, humidity, light levels, and occupancy levels. In an autonomous vehicle data-driven cyber-physical system, sensors may monitor the vehicle's speed, acceleration, and proximity to other vehicles or obstacles. The data collected by these sensors is processed in real-time to generate insights and inform decision-making. In a smart building data-driven cyber-physical system, real-time data processing can enable the building's heating, ventilation, and air conditioning (HVAC) system to adjust automatically to maintain optimal temperature and humidity levels. In an autonomous vehicle data-driven cyber-physical system, real-time data processing can enable the vehicle to adjust its speed and direction to avoid collisions with other vehicles or obstacles.

Real-time processing plays a pivotal role in DDCPS, demanding sophisticated algorithms and substantial computational resources for swift data analysis and interpretation. This often necessitates the application of machine learning and artificial intelligence techniques to discern patterns, anomalies, and make predictions regarding future events. In DDCPS, real-time processing extends beyond mere data transmission and storage, serving as a means to intelligently harness the wealth of data available.

DDCPS constitute a multidisciplinary field that integrates physical systems with computational systems, interconnected through sensors and communication networks. These systems collect diverse environmental data, encompassing variables such as temperature, humidity, pressure, and location, sourced from domains like industrial equipment, transportation systems, and energy management systems. Achieving real-time processing within DDCPS is essential for timely decision-making, anomaly detection, and the optimization of system performance.

2.2.6 Real-Time Decision-Making

DDCPS use data generated by physical sensors and computational models to make decisions. Real-time decision-making is a key characteristic of DDCPS that enables them to make decisions based on real-time data collected from physical systems. This decision-making approach involves collecting, processing, and analyzing large amounts of data from sensors and other sources in order to derive insights that inform decision-making. The acquisition and subsequent conversion of data into actionable insights are fundamental for the creation of immediate metrics. These real-time metrics afford businesses a competitive edge, as they amplify computation power and expedite both decision-making processes and failure detection. Real-time decision-making in DDCPS requires advanced data analytics tools and techniques, including machine learning and artificial intelligence algorithms. These tools enable DDCPS to automatically detect patterns in data, make predictions, and identify anomalies that may indicate potential issues with physical systems.

In DDCPS, real-time decision-making is used to optimize the performance of physical systems, improve efficiency, and enhance safety [47]. For instance, in an autonomous vehicle data-driven cyber-physical system, instant decisions can enhance route optimization and adapt vehicle speed according to traffic dynamics and road states. In the context of industrial DDCPS, real-time decision-making plays a pivotal role in identifying equipment malfunctions and taking prompt actions to avert accidents while reducing operational interruptions.

2.3 Components of DDCPS

2.3.1 *Sensing Components*

Sensing components in DDCPS encompass a wide range of devices and technologies that enable the system to perceive and capture data from the surrounding physical environment [48]. These components are designed to gather information from various sources, including physical measurements, environmental conditions, and interactions. One of the key aspects of sensing components is their ability to detect and measure physical parameters accurately and reliably. This involves employing different types of sensors, such as temperature sensors, pressure sensors, humidity sensors, accelerometers, gyroscopes, proximity sensors, and optical sensors, among others. Each sensor type specializes in detecting specific physical quantities or phenomena, allowing for comprehensive data collection. Furthermore, sensing components can also incorporate advanced techniques such as computer vision and image processing algorithms. These techniques enable the system to perceive visual information, recognize objects, track movements, and extract relevant features from images or video streams. To ensure seamless integration with the CPS, sensing components are often designed to be energy-efficient and capable of real-time data acquisition. They may employ wireless communication protocols to transmit data to the computational components for further analysis and decision-making. This enables the system to continuously monitor the physical world, providing a timely and accurate representation of the environment. The data captured by sensing components serves as the foundation for data-driven decision-making and control in CPS. It provides valuable insights into the behavior of the physical system, facilitating adaptive and intelligent responses to changing conditions. By leveraging the data from sensing components, CPS can optimize energy consumption, improve operational efficiency, enhance safety, and enable innovative applications in various domains such as smart cities, industrial automation, healthcare, and transportation. Overall, sensing components form an integral part of DDCPS, enabling the system to sense, perceive, and interact with the physical world [49]. Their accurate and timely data acquisition capabilities lay the foundation for intelligent decision-making and control, driving the advancements and applications of CPS.

2.3.2 *Computational Components*

Computational components play a crucial role in DDCPS by processing and analyzing the data collected from the sensing components. These components encompass various computing devices, algorithms, and software systems that enable intelligent decision-making and control actions based on the captured data. The computational components receive the raw data from the sensing components and apply a range of computational techniques to transform and extract meaningful information from it.

This includes tasks such as data pre-processing, feature extraction, pattern recognition, statistical analysis, machine learning, and data fusion. By leveraging these techniques, computational components enable the DDCPS to understand and interpret the data in real-time, facilitating insights and knowledge generation [50]. One of the key aspects of computational components is their ability to handle large volumes of data and perform complex computations efficiently [51]. This often involves utilizing high-performance computing architectures, parallel processing, distributed computing, and cloud computing resources. These technologies ensure the scalability, speed, and accuracy required for real-time analysis and decision-making in DDCPS. Computational components also involve the development and implementation of software systems and algorithms tailored to the specific DDCPS application [52]. This may include developing models, simulations, optimization algorithms, control algorithms, and decision support systems. These software components enable the DDCPS to optimize operations, predict future states, detect anomalies, and make informed decisions based on the analyzed data. The integration of computational components with sensing components forms a closed-loop feedback system in DDCPS. The computational components continuously analyze the data from the sensing components, extract valuable insights, and generate control signals or recommendations. These control signals are then communicated to the communication and control components to enable the system to adapt, adjust, or automate physical processes accordingly.

2.3.3 Communication Components

Communication components are an integral part of DDCPS, facilitating the seamless exchange of data and information between different system components. These components encompass various technologies, protocols, and networking infrastructure that enable reliable and efficient communication within the CPS ecosystem. The primary function of communication components is to enable the transfer of data between the sensing, computational, and control components of DDCPS. They establish the necessary connections and channels for transmitting data, commands, and feedback across the system. This enables real-time collaboration, coordination, and synchronization between different components, contributing to the overall functionality and performance of the DDCPS. Communication components employ a range of technologies and protocols, including wired and wireless communication [53]. On the one hand, wired communication technologies, such as Ethernet or fieldbus systems, provide reliable and high-bandwidth connections within localized areas. On the other hand, wireless communication technologies, such as Wi-Fi, Bluetooth, or cellular networks, enable flexible and mobile connectivity over larger distances [54]. To ensure the secure and efficient transfer of data, communication components often incorporate encryption, authentication, and error detection mechanisms. These measures protect the integrity and confidentiality of the transmitted data, safeguarding the DDCPS against unauthorized access and tampering [55]. In addition to inter-component communication, communication components also enable connectivity with external systems or networks. This allows DDCPS to integrate with external data sources, cloud platforms, or other DDCPS instances, enabling collaborative decision-making and sharing of resources. The choice of communication components and protocols depends on various factors such as the system's requirements, scalability, reliability, latency, bandwidth, and security. It is crucial to carefully design and configure the communication infrastructure to meet the specific needs of the DDCPS application. Overall, communication components form a vital backbone of DDCPS, enabling seamless and reliable data exchange between different system components, fostering real-time collaboration, and facilitating integration with external systems.

2.3.4 Control Components

Control components play a vital role in DDCPS by enabling the system to regulate and manipulate physical processes based on the data analysis and decision-making outcomes. These components encompass algorithms, controllers, actuators, and other mechanisms that ensure the system's behavior aligns with desired objectives and requirements. The control components receive inputs from the computational and sensing components, which provide insights and information about the current state of the physical system. Based on this input, control algorithms and strategies are employed to generate control signals that dictate the actions of actuators, influencing the physical processes in real-time [56]. Control components can encompass various types of control techniques, ranging from traditional control methods such as proportional-integral-derivative (PID) control to more advanced approaches such as model predictive control (MPC), fuzzy logic control, or adaptive control [57]. The selection of the control technique depends on the system's complexity, dynamics, and desired performance criteria. Actuators are a crucial part of control components as they convert control signals into physical actions. These can include motors, valves, pumps, switches, or other devices that influence the physical system. Actuators are responsible for manipulating variables such as speed, position, temperature, flow rate, or pressure to achieve the desired control objectives. In DDCPS, control components often leverage real-time data analysis and feedback loops to continuously monitor and adjust the system's behavior [58]. By integrating sensing, computational, and communication components, control components enable closed-loop control systems that can adapt and respond to changing conditions in real-time. The design and implementation of control components involve considerations such as stability, performance, robustness, safety, and energy efficiency. Control engineers utilize mathematical models, system identification techniques, and simulation tools to develop and optimize control algorithms that meet the system's requirements.

2.4 Examples of DDCPS in Different Industries

Automobiles, aeroplanes, the electrical grid, manufacturing facilities, medical systems, and buildings are just a few examples of the many engineered systems on which modern civilization relies heavily. DDCPS have the potential to revolutionise how we interact with, operate, and design them.

2.4.1 Smart Grids

Reliable access to energy infrastructure serves as the bedrock of our industrialized societies. Traditionally, the electric grid has been characterized by a relatively small number of high-volume production facilities on one end and a vast array of generally low-to-medium volume consumer installations on the other. This grid operates with a demand that varies by a factor of four between its daily lows and peaks [59]. However, the landscape of energy production is rapidly evolving, with a growing reliance on renewable energy sources. These renewable sources are typically harnessed from a larger number of facilities, each with volatile output volumes that often do not align with the prevailing consumption patterns. Consequently, the conventional asymmetric and centralized management model of the electric grid is becoming progressively inadequate [60].

The new technical emphasis in DDCPS for a smart grid offers a comprehensive solution by enabling decentralized and cooperative coordination of both technical and organizational processes [61]. The transformation extends from controlling individual photovoltaic installations to managing the intricacies of energy billing and trading. This shift toward decentralization empowers the creation of microgrids, which are essentially local grids capable of functioning independently or in conjunction with the main grid. In parallel, automation technologies, including the use of smart meters at a granular level, play a pivotal role in monitoring and predicting energy consumption and renewable energy production. The fine-grained oversight facilitates the reliable short-term balancing of supply and demand. Moreover, integrating intelligent devices and managed installations, such as low-volume energy buffers like batteries, supports a transition from a supply-centric to a demand-side management approach for the grid. The highly automated control mechanisms inherent in DDCPS enable the scalability of these processes to accommodate a growing number of participants. The scalability ensures that the system can adapt to varying needs seamlessly. Additionally, the life-cycle integration feature allows for continuous updates and enhancements to the highly automated control system across the entirety of the smart grid. Inefficiencies in local control can be readily identified and replaced, while additional data and distributed sensing capabilities can be integrated modularly.

In a world, where further development and utilization of technological and organizational advancements are pursued, enhancing the overall performance and efficiency of smart grids within the framework of DDCPS presents additional opportunities for tailored solutions and fine-tuning by both manufacturers and users of DDCPS-based smart grids.

2.4.2 *Agriculture*

The research in the field of DDCPS has immense potential in contributing towards the growth and progress of agriculture [62–64]. DDCPS offer various solutions for enhancing the overall efficiency and productivity of food production and consumption through the integration of innovative technologies such as precision agriculture, intelligent water management, and efficient food distribution. With the increasing demand for food globally, the importance of maximizing agricultural productivity has become more pressing than ever. Fortunately, DDCPS research provides us with a solution to this challenge by utilizing cutting-edge technologies to optimize food production, minimize waste, and reduce costs. By adopting precision agriculture, farmers can obtain critical data about crop growth patterns and make informed decisions about resource utilization, thereby increasing their yield and profitability. Moreover, the intelligent management of water resources through DDCPS can improve the efficiency of water usage in agriculture. By utilizing real-time data and advanced analytics, farmers can optimize irrigation scheduling, reduce water loss, and enhance crop quality. This not only saves water resources but also helps to protect the environment.

Agricultural operations are intricate and multifaceted. Caramihai and Dumitrache suggest a DDCPS control methodology that utilizes agent-based behavior to ensure agricultural proactivity in response to environmental and market fluctuations [65]. They contend that implementing real-time control services in farming can aid in addressing the growing disparity between food demand and supply. Mehdipour et al. have presented a comprehensive technological approach to manage rodents using DDCPS [66]. Their work has led to the development of an advanced "smart pest control" solution (SPeC), which includes a "rat detection system" (RDS) designed to monitor rat populations in agricultural areas [67]. By implementing this system, significant cost savings can be achieved in rodent control measures, as well as a reduction in crop losses and environmental pollution.

In essence, DDCPS research offers a comprehensive approach towards enhancing food production and consumption efficiency in agriculture. The potential of DDCPS for revolutionizing the industry is enormous, and by implementing approaches combined with DDCPS can lead to a sustainable and secure future for the agricultural sector.

2.4.3 Healthcare

DDCPS have emerged as a crucial player in the healthcare industry, enabling us to enhance our ability to provide high-quality patient care. A wide range of research studies has been conducted in the field of healthcare, with a particular focus on the development of smart sensor systems that can monitor and provide real-time updates on patients' health conditions and provide alerts as necessary [68]. In addition, the research also emphasizes the design and development of tele-medicine systems that can provide remote healthcare services to patients, thereby making healthcare more accessible and convenient. Furthermore, the use of semi-autonomous tele-operated home service robots is also being explored, which can aid patients with their daily physical activities, thereby improving their quality of life. By investing in DDCPS research, we can leverage technology to improve healthcare outcomes and provide more personalized, efficient, and effective care to patients.

DDCPS have found diverse applications within the healthcare sector, ranging from media devices designed to enhance the efficiency of medical procedures and surgeries, to remote services grounded on data acquisition [69]. A substantial shift in automation levels has occurred, transitioning from primarily passive devices, which were under human control, to devices enabled by modern information technology [70]. The integration of medical cyber physical systems (MCPS) has become more prevalent in hospitals due to their ability to offer constant, high-quality care for patients [71]. The design of MCPS must be approached with caution as there are several challenges to consider. These challenges include the need for context-aware intelligence to ensure that the system functions in accordance with the patient's specific needs, autonomy to enable MCPS to operate independently without human intervention, and security and privacy to protect patient data and information. Additionally, device certifiability is also crucial to ensure that MCPS adhere to strict safety regulations. Despite these challenges, the benefits of MCPS make them a valuable tool for enhancing the quality of care in hospitals, and ongoing research in this field aims to overcome these obstacles and develop safe and effective MCPS.

A MCPS includes two kinds of devices: monitoring devices and actuation devices as Fig. 2.7 shows. Monitoring devices offer various forms of patient data that are essential to the clinic, like sensors and bedside heart rate and oxygen level monitors. Actuation devices offer data on actuation therapy-ventilators and immersion pumps, for example-that can alter the patient's physiological condition.

Interoperability of medical devices offers the potential to lower health care costs while enhancing patient safety and treatment outcomes. In order to accomplish interoperability, medical devices (software, applications) and other equipment share the same information model and communication protocol. Any compatible data, from any source—including the manufacturer or a particular device type—can be used with them. A system is referred to as a distributed device if it makes use of several kinds of medical equipment. User requirements are utilized for validation, whereas technical requirements are employed for verification. MCPS or healthcare systems include intelligent operating rooms and hospitals, image-guided therapy and surgery,

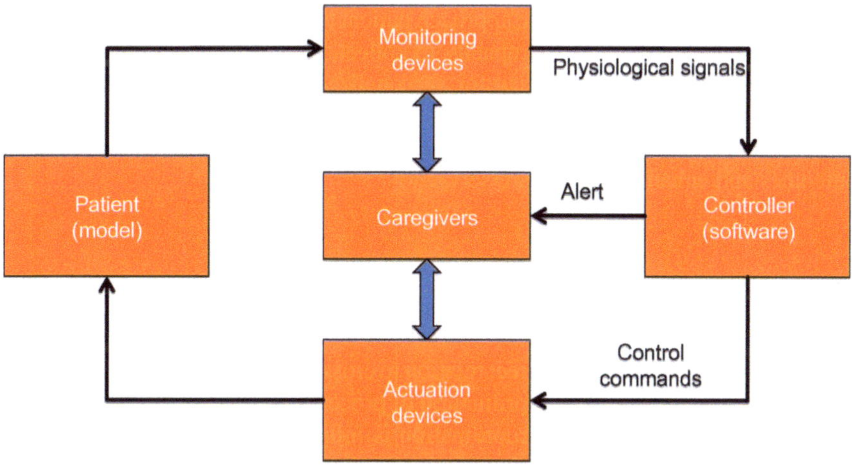

Fig. 2.7 In MCPS, a patient is regarded as a model, and controller receives physiological signals via network to send alert to caregivers. Caregivers receive alert to interact with monitoring devices and actuation devices. Moreover, controller will also give control commands via network to actuation devices

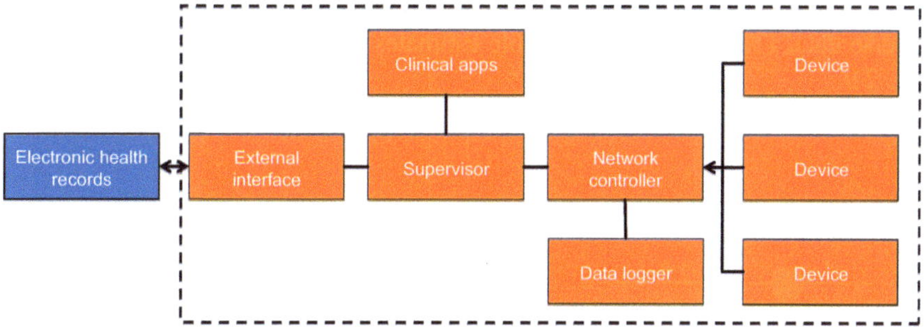

Fig. 2.8 Applications that operate on a supervisor and communicate with medical equipment are used by clinicians. Through the use of adapters or built-in equipment interfaces, patient-linked medical devices and other equipment are connected to the system. Similar to an airplane's black box, the network controller connects the devices to the Supervisor and transmits data to the data logger

fluid flow control for pharmaceuticals and biological testing, and the creation of both physical and neural prosthetics as Fig. 2.8.

Newly developed healthcare devices and equipment now possess the capability to actively manage essential physiological processes and functions. These capabilities are built upon embedded computing, sensing, modeling, and communication, and their deep integration with physical elements and processes. The high level of integration facilitates a new generation of DDCPS that are able to reach unprecedented levels of functionality, adaptability, and effectiveness, far surpassing what was achievable with the traditional passive systems. These DDCPS healthcare devices are transforming not only how we approach patient treatment, but also how we gather and utilize medical data. Their capacity to collect, analyze, and act upon real-time data is creating new opportunities for personalized, effective healthcare delivery. The predictive nature of these systems is paving the way for preventative care and early detection of potential health issues, ultimately contributing to improved patient outcomes and enhanced healthcare services.

2.4.4 Intelligent Transportation

Transportation systems have a direct relation to cross-domain integration [72]. Transportation services require the coordination of processes across sectors like logistics, automotive, and rail [73]. The coordination is influenced both by the corresponding vehicles and infrastructural components (e.g., roads and communications), with differences including speed, capacity, cost, and governance [74].

Intelligent transportation systems (ITS) embody DDCPS driven by big data [75]. These systems and infrastructures are responsible for effectively managing extensive datasets. The datasets incorporate a diverse range of traffic information elements, including geographical location, speed, count, and behavior patterns. By aggregating diverse data, ITS can generate reliable estimates of traffic conditions as shown in Fig. 2.9.

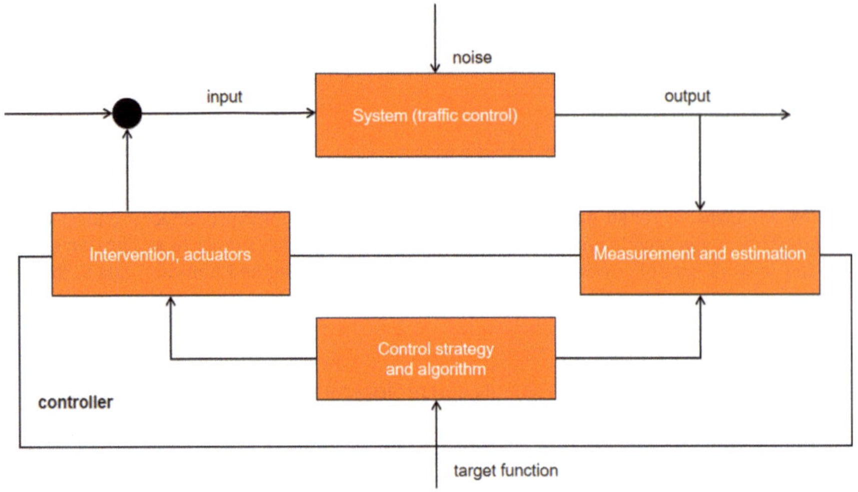

Fig. 2.9 An intelligent transportation system receives an input and gives an output. The system includes traffic control, which will be influenced by noise, intervention, actuators, measurement and estimation. Moreover, on the basis of measurement and estimation, the controller comprises control strategy and algorithm to control actuators

While most DDCPS domains share similarities in the challenges they face, there are several unique problems linked to transportation DDCPS [74]. The customization and variability of system architecture weaves a level of complexity rarely observed. The ultra-competitive global landscape mandates ever-evolving requirements for enhanced capabilities, resulting in the need to rapidly adapt systems. The changing landscape leads to an increased importance of cross-cutting aspects such as safety and security; and the evolving need for life-cycle integration to allow for pattern analysis of accident statistics, continuous updates to vehicles to remove defects or avoid unsuitable driver behavior, etc. In addition to the competition in the pursuit of vehicle consumers, the global competition at the component level is equally fierce, resulting in a diverse and constantly evolving set of component suppliers that must provide products to be integrated into the whole.

While many DDCPS domains encounter common challenges, transportation DDCPS presents a unique set of issues [74]. Its intricate system architecture, characterized by customization and variability, introduces a level of complexity rarely seen elsewhere. In this ultra-competitive global landscape, the demand for enhanced capabilities continually evolves, necessitating rapid system adaptation. The dynamic nature of this field accentuates the significance of cross-cutting elements, such as safety and security. Furthermore, the evolving landscape underscores the growing importance of life-cycle integration. This integration facilitates pattern analysis of accident statistics and enables ongoing updates to vehicles to eliminate defects and prevent inappropriate driver behavior, among other benefits. Beyond the fierce competition for vehicle consumers, there is a global competition at the component level. This competition results in a diverse and perpetually changing pool of component suppliers, each striving to provide products that seamlessly integrate into the larger system.

Intelligent transportation systems (ITS) embody DDCPS driven by big data [75]. These systems and infrastructures are responsible for effectively managing extensive datasets. The datasets incorporate a diverse range of traffic information elements, including geographical location, speed, count, and behavior patterns. By aggregating diverse data, ITS can generate reliable estimates of traffic conditions.

DDCPS leverage large-scale data analysis to identify traffic flow patterns and predict congestion and traffic conditions, which enables intelligent optimization of traffic signals and routing algorithms, reducing congestion and travel times. DDCPS integrate real-time data from various sources, including sensors and cameras, to monitor traffic conditions. In case of accidents or incidents, the system can promptly detect and alert authorities, facilitating faster response times. By analyzing historical and real-time data, DDCPS can offer dynamic route planning to drivers. It takes into account current traffic conditions and suggests the most efficient routes to reach destinations, enhancing overall travel efficiency.

DDCPS address the challenge of integrating diverse components and systems within the intelligent transportation framework by facilitating seamless communication between vehicles, infrastructure, and control centers, ensuring effective coordination. DDCPS handle massive data streams from various sources, enabling quick decision-making and responsive actions. ITS must adapt swiftly to changing scenarios, such as accidents, road closures, or sudden traffic spikes. DDCPS employ real-time data to adjust traffic management strategies and ensure optimal flow. With extensive data collection comes the responsibility to ensure data security and protect user privacy. DDCPS employ robust encryption and authentication measures to safeguard sensitive transportation-related data. As urban populations grow and transportation networks expand, DDCPS must scale to accommodate increased data volume and user demands. Scalable architectures and cloud-based solutions help addressing this challenge.

2.4.5 *Smart Manufacturing*

The concept of intertwining the digital and physical realms is increasingly gaining acceptance among manufacturers [76]. Several strategic initiatives have emerged in this regard, including Germany's Industry 4.0, the Industrial Internet in the United States, and the Made in China 2025 movement [77]. These programs advocate for the integration of contemporary information technologies within the manufacturing sector, propelling the evolution of smart manufacturing. Smart manufacturing's objective is to transform data collected throughout the product life cycle into actionable intelligence, positively influencing various facets of manufacturing [78]. The present-day manufacturing industry is witnessing an unprecedented surge in the volume of data generated by manufacturing systems [79]. Systematic and computational scrutiny of this manufacturing data will pave the way for more insightful decision-making, thereby augmenting the effectiveness of smart manufacturing [80]. Thus, it can be stated that data-driven manufacturing is a prerequisite for the realization of smart manufacturing. Accordingly, DDCPS are emerging as a pivotal tool to bolster manufacturing competitiveness [81]. It's evident that manufacturers are increasingly acknowledging the strategic significance of data in their operations.

The manufacturing sector across the globe is currently grappling with an array of challenges such as evolving customer demands, escalating labor expenses, skyrocketing costs of energy resources, environmental contamination, and increasing uncertainties [82]. In an effort to tackle these problems, multinational manufacturing corporations and industrialized nations are pioneering the progression of the Smart manufacturing system (SMS) [83].

The application of DDCPS in manufacturing facilities facilitates an enhanced product development process and comprehensive control over the production system [84]. This enhancement and control are achieved through the real-time exchange of all necessary production information, leveraged by IoT.

Industry 4.0 utilizes cutting-edge technologies and intelligence to drive the digitization of the manufacturing industry. The smart factory is a central concept in Industry 4.0, which refers to a manufacturing system that integrates processes and operations through DDCPS [85]. The smart factory is a crucial element in Industry 4.0

initiative, which aims to achieve intelligent manufacturing through interconnected production systems and vertical integration of production processes [86].

Utilizing robotics, automation, embedded systems, and information systems, the smart factory plays a vital role in achieving Industry 4.0 objectives. This concept marks a transformation from conventional to intelligent manufacturing, paving the way for digital manufacturing and digital twin models. To achieve this transition, emerging technologies such as DDCPS are essential [87]. The smart factory utilizes advanced technology and manufacturing models to achieve self-optimized performance, adaptability to changing circumstances, and real-time learning, enabling highly flexible, efficient, and customizable production processes that operate autonomously.

To institute these DDCPS in manufacturing enterprises, it's crucial to integrate legacy systems and engineering tools with cloud computing and IoT technologies. Moreover, these advanced systems must empower real-time information exchanges across various levels of business operations, from the shop-floor level to the strategic business level. This holistic connectivity will thus enable optimized production processes and improved decision-making capabilities.

2.5 Challenges of DDCPS

DDCPS have far higher economic and societal potential than has been recognised, and significant investments are being made globally to advance the technology. At the same time, DDCPS encounter many challenges arising from their own characteristics as Fig. 2.10.

DDCPS exhibit a strong interdependence between the physical components and the underlying information and communication layers. This interdependence results in unique security vulnerabilities that are distinct from those found in traditional networked systems [88]. The physical components of such systems introduce safety and reliability criteria that are qualitatively different from those in general-purpose computing, which poses significant hurdles. Additionally, physical and object-oriented software components differ fundamentally from one another. Traditional abstractions built on method calls and threads are ineffective [48].

Cyber attacks and the potential damage caused by malfunctions and failures of physical control systems are coupled in DDCPS [89]. By combining safety-critical control systems with information systems that are security-critical, we have been placing computer systems in a risk environment that is greater than the sum of the hazards of each of their individual components [90]. For instance, industrial control systems have generally been isolated from public networks and are not vulnerable to Internet-based cyber attacks. As a result, physical systems are typically more vulnerable than information systems because they were never intended to handle such remote attacks. Furthermore, the adverse effects of cyber attacks and malfunctions are

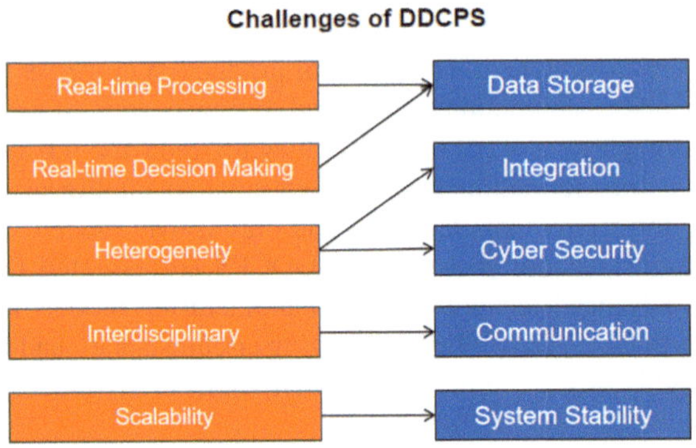

Fig. 2.10 The challenges that DDCPS face can be categorized into five groups, which are data storage, integration, communication, cyber security, system stability individually. Specifically, the real-time processing and real-time decision-making capabilities of DDCPS lead to challenges of data storage. The heterogeneity of DDCPS presents integration challenges. The interdisciplinary nature of DDCPS involves experts from various fields, which can lead to communication challenges. The heterogeneity of DDCPS can lead to significant challenges in terms of cyber security. While scalability is a desirable feature, it can introduce challenges related to system stability

worsened by the failure of cyber system designers to account for potentially harmful failures, as well as the omission of potential cyber attack pathways to control systems by control system designers.

2.5.1 Data Storage

DDCPS generate massive amounts of data, including sensor readings, control signals, and system logs. Storing and managing such large volumes of data requires scalable storage solutions capable of handling high data ingestion rates [91]. Data in real-time or near real-time, require storage systems that can handle high data ingestion rates and provide low-latency access [32]. Storage systems need to support different data formats and provide mechanisms for efficient indexing and querying across these diverse data types [92]. Traditional storage architectures may struggle to keep up with the rapid data flow. DDCPS often require integrating data storage with analytics frameworks for real-time or batch processing. Storage systems need to provide seamless integration with analytics tools, such as distributed processing frameworks, to enable efficient data processing and analysis [93]. As DDCPS operate in critical domains where system failures can have severe consequences, Storage systems must be fault-tolerant and reliable, ensuring data durability and availability in the presence of hardware or software failures.

DDCPS rely on accurate and trustworthy data for decision-making. Ensuring data veracity involves addressing issues like data integrity, security, privacy, and provenance. Storage systems need to provide mechanisms for data validation, authentication, and access control. Some data may be required for real-time analysis and decision-making, while other data may need to be stored for long-term analysis, compliance, or audit purposes.

Real-time processing in DDCPS involves the continuous collection, transmission, and analysis of data from various sensors and devices in the physical world. This data is often generated at high frequencies and must be processed quickly to provide timely insights and control actions. This real-time processing generates a significant amount of data at a rapid pace. The real-time processing capability of DDCPS leads to challenges of data storage contain data volume, data velocity, data latency. To be specific, the high frequency of data generation results in a massive volume of data being produced, which requires substantial storage capacity. Data is generated and processed at a high velocity, necessitating efficient data ingestion and storage mechanisms to keep up with the pace of incoming data. Real-time processing demands low latency in data storage and retrieval to ensure that decisions and actions can be taken promptly. Any delays in data storage can lead to suboptimal or even unsafe system behavior.

Real-time decision-making in DDCPS involves analyzing incoming data streams to make immediate decisions that affect the physical processes or systems being monitored. These decisions often rely on historical data, predictive models, or predefined rules. Challenges related to data storage caused by real-time decision-making include data accessibility, data integration, data retention. Specifically, decision-making algorithms require rapid access to historical data for comparison, pattern recognition, and forecasting. Efficient data retrieval becomes crucial for timely decision-making. Real-time decisions may involve combining data from various sources and time frames. Effective data integration mechanisms must be in place to support these decision processes. Storing historical data for future analysis, auditing, or compliance purposes is essential. However, this can result in the accumulation of large volumes of data over time, posing challenges in terms of long-term storage and data management.

In a word, the rapid data generation and the need for quick access in DDCPS demand robust data storage solutions to handle the volume, velocity, and timely access requirements, especially for supporting real-time processing and decision-making.

2.5.2 Integration

As DDCPS grow more complex and include multiple subsystems and components, the issue of heterogeneity becomes more pronounced. Heterogeneity in DDCPS may include diversity in system components, protocols, data structures, and more.

The diversity and differences in the characteristics, capabilities, and behaviors of components in DDCPS lead to integration challenges.

One major challenges in DDCPS are the integration of heterogeneous systems. Different subsystems or components may have been developed independently using different standards, protocols, and architectures. This can make it difficult to integrate them into a single DDCPS, and may require significant efforts in terms of interoperability and compatibility [94].

To be more specific, challenges in the integration of heterogeneous systems can be reflected in the following aspects: data integration, communication protocols, hardware and software differences, diverse sensors and actuators, scalability and evolution, fault tolerance and resilience. DDCPS components may generate data in various formats and with different data rates. For instance, sensors might produce data in different units or communication protocols. Integrating this heterogeneous data into a unified format that can be processed and analyzed cohesively can be challenging [95]. Cyber components may use different communication protocols, which may not be compatible. Ensuring seamless communication between these components requires bridging or translating protocols, adding complexity to the integration process. The physical components in DDCPS may have varying hardware specifications and run different software. Integrating software applications across these diverse hardware platforms can be challenging, as they may require custom adaptations or interfaces. Sensors and actuators in DDCPS may have different measurement accuracies, response times, and capabilities. Ensuring that these diverse sensors and actuators work together effectively and produce coherent control actions can be a significant integration challenge. DDCPS are often designed to be scalable and adaptable to evolving requirements. Managing the heterogeneity of components while adding or modifying them over time can be a complex task, as it requires maintaining compatibility and coherence in the system. Heterogeneity can also impact fault tolerance and system resilience. Ensuring that the system can gracefully handle failures or unexpected behaviors in various components, especially when they differ significantly, is a challenge in DDCPS integration.

2.5.3 Communication

Reliable communication is vital for DDCPS [96]. The time-sensitive nature of DDCPS demands low-latency communication. Real-time or near-real-time responses are often necessary, particularly in control systems like automated vehicles or industrial automation, where delays can lead to hazardous situations [97]. The nature of these systems often means they're used in critical applications where any disruption in data flow can lead to system failure, compromising safety or causing considerable financial losses. As such, it is crucial to develop communication protocols with robust error detection and correction features.

With the sheer amount of data generated, transmitted, and stored by DDCPS, secure communication channels are critical to protect against unauthorized access and data breaches [98]. This involves encrypting data in transit, managing access controls, and regular security updates. Privacy is another concern because DDCPS often handle sensitive information, necessitating measures to anonymize data and protect user identities.

DDCPS are characterized by their interdisciplinary nature, which means they draw upon knowledge and expertise from multiple fields such as computer science, engineering, data science, and domain-specific domains like healthcare or transportation. This interdisciplinary aspect leads to several communication challenges within DDCPS: diverse terminology, varying Goals and objectives, interdisciplinary collaboration, complex problem-solving, decision-making. To be specific, different disciplines often use their own jargon and terminology. When experts from various fields collaborate on a DDCPS project, misunderstandings can occur due to the different meanings attached to similar terms. Effective communication requires reconciling these differences and establishing a common language. Each discipline may have its own priorities and objectives. Engineers may focus on system reliability, while data scientists prioritize predictive modeling. These varying goals can lead to misalignment and conflicts in decision-making and resource allocation. Effective collaboration among experts from diverse backgrounds can be challenging. Communication barriers, differences in problem-solving approaches, and a lack of shared understanding can hinder the integration of expertise and knowledge from different fields. DDCPS often deal with complex, multifaceted problems that require input and insights from multiple disciplines. Coordinating these inputs and arriving at solutions that satisfy the requirements of all involved parties can be a significant communication challenge. In interdisciplinary teams, decisions need to be made collectively. Disagreements or miscommunication during decision-making processes can lead to delays and suboptimal solutions.

2.5.4 Cyber Security

DDCPS leverage large amounts of data and advanced machine learning algorithms to control and optimize physical systems. However, the integration of data-driven approaches into CPS poses significant challenges for network security [99]. DDCPS rely heavily on data to function properly, so any data tampering or theft can have serious consequences [100]. Unauthorised access can lead to manipulated data, which can, in turn, compromise the performance and safety of the system. DDCPS are vulnerable to various types of cyber attacks, including data breaches, malware, denial-of-service attacks, and even more sophisticated advanced persistent threats (APTs). The interconnectedness of these systems makes them an attractive target for cyber criminals. Moreover, DDCPS often involve IoT devices. However, many of these devices lack robust built-in security features, making them a weak link in the system's security [101]. Machine learning algorithms that DDCPS rely on can be vulnerable to adversarial attacks, where attackers deliberately input deceptive data to manipulate the system's output. DDCPS require constant and reliable communication between components. However, this communication can be intercepted or disrupted by malicious actors, affecting the system's performance.

Heterogeneity in the context of DDCPS refers to the presence of diverse components, technologies, and data sources within the system. This heterogeneity can lead to several cybersecurity challenges: diverse interfaces and protocols, vulnerability variability, interoperability issues, data divergence, monitoring and detection complexity, resource constraints. Heterogeneous DDCPS often incorporate various sensors, actuators, and communication protocols. This diversity makes it challenging to establish consistent security mechanisms across the system, as each component may require different security measures and may be vulnerable to unique threats. Different components of a DDCPS may have varying levels of security vulnerabilities. Some may be well-protected, while others may lack robust security measures. Cyber attackers can exploit the weakest links in the system, making it essential to identify and address vulnerabilities across the heterogeneous landscape. Heterogeneous DDCPS components may not seamlessly communicate or share data. Ensuring secure interoperability among these components becomes a complex task. Any security gap in the integration points can be exploited by cybercriminals. Heterogeneous data sources may produce data with varying formats, accuracy levels, and update rates. Handling and securing this diverse data can be challenging, especially when trying to maintain data integrity and confidentiality. The diverse nature of DDCPS components makes it challenging to establish a unified and effective cybersecurity monitoring and detection system. Anomalies or security breaches may manifest differently in various components, requiring a comprehensive approach to threat detection and response. Some components of a DDCPS may have limited computing power and memory, making it difficult to implement robust security measures. This resource constraint can be exploited by attackers to compromise these vulnerable components.

2.5.5 System Stability

The system stability of DDCPS refers to the ability of the system's various components, including physical and computational elements, to maintain a controllable and predictable state during its operation, ensuring the achievement of expected performance and functionality. System stability is a critical characteristic essential for ensuring the reliability, availability, and security of DDCPS. In DDCPS, data-driven models are often used for prediction and control. However, developing reliable and accurate data-driven models for complex and dynamic CPS can be challenging, and model inaccuracies can lead to system instability [102].

The system's stability could be compromised if the data used for decision making is not accurate or reliable, leading to erratic behavior and undesirable outcomes [103]. DDCPS operate in real-time, and therefore, data must be processed and actions must be decided upon within a strict time frame [104]. If the data processing capabilities of the DDCPS are not adequate, it could lead to delays, instability, and potential system failures. The accuracy and reliability of DDCPS are also heavily dependent on the quality of the data input. Like any other system reliant on digital data, DDCPS are subject to potential cyber security threats. Attacks could destabilize the system and potentially cause dangerous situations, especially in critical infrastructure systems like power grids or health monitoring systems [105]. DDCPS often comprise various subsystems from different vendors. Ensuring these subsystems can communicate effectively and work together to maintain system stability is a complex task. A lack of standardization in DDCPS designs can increase this challenge.

While scalability is an important feature for accommodating growth and flexibility, it can introduce challenges related to system stability in DDCPS for several reasons: increased complexity, data overload, resource allocation, network congestion, interoperability. As DDCPS scale up, they often become more complex with additional components, sensors, and computational resources. Managing this increased complexity can be challenging, leading to potential points of failure that can affect system stability. Scalability often means handling larger volumes of data from various sources. Managing and processing this data efficiently can be demanding, and if not handled properly, it can lead to bottlenecks, delays, or data loss, impacting the stability of the data-driven components of the system. Scaling may require the allocation of additional computational resources, such as servers or cloud resources. Ensuring proper resource allocation and management becomes critical to maintain system stability, as inadequate resource provisioning can lead to performance issues or even system crashes. In DDCPS, data is frequently transmitted between physical and computational components through networks. Scalability can result in increased network traffic, potentially causing congestion, packet loss, or latency issues that affect data transmission and system response times. As DDCPS scale, they may need to interface with a wider range of devices, protocols, and data formats. Ensuring interoperability between different components and systems can be complex and may introduce compatibility issues that affect system stability.

To address these challenges, DDCPS designers and operators must carefully plan for scalability, considering the potential impact on system stability. This may involve optimizing algorithms, ensuring adequate resource provisioning, implementing effective monitoring and fault-tolerance mechanisms, and conducting thorough testing to validate the system's stability under varying loads and conditions. Balancing scalability with system stability is crucial to achieving the desired performance and reliability in DDCPS.

2.6 Summary

We focus on understanding both CPS and DDCPS through thorough definitions and diverse perspectives. The subsequent exploration narrows in on the distinct features that characterize DDCPS. To offer a comprehensive perspective, we delve into the intricate building blocks of DDCPS, highlighting the crucial roles played by sensing, computation, communication, and control within this framework. Bridging the gap between theoretical concepts and practical implementations, a range of real-world applications spanning various domains are showcased, demonstrating the remarkable achievements made possible by harnessing CPS. Finally, we conclude by shedding light on the impending challenges that DDCPS are anticipated to face in the future.

Competing Interests

1. The research is supported by the National Key Research and Development Project under Grant 2023YFB3307300, the National Natural Science Foundation of China under Grants 62373014, 62422301, 62125301, and 92467205.
2. The authors have no conflicts of interest to declare that are relevant to the content of this chapter.

References

1. R. Baheti and H. Gill, "Cyber-physical systems," *The impact of control technology*, vol. 12, no. 1, pp. 161–166, 2011.
2. E. A. Lee, "Cps foundations," in *Proceedings of the 47th design automation conference*, pp. 737–742, 2010.
3. S. M. Dibaji, M. Pirani, D. B. Flamholz, A. M. Annaswamy, K. H. Johansson, and A. Chakrabortty, "A systems and control perspective of cps security," *Annual reviews in control*, vol. 47, pp. 394–411, 2019.
4. B. Dafflon, N. Moalla, and Y. Ouzrout, "The challenges, approaches, and used techniques of cps for manufacturing in industry 4.0: A literature review," *The International Journal of Advanced Manufacturing Technology*, vol. 113, pp. 2395–2412, 2021.
5. Y. Lu, "Cyber physical system (cps)-based industry 4.0: A survey," *Journal of Industrial Integration and Management*, vol. 2, no. 03, p. 1750014, 2017.
6. F.-Y. Wang, "The emergence of intelligent enterprises: From cps to cpss," *IEEE Intelligent Systems*, vol. 25, no. 4, pp. 85–88, 2010.
7. S. Kim and S. Park, "Cps (cyber physical system) based manufacturing system optimization," *Procedia computer science*, vol. 122, pp. 518–524, 2017.
8. D. Antonioli and N. O. Tippenhauer, "Minicps: A toolkit for security research on cps networks," in *Proceedings of the First ACM workshop on cyber-physical systems-security and/or privacy*, pp. 91–100, 2015.
9. S. Weyer, T. Meyer, M. Ohmer, D. Gorecky, and D. Zühlke, "Future modeling and simulation of cps-based factories: an example from the automotive industry," *Ifac-Papersonline*, vol. 49, no. 31, pp. 97–102, 2016.
10. J. Jamaludin and J. M. Rohani, "Cyber-physical system (cps): State of the art," in *2018 International Conference on Computing, Electronic and Electrical Engineering (ICE Cube)*, pp. 1–5, IEEE, 2018.
11. K.-D. Kim and P. Kumar, "An overview and some challenges in cyber-physical systems," *Journal of the Indian Institute of Science*, vol. 93, no. 3, pp. 341–352, 2013.
12. W. Wolf, "Cyber-physical systems," *Computer*, vol. 42, no. 03, pp. 88–89, 2009.
13. E. A. Lee, "Cyber-physical systems-are computing foundations adequate," in *Position paper for NSF workshop on cyber-physical systems: research motivation, techniques and roadmap*, vol. 2, pp. 1–9, Austin, TX, 2006.
14. E. A. Lee, "The past, present and future of cyber-physical systems: A focus on models," *Sensors*, vol. 15, no. 3, pp. 4837–4869, 2015.

15. A. V. Jha, B. Appasani, A. N. Ghazali, P. Pattanayak, D. S. Gurjar, E. Kabalci, and D. Mohanta, "Smart grid cyber-physical systems: Communication technologies, standards and challenges," *Wireless Networks*, vol. 27, pp. 2595–2613, 2021.
16. J. Lee, H. D. Ardakani, S. Yang, and B. Bagheri, "Industrial big data analytics and cyber-physical systems for future maintenance & service innovation," *Procedia cirp*, vol. 38, pp. 3–7, 2015.
17. A. Darwish and A. E. Hassanien, "Cyber physical systems design, methodology, and integration: the current status and future outlook," *Journal of Ambient Intelligence and Humanized Computing*, vol. 9, pp. 1541–1556, 2018.
18. V. Lesch, M. Züfle, A. Bauer, L. Iffländer, C. Krupitzer, and S. Kounev, "A literature review of iot and cps—what they are, and what they are not," *Journal of Systems and Software*, vol. 200, p. 111631, 2023.
19. K. Wan, K. Man, and D. Hughes, "Specification, analyzing challenges and approaches for cyber-physical systems (cps).," *Engineering Letters*, vol. 18, no. 3, 2010.
20. N. Jazdi, "Cyber physical systems in the context of industry 4.0," in *2014 IEEE international conference on automation, quality and testing, robotics*, pp. 1–4, IEEE, 2014.
21. J. Shi, J. Wan, H. Yan, and H. Suo, "A survey of cyber-physical systems," in *2011 international conference on wireless communications and signal processing (WCSP)*, pp. 1–6, IEEE, 2011.
22. Y. Tan, S. Goddard, and L. C. Pérez, "A prototype architecture for cyber-physical systems," *SIGBED Rev.*, vol. 5, p. 26, 2008.
23. Y. Tan, S. Goddard, and L. C. Perez, "A prototype architecture for cyber-physical systems," *ACM Sigbed Review*, vol. 5, no. 1, pp. 1–2, 2008.
24. R. Rajkumar, I. Lee, L. Sha, and J. Stankovic, "Cyber-physical systems: the next computing revolution," in *Proceedings of the 47th design automation conference*, pp. 731–736, 2010.
25. A. Taherkordi and F. Eliassen, "Towards independent in-cloud evolution of cyber-physical systems," in *2014 IEEE International Conference on Cyber-Physical Systems, Networks, and Applications*, pp. 19–24, IEEE, 2014.
26. X. Liu, J. Cao, Y. Yang, and S. Jiang, "Cps-based smart warehouse for industry 4.0: A survey of the underlying technologies," *Computers*, vol. 7, no. 1, p. 13, 2018.
27. D.-Y. Kim, S. Kim, H. Hassan, and J. H. Park, "Radio resource management for data transmission in low power wide area networks integrated with large scale cyber physical systems," *Cluster Computing*, vol. 20, pp. 1831–1842, 2017.
28. F.-Y. Wang and D. Liu, *Networked control systems*. Springer, 2008.
29. H. Sandberg, S. Amin, and K. H. Johansson, "Cyberphysical security in networked control systems: An introduction to the issue," *IEEE Control Systems Magazine*, vol. 35, no. 1, pp. 20–23, 2015.
30. F. Farivar, M. S. Haghighi, A. Jolfaei, and S. Wen, "On the security of networked control systems in smart vehicle and its adaptive cruise control," *IEEE Transactions on Intelligent Transportation Systems*, vol. 22, no. 6, pp. 3824–3831, 2021.
31. G. Ravikumar, B. Hyder, and M. Govindarasu, "Efficient modeling of iec-61850 logical nodes in ieds for scalability in cps security testbed," in *2020 IEEE/PES Transmission and Distribution Conference and Exposition (T&D)*, pp. 1–5, IEEE, 2020.
32. X. Zheng, C. Julien, R. Podorozhny, F. Cassez, and T. Rakotoarivelo, "Efficient and scalable runtime monitoring for cyber–physical system," *IEEE Systems Journal*, vol. 12, no. 2, pp. 1667–1678, 2016.
33. G. Ravikumar, B. Hyder, and M. Govindarasu, "Efficient modeling of hil multi-grid system for scalability & concurrency in cps security testbed," in *2019 North American Power Symposium (NAPS)*, pp. 1–6, IEEE, 2019.
34. M. Govindarasu and C. Liu, "Cyber physical security testbed for the smart grid: fidelity, scalability, remote access, and federation," in *National CPS Energy Workshop*, 2013.
35. L. Zhang, S. Li, L. Wihl, M. Kazemtabrizi, S. Q. Ali, J.-N. Paquin, and S. Labbé, "Cybersecurity study of power system utilizing advanced cps simulation tools," in *Proceedings of the 2019 PAC World Americas Conference, Raleigh, NC, USA*, pp. 19–22, 2019.

36. F. De Rango, G. Potrino, M. Tropea, and P. Fazio, "Energy-aware dynamic internet of things security system based on elliptic curve cryptography and message queue telemetry transport protocol for mitigating replay attacks," *Pervasive and Mobile Computing*, vol. 61, p. 101105, 2020.
37. S. Vinoski, "Advanced message queuing protocol," *IEEE Internet Computing*, vol. 10, no. 6, pp. 87–89, 2006.
38. J. Sztipanovits, X. Koutsoukos, G. Karsai, N. Kottenstette, P. Antsaklis, V. Gupta, B. Goodwine, J. Baras, and S. Wang, "Toward a science of cyber–physical system integration," *Proceedings of the IEEE*, vol. 100, no. 1, pp. 29–44, 2011.
39. P. Derler, E. A. Lee, and A. S. Vincentelli, "Modeling cyber–physical systems," *Proceedings of the IEEE*, vol. 100, no. 1, pp. 13–28, 2011.
40. J. Sztipanovits, "Composition of cyber-physical systems," in *14th Annual IEEE International Conference and Workshops on the Engineering of Computer-Based Systems (ECBS'07)*, pp. 3–6, IEEE Computer Society, 2007.
41. V. Volkova, G. Gorelova, and N. Pankratova, "The development of the cyberphysical system concept on base of the interdisciplinary theories," in *2020 IEEE 2nd International Conference on System Analysis & Intelligent Computing (SAIC)*, pp. 1–6, IEEE, 2020.
42. J. Luo, S. Green, P. Feghali, G. Legrady, and Ç. K. Koç, "Reinforcement learning and trustworthy autonomy," *Cyber-Physical Systems Security*, pp. 191–217, 2018.
43. N. Wells, R. Johnson, and S. Salyer, "Interdisciplinary collaboration," *Clinical Nurse Specialist*, vol. 12, no. 4, pp. 161–168, 1998.
44. S. Roy and S. K. Das, *Principles of Cyber-Physical Systems: An Interdisciplinary Approach*. Cambridge University Press, 2020.
45. A. Fischer, S. Greiff, and J. Funke, "The process of solving complex problems," *Journal of Problem Solving*, vol. 4, no. 1, pp. 19–42, 2011.
46. A. Platzer, "Teaching cps foundations with contracts," 2013.
47. P. Morella, M. P. Lambán, J. A. Royo, and J. C. Sánchez, "The importance of implementing cyber physical systems to acquire real-time data and indicators," *J*, vol. 4, no. 2, pp. 147–153, 2021.
48. E. A. Lee, "Cyber physical systems: Design challenges," in *2008 11th IEEE international symposium on object and component-oriented real-time distributed computing (ISORC)*, pp. 363–369, IEEE, 2008.
49. H. Song, D. B. Rawat, S. Jeschke, and C. Brecher, *Cyber-physical systems: foundations, principles and applications*. Morgan Kaufmann, 2016.
50. L. Sha, T. Abdelzaher, K.-E. Årzén, A. Cervin, T. Baker, A. Burns, G. Buttazzo, M. Caccamo, J. Lehoczky, and A. K. Mok, "Real time scheduling theory: A historical perspective," *Real-time systems*, vol. 28, pp. 101–155, 2004.
51. T. Ngo, "Data mining: practical machine learning tools and technique, by ian h. witten, eibe frank, mark a. hell," *ACM SIGSOFT Software Engineering Notes*, vol. 36, no. 5, pp. 51–52, 2011.
52. J. Lee, B. Bagheri, and H.-A. Kao, "A cyber-physical systems architecture for industry 4.0-based manufacturing systems," *Manufacturing letters*, vol. 3, pp. 18–23, 2015.
53. H. Geng, *Internet of things and data analytics handbook*. John Wiley & Sons, 2017.
54. Y.-C. Hu, A. Perrig, and D. B. Johnson, "Rushing attacks and defense in wireless ad hoc network routing protocols," in *Proceedings of the 2nd ACM workshop on Wireless security*, pp. 30–40, 2003.
55. S. Li, L. D. Xu, and S. Zhao, "The internet of things: a survey," *Information systems frontiers*, vol. 17, pp. 243–259, 2015.
56. G. F. Franklin, J. D. Powell, A. Emami-Naeini, and J. D. Powell, *Feedback control of dynamic systems*, vol. 4. Prentice hall Upper Saddle River, 2002.
57. G. Goodwin, S. Graebe, and M. Salgado, "Control system design prentice-hall," *New Jersey*, 2001.
58. K. J. Åström and R. M. Murray, *Feedback systems: an introduction for scientists and engineers*. Princeton university press, 2021.

59. P. Palensky and F. Kupzog, "Smart grids," *Annual Review of Environment and Resources*, vol. 38, pp. 201–226, 2013.
60. X. Yu, C. Cecati, T. Dillon, and M. G. Simoes, "The new frontier of smart grids," *IEEE Industrial Electronics Magazine*, vol. 5, no. 3, pp. 49–63, 2011.
61. H. Li, A. Dimitrovski, J. B. Song, Z. Han, and L. Qian, "Communication infrastructure design in cyber physical systems with applications in smart grids: A hybrid system framework," *IEEE Communications Surveys & Tutorials*, vol. 16, no. 3, pp. 1689–1708, 2014.
62. S. H. Ahmed, G. Kim, and D. Kim, "Cyber physical system: Architecture, applications and research challenges," *2013 IFIP Wireless Days (WD)*, pp. 1–5, 2013.
63. P. Guo, P. O. Dusadeerungsikul, and S. Y. Nof, "Agricultural cyber physical system collaboration for greenhouse stress management," *Computers and electronics in agriculture*, vol. 150, pp. 439–454, 2018.
64. I. Haris, A. Fasching, L. Punzenberger, and R. Grosu, "Cps/iot ecosystem: Indoor vertical farming system," in *2019 IEEE 23rd International Symposium on Consumer Technologies (ISCT)*, pp. 47–52, IEEE, 2019.
65. S. I. Caramihai and I. Dumitrache, "Agricultural enterprise as a complex system: A cyber physical systems approach," in *2015 20th International Conference on Control Systems and Computer Science*, pp. 659–664, IEEE, 2015.
66. F. Mehdipour, K. C. Nunna, and K. J. Murakami, "A smart cyber-physical systems-based solution for pest control (work in progress)," in *2013 IEEE International Conference on Green Computing and Communications and IEEE Internet of Things and IEEE Cyber, Physical and Social Computing*, pp. 1248–1253, IEEE, 2013.
67. F. Mehdipour, "Smart field monitoring: An application of cyber-physical systems in agriculture (work in progress)," in *2014 IIAI 3rd International Conference on Advanced Applied Informatics*, pp. 181–184, IEEE, 2014.
68. S. A. Haque, S. M. Aziz, and M. Rahman, "Review of cyber-physical system in healthcare," *international journal of distributed sensor networks*, vol. 10, no. 4, p. 217415, 2014.
69. Y. Zhang, M. Qiu, C.-W. Tsai, M. M. Hassan, and A. Alamri, "Health-cps: Healthcare cyber-physical system assisted by cloud and big data," *IEEE Systems Journal*, vol. 11, no. 1, pp. 88–95, 2015.
70. H. K. Smalley and P. Keskinocak, "Automated medical resident rotation and shift scheduling to ensure quality resident education and patient care," *Health care management science*, vol. 19, pp. 66–88, 2016.
71. I. Lee and O. Sokolsky, "Medical cyber physical systems," in *Proceedings of the 47th design automation conference*, pp. 743–748, 2010.
72. Z. Zhou, Y. Li, J. Li, K. Yu, G. Kou, M. Wang, and B. B. Gupta, "Gan-siamese network for cross-domain vehicle re-identification in intelligent transport systems," *IEEE Transactions on Network Science and Engineering*, 2022.
73. C. Csiszár and A. Zarkeshev, "Demand-capacity coordination method in autonomous public transportation," *Transportation research procedia*, vol. 27, pp. 784–790, 2017.
74. D. B. Rawat, C. Bajracharya, and G. Yan, "Towards intelligent transportation cyber-physical systems: Real-time computing and communications perspectives," in *SoutheastCon 2015*, pp. 1–6, IEEE, 2015.
75. Y. Lin, P. Wang, and M. Ma, "Intelligent transportation system (its): Concept, challenge and opportunity," in *2017 IEEE 3rd international conference on big data security on cloud (bigdatasecurity), IEEE international conference on high performance and smart computing (hpsc), and IEEE international conference on intelligent data and security (ids)*, pp. 167–172, IEEE, 2017.
76. F. Tao, Q. Qi, A. Liu, and A. Kusiak, "Data-driven smart manufacturing," *Journal of Manufacturing Systems*, vol. 48, pp. 157–169, 2018.
77. F. Tao and Q. Qi, "New it driven service-oriented smart manufacturing: framework and characteristics," *IEEE Transactions on Systems, Man, and Cybernetics: Systems*, vol. 49, no. 1, pp. 81–91, 2017.

78. P. O'Donovan, K. Leahy, K. Bruton, and D. T. O'Sullivan, "An industrial big data pipeline for data-driven analytics maintenance applications in large-scale smart manufacturing facilities," *Journal of Big Data*, vol. 2, no. 1, pp. 1–26, 2015.
79. S. Yin and O. Kaynak, "Big data for modern industry: challenges and trends [point of view]," *Proceedings of the IEEE*, vol. 103, no. 2, pp. 143–146, 2015.
80. G. Shao, S.-J. Shin, and S. Jain, "Data analytics using simulation for smart manufacturing," in *Proceedings of the Winter Simulation Conference 2014*, pp. 2192–2203, IEEE, 2014.
81. A. Kusiak, "Smart manufacturing must embrace big data," *Nature*, vol. 544, no. 7648, pp. 23–25, 2017.
82. H. S. Kang, J. Y. Lee, S. Choi, H. Kim, J. H. Park, J. Y. Son, B. H. Kim, and S. D. Noh, "Smart manufacturing: Past research, present findings, and future directions," *International journal of precision engineering and manufacturing-green technology*, vol. 3, pp. 111–128, 2016.
83. P. Zheng, H. Wang, Z. Sang, R. Y. Zhong, Y. Liu, C. Liu, K. Mubarok, S. Yu, and X. Xu, "Smart manufacturing systems for industry 4.0: Conceptual framework, scenarios, and future perspectives," *Frontiers of Mechanical Engineering*, vol. 13, pp. 137–150, 2018.
84. S. Choi, G. Kang, C. Jun, J. Y. Lee, and S. Han, "Cyber-physical systems: A case study of development for manufacturing industry," *International Journal of Computer Applications in Technology*, vol. 55, no. 4, pp. 289–297, 2017.
85. M. Ryalat, H. ElMoaqet, and M. AlFaouri, "Design of a smart factory based on cyber-physical systems and internet of things towards industry 4.0," *Applied Sciences*, vol. 13, no. 4, p. 2156, 2023.
86. S. Wang, J. Wan, D. Li, and C. Zhang, "Implementing smart factory of industrie 4.0: an outlook," *International journal of distributed sensor networks*, vol. 12, no. 1, p. 3159805, 2016.
87. M. M. Mabkhot, A. M. Al-Ahmari, B. Salah, and H. Alkhalefah, "Requirements of the smart factory system: A survey and perspective," *Machines*, vol. 6, no. 2, p. 23, 2018.
88. A. Sanjab and W. Saad, "On bounded rationality in cyber-physical systems security: Game-theoretic analysis with application to smart grid protection," in *2016 Joint Workshop on Cyber-Physical Security and Resilience in Smart Grids (CPSR-SG)*, pp. 1–6, IEEE, 2016.
89. C. W. Axelrod, "Managing the risks of cyber-physical systems," in *2013 IEEE Long Island Systems, Applications and Technology Conference (LISAT)*, pp. 1–6, IEEE, 2013.
90. Y. Ashibani and Q. H. Mahmoud, "Cyber physical systems security: Analysis, challenges and solutions," *Computers & Security*, vol. 68, pp. 81–97, 2017.
91. A. J. Jara, D. Genoud, and Y. Bocchi, "Big data for cyber physical systems: an analysis of challenges, solutions and opportunities," in *2014 Eighth International Conference on Innovative Mobile and Internet Services in Ubiquitous Computing*, pp. 376–380, IEEE, 2014.
92. V. Jirkovskỳ, M. Obitko, and V. Mařík, "Understanding data heterogeneity in the context of cyber-physical systems integration," *IEEE Transactions on Industrial Informatics*, vol. 13, no. 2, pp. 660–667, 2016.
93. B. Li, Y. Wu, J. Song, R. Lu, T. Li, and L. Zhao, "Deepfed: Federated deep learning for intrusion detection in industrial cyber–physical systems," *IEEE Transactions on Industrial Informatics*, vol. 17, no. 8, pp. 5615–5624, 2020.
94. J. Tavčar and I. Horvath, "A review of the principles of designing smart cyber-physical systems for run-time adaptation: Learned lessons and open issues," *IEEE Transactions on Systems, Man, and Cybernetics: Systems*, vol. 49, no. 1, pp. 145–158, 2018.
95. A. Zanella, N. Bui, A. Castellani, L. Vangelista, and M. Zorzi, "Internet of things for smart cities," *IEEE Internet of Things journal*, vol. 1, no. 1, pp. 22–32, 2014.
96. L. Da Xu, W. He, and S. Li, "Internet of things in industries: A survey," *IEEE Transactions on industrial informatics*, vol. 10, no. 4, pp. 2233–2243, 2014.
97. A. Al-Fuqaha, M. Guizani, M. Mohammadi, M. Aledhari, and M. Ayyash, "Internet of things: A survey on enabling technologies, protocols, and applications," *IEEE communications surveys & tutorials*, vol. 17, no. 4, pp. 2347–2376, 2015.
98. Z.-S. Hou and Z. Wang, "From model-based control to data-driven control: Survey, classification and perspective," *Information Sciences*, vol. 235, pp. 3–35, 2013.

99. R. Alguliyev, Y. Imamverdiyev, and L. Sukhostat, "Cyber-physical systems and their security issues," *Computers in Industry*, vol. 100, pp. 212–223, 2018.
100. G. Wu, J. Sun, and J. Chen, "A survey on the security of cyber-physical systems," *Control Theory and Technology*, vol. 14, no. 1, pp. 2–10, 2016.
101. W. Shi, J. Cao, Q. Zhang, Y. Li, and L. Xu, "Edge computing: Vision and challenges," *IEEE internet of things journal*, vol. 3, no. 5, pp. 637–646, 2016.
102. F. Hussain, R. Hussain, S. A. Hassan, and E. Hossain, "Machine learning in iot security: Current solutions and future challenges," *IEEE Communications Surveys & Tutorials*, vol. 22, no. 3, pp. 1686–1721, 2020.
103. Y. Jiang, S. Yin, and O. Kaynak, "Data-driven monitoring and safety control of industrial cyber-physical systems: Basics and beyond," *IEEE Access*, vol. 6, pp. 47374–47384, 2018.
104. H. El-Sayed, S. Sankar, M. Prasad, D. Puthal, A. Gupta, M. Mohanty, and C.-T. Lin, "Edge of things: The big picture on the integration of edge, iot and the cloud in a distributed computing environment," *ieee access*, vol. 6, pp. 1706–1717, 2017.
105. J. Liu, J. Wan, Q. Wang, P. Deng, K. Zhou, and Y. Qiao, "A survey on position-based routing for vehicular ad hoc networks," *Telecommunication Systems*, vol. 62, pp. 15–30, 2016.

Chapter 3
Data Collection in Cyber Physical Systems

Abstract Data collection is a necessary technology in the domain of cyber-physical systems (CPS). The selection of appropriate sensors and devices hinges upon a multitude of factors encompassing resolution, accuracy, power efficiency, compatibility, as well as cost considerations. These factors must be understood to create robust CPS that meet specific requirements. CPS generate large volumes of data, characterized by volume, velocity, and variety. Processing and analyzing this data can be challenging but offer significant benefits. Achieving real-time responsiveness and minimizing latency in CPS requires considering factors such as edge computing, predictive analytics, data compression, prioritized communication, and energy-efficient designs. Evaluating CPS performance involves factors like response time, predictability, meeting deadlines, system dynamics, network communication impact, safety requirements, and fault tolerance. Addressing data quality and reliability challenges is important for accurate and consistent data collection in CPS. Ongoing advancements in data collection techniques continue to improve the accuracy and dependability of CPS in various applications.

This chapter begins with the exploration of the devices involved, as they form the foundation of data collection in CPS. Then it delves into the types of data collected, highlighting the complexity and diversity of information captured. Next, the presentation is real-time responsiveness and latency, which are important determinants of system efficiency. The final section describes data quality and reliability, emphasizing their role in ensuring system reliability.

3.1 Sensors and Auxiliary Components

Sensors and devices assume a pivotal role in the context of cyber-physical systems (CPS), serving as conduits for data acquisition, device status monitoring, and system control [1]. A diverse array of sensor types, including those for temperature, pressure, motion, and illuminance, as well as various device categories encompassing

F. Li et al., *Data-Driven Cyber Physical Systems*,
https://doi.org/10.1007/978-981-96-8709-1_3

actuators, transducers, and data storage components, contribute to the comprehensive landscape. The process of selecting apt sensors and devices rests upon a nuanced interplay of factors such as resolution, accuracy, power consumption, compatibility, and cost. This requisite understanding empowers the creation of resilient systems aligned with specific requisites, thus unleashing CPS's transformative potential across diverse industries.

A prerequisite for sensor and device selection lies in an in-depth comprehension of distinct typologies, ranging from temperature, humidity, and pressure sensors to accelerometers, gyroscopes, and magnetometers [2]. This grasp of typological diversity is essential in facilitating a comprehensive grasp of the integral role played by these constituents within CPS.

The judicious selection of sensors and devices mandates a comprehensive evaluation encompassing resolution, accuracy, power efficiency, form factor, dependability, availability, financial outlay, and interoperability [3, 4]. The evaluation process further demands an exploration of diverse data collection scenarios and ensuing applications, each bespeaking unique sensor and device demands tailored to their distinct operational exigencies. Notably, industrial automation settings may necessitate high-precision temperature and pressure sensors, while light and humidity sensors find primacy within smart home environments [5]. By orchestrating a meticulous analysis of these multifarious considerations, an informed and discerning selection of sensors and devices ensues, culminating in a superlative manifestation of CPS.

The realization of efficacious CPS design and implementation is inexorably linked to an astute grasp of the intricate facets underpinning sensors and devices. The meticulous curation of appropriate sensors and devices stands as a lynchpin in engendering peak system performance and operational effectiveness. Thus, a thorough assimilation of factors governing sensors and ancillary components, coupled with a nuanced understanding of sensor and device typologies, emerges as a preeminent prerequisite. This holistic approach begets the formulation of CPS that seamlessly align with the idiosyncratic prerequisites characterizing distinct applications.

3.1.1 Type of Sensor and Auxiliary Components

A fundamental aspect in the realm of CPS is gaining a comprehensive understanding of the sensors and devices that can be effectively utilized. Sensors such as temperature sensors, humidity sensors, pressure sensors, and illuminance sensors, as well as motion and orientation sensors like accelerometers, gyroscopes, and magnetometers, play a significant role in monitoring and assessing environmental conditions. Actuators, transducers, and data storage devices also contribute to the overall functionality of a CPS. In CPS, the interplay of sensors and auxiliary components needs to be considered as they interact with the physical world, collect data, and enable real-time monitoring [6]. By carefully considering the available options, understanding their capabilities, and thoughtfully integrating them into the system, designers can optimize the performance of the CPS. Comprehensive understanding and thoughtful utilization of sensors and devices are essential for designing CPS that effectively operate within specific application contexts and meet the system's requirements. The schematic diagram of the various sensors and auxiliary components used for data collection in the CPS is shown in Fig. 3.1.

Environmental monitoring sensors are one category of sensors commonly used in CPS [7]. These sensors are designed to monitor and assess various environmental conditions, providing important data for system operation. Temperature sensors, for example, enable the measurement of ambient temperature, allowing the system to respond accordingly and make informed decisions in applications such as climate control and thermal management [8]. Moreover, temperature sensors have the capability of detecting temperature variations with high accuracy, which makes them particularly useful in industries such as food, pharmaceutical, and chemical processing. Humidity sensors detect the moisture content in the air, providing valuable insights for environments where maintaining optimal humidity levels is essential, such as greenhouses or data centers. In addition, humidity sensors can detect changes in humidity levels, which can help preventing problems such as mold growth, corrosion, and damage to electronic equipment. Pressure sensors, on the other hand, measure the pressure of gases or liquids, finding applications in areas such as industrial processes, weather monitoring, or even medical devices like blood pressure monitors [9]. For example, in the oil and gas industry, pressure sensors are used to

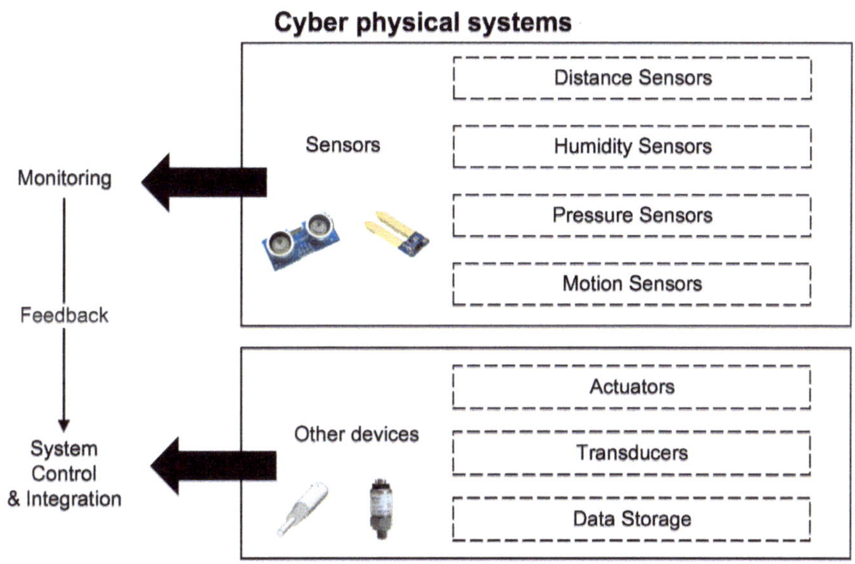

Fig. 3.1 Data are collected in CPS through various sensors and auxiliary components. These include sensors, cameras, actuators and communication modules. The sensors are used to monitor the system and provide feedback to the system, while the auxiliary components receive the feedback and control the system

monitor pipeline pressure to prevent leaks and ensure safe operations. Illuminance sensors detect ambient light levels, enabling the system to automatically adjust lighting systems or implement energy conservation efforts based on the available natural light [10]. Additionally, illuminance sensors can be used to detect the presence or absence of light in applications such as security systems.

Motion and orientation sensors provide physical data to the CPS. Accelerometers are used to measure acceleration forces, allowing the system to detect a range of movements, changes in velocity, and even gestures [11]. For instance, they are used in tracking movements and in activity-tracking devices to monitor physical activity. Similarly, they find applications in-vehicle navigation systems, where they help detecting the vehicle's movements. Gyroscopes sense angular motion and rotational orientation, providing critical data for a range of applications, from robotics, to unmanned aerial vehicles (UAVs) [12] to virtual reality (VR) systems [13]. By measuring rotation and angular velocity, they enable precise movement control and orientation tracking, facilitating the development of highly accurate and responsive systems. Magnetometers detect changes in magnetic fields and find applications in compasses, navigation systems, and even medical devices like magnetic resonance imaging (MRI) machines [14]. Moreover, magnetometers can be used to detect the presence of metal, which is useful in applications such as security systems [15]. With the increasing demand for smart and connected devices, the importance of motion and orientation sensors in CPS is set to grow, enabling the development of more sophisticated and responsive systems.

In addition to sensors, there are several auxiliary components that contribute to the functionality of CPS. Actuators, for instance, play a critical role in converting electrical signals into physical actions, which in turn enables the control of mechanical systems or robotic movements [16]. Actuators come in many different types, such as hydraulic, pneumatic, or electric, and are used in a variety of applications, including industrial automation, robotics, and aerospace engineering. For example, in an industrial automation system, actuators may be used to control robotic arms or manipulate valves to regulate fluid flow [17]. Another important device in CPS is the transducer, which converts one form of energy into another [18]. This facilitates data transfer and processing within the system. Transducers are commonly used to convert physical signals into electrical signals or vice versa, allowing for seamless integration of different components. There are many different types of transducers, including pressure transducers, temperature transducers, and flow meters, to name just a few. Each type of transducer plays an important role in sensing and measuring different types of physical phenomena. Data storage components are also an essential part of the CPS. Solid-state drives and cloud storage systems, for instance, provide the means to store and manage the vast amount of data generated by the CPS. They are critical for storing historical data, performing real time analysis, and enabling future data-driven decision-making processes. With advances in machine learning and artificial intelligence, the importance of data storage and management in CPS is only likely to grow in the future.

Effective data collection, device monitoring, and seamless system control are essential aspects of modern CPS [19]. With the increasing popularity of the Internet of Things (IoT) and smart devices, there is a growing demand for reliable and efficient CPS solutions. Achieving optimal performance, reliability, and efficiency in CPS requires carefully selecting the sensors and devices used, each of which has unique strengths and weaknesses in CPS. Each sensor and device in a CPS has its unique strengths and weaknesses. Choosing the right combination of sensors and devices can significantly enhance the overall performance and efficiency of the system. For example, sensors can range from simple temperature sensors to advanced image recognition cameras, and devices can include actuators, controllers, and communication modules. The selection should align with the specific requirements and objectives of the CPS application. Moreover, as sensor technology continues to advance and CPS applications evolve, the availability and diversity of sensors and devices will continue to expand. This ongoing development provides CPS with a broader array of options to tailor solutions to specific needs and leverage new innovations. This adaptability enables the creation of more sophisticated and efficient CPS solutions, ensuring that they can effectively meet the evolving demands of various industries and applications.

In order to design a robust and effective CPS, it requires an understanding of the characteristics and interrelationships of different types of sensors and auxiliary components. Environmental sensors, such as temperature, humidity, pressure, and illuminance sensors, enable the monitoring of environmental conditions and inform system decisions accordingly. Motion and orientation sensors, including accelerometers, gyroscopes, and magnetometers, detect device movement and orientation changes, allowing for precise control and tracking. Additionally, actuators, transducers, and data storage components play important roles in achieving system functionality and data management. The selection of appropriate sensors and devices depends on various factors, such as the specific application requirements, measurement accuracy, power consumption, compatibility, and cost. These considerations will be further explored in the next subsection. By carefully evaluating these factors and making informed decisions, designers can create CPS solutions that meet the needs of different industries and applications. Moreover, it is important to note that the use of sensors and devices is not limited to CPS applications, as they can be used in various other domains such as healthcare, environmental monitoring, and transportation. Therefore, it is important to keep up-to-date with the latest advancements in sensor technology and explore their potential applications in different fields.

3.1.2 Factors for Selecting Sensors and Auxiliary Components

When selecting sensors and auxiliary components for CPS, several key factors must be considered. One of the most important considerations is the measurement range and accuracy of the sensors, as they need to align with the specific requirements of

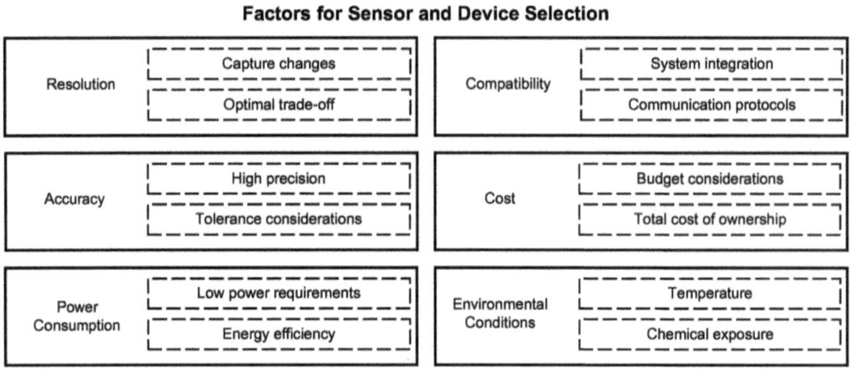

Fig. 3.2 Factors to consider when selecting sensors and auxiliary components in a CPS include accuracy, response time, reliability, power consumption, and cost

the system [3]. Power consumption and physical size of sensors and devices also need to be taken into account, as these factors play a significant role in determining the system's energy efficiency and overall integration performance. Moreover, the reliability, availability, and cost of the sensors and devices must be considered [4]. Another factor to consider when selecting sensors and auxiliary components for CPS is the evaluation of their interfaces and compatibility. This is important to ensure seamless communication and smooth integration with other components of the system. By evaluating the sensor's interfaces and compatibility, you can ensure that they work seamlessly and effectively within the framework of CPS.

The selection of sensors and devices in CPS is a multifaceted process that requires a comprehensive understanding of the system's requirements and constraints [20]. Factors such as resolution, accuracy, power consumption, compatibility, cost, and environmental conditions all contribute to the decision-making process. By thoroughly evaluating these factors, informed decisions can be made to meet the specific needs of the CPS. This ensures that the chosen sensors and devices are capable of meeting the system's objectives and operating effectively in real-world applications. Ultimately, the careful consideration of these factors contributes to the successful implementation and optimization of the CPS, enhancing its overall functionality, reliability, and efficiency. The key factors to consider when selecting sensors and auxiliary components in a CPS are shown in Fig. 3.2.

Resolution

When selecting sensors for CPS, there are several factors to consider in addition to resolution. For example, the range of the sensor, or the minimum and maximum values it can measure, is also an important consideration. Additionally, the accuracy of the sensor itself is a key factor to ensure that the data collected is reliable. Another factor to consider is the response time of the sensor, which refers to how quickly it can detect changes and provide updated measurements. In applications such as precision manufacturing or real-time monitoring systems, high-resolution sensors are particularly important, as they provide more detailed measurement results and can capture smaller changes [21]. However, choosing a resolution that is too high may cause some problems. For example, high-resolution sensors produce larger datasets, which can lead to challenges in processing and storing data [22]. Additionally, high-resolution sensors tend to be more expensive, which may increase system costs. Therefore, all of these factors must be considered thoroughly when selecting a sensor for CPS. By taking into account the application requirements, system resources, cost factors, and other relevant considerations, the appropriate resolution can be selected to ensure that the chosen sensor provides the necessary measurement accuracy and effectively supports the operation and application of CPS.

Accuracy

Accuracy is one of the factors to be considered when collecting data. It can affect the reliability and precision of the data. On the one hand, the required level of accuracy depends on the specific application and its associated tolerances [23]. For example, in critical applications such as healthcare, manufacturing or self-driving cars, high-precision sensors ensure accurate measurements and reliable system performance. This is because even slight errors in measurement can lead to serious consequences. On the other hand, in some applications where a lower level of accuracy is acceptable, cost-effective sensors with slightly reduced precision can be chosen. However, it is important to strike a balance between the desired accuracy level and the cost implications [24]. This is attributed to the fact that opting for sensors of excessively high precision can incur elevated costs and may not be necessary for the application. By optimizing the overall system performance and budgetary considerations, the best choice of sensors can be made to meet the desired level of accuracy within the given budget.

Power consumption

CPS need to consider power consumption in battery-powered or energy-constrained [25]. CPS are increasingly being used in a wide range of applications, from smart homes to autonomous vehicles, and they rely on sensors to collect data and control the system. However, using sensors with high energy requirements can rapidly deplete the system's energy resources, which can lead to frequent battery replacements, shorter operational lifetimes, or the need for additional power sources. Therefore, it is important to choose sensors and devices with low power consumption to optimize energy efficiency, prolong battery life, and ensure uninterrupted system operation. Moreover, energy-efficient sensors offer not only these benefits but also contribute to sustainability by minimizing power consumption and reducing environmental impact [26]. The use of energy-efficient sensors can help to reduce the carbon footprint of CPS, which is an important aspect of sustainable development. Employing energy-efficient sensors enables uninterrupted CPS operation while mitigating energy source requirements. This exerts a notable environmental influence by curbing non-renewable energy consumption and fostering sustainability. Hence, it's imperative to factor in sensor energy consumption during CPS design for enduring sustainability.

Compatibility

Compatibility is a common factor influencing seamless integration and effective communication within the CPS [27]. It is essential that sensors and devices are compatible with the overall system architecture, communication protocols, and interfaces. Compatibility ensures that the various components of the CPS can communicate and share data efficiently, thus enabling coordinated operations and seamless interoperability. When selecting sensors and devices, several factors must be considered to ensure compatibility with the existing infrastructure. These factors include data formats, communication protocols, connectivity options, and compatibility with existing infrastructure [28]. By choosing compatible components, designers can streamline the system integration process, minimize compatibility issues, and enhance the overall performance and reliability of the CPS. Moreover, it is important to note that ensuring compatibility is not a one time task. It is an ongoing process that requires constant monitoring and updating. As CPS evolve and new technologies emerge, the compatibility of existing components needs to be evaluated and necessary adjustments need to be implemented. This not only helps to improve the overall performance of CPS but also ensures its long-term sustainability and adaptability to changing requirements.

Cost

Cost becomes an imperative consideration when selecting sensors and devices, particularly when operating under financial constraints. One of the most significant considerations when it comes to cost is the technology used, which can significantly impact the price of the sensors and devices [29]. Other factors, such as performance, brand, and volume, can also affect the cost. In order to achieve the desired functionalities and performance, designers must balance the system's budgetary limitations with the cost of the sensors and devices. Additionally, it is imperative to consider not only the upfront cost but also the maintenance and replacement costs over the system's lifecycle. One way to balance cost-effectiveness with performance is to evaluate multiple options. This can help designers make informed decisions that are aligned with the financial considerations of the CPS project. Evaluating the cost-effectiveness of different sensors and devices can provide designers with a better understanding of the long-term costs and benefits of each option. By doing so, they can make informed decisions that will not compromise the overall performance of the system. It is essential to consider all relevant factors and to weigh them carefully when making a decision, in order to ensure that the CPS project is cost-effective and meets the required performance standards.

Environmental conditions

Furthermore, when selecting sensors and devices, it is important to consider the various environmental conditions in which they will be used. These conditions can be extremely diverse and can pose unique challenges. For example, sensors and devices may be exposed to extreme temperatures, both high and low, as well as high levels of humidity. They may also need to function properly in environments with intense vibrations or harsh chemicals [30]. Consequently, when evaluating different sensors and devices, it is important to assess their ability to withstand and reliably operate in these conditions. One way to ensure that the sensors and devices selected can handle these challenging environments is to use environmental ratings, such as ingress protection ratings [31]. These ratings can provide valuable guidance on the level of protection a device can offer against a variety of environmental factors, such as dust and moisture. By taking into account these ratings and selecting components with appropriate ratings, it is possible to ensure that the chosen components are suitable for the intended application environment. This can help to achieve reliable and effective performance, even in the most demanding conditions. Moreover, it is important to note that selecting components based on environmental ratings is just one aspect of the selection process. Other factors, such as cost, functionality, and ease of use, must also be considered to make the most informed decision possible.

Careful consideration of factors such as resolution, accuracy, power consumption, compatibility, cost, and environmental conditions ensures that the right sensors and devices are selected for CPS. By evaluating these factors, designers can optimize the system's performance and functionality, enabling reliable data collection and seamless system control. However, selecting the right sensors and devices is only the first step. The next chapter will explore various scenarios for data collection with sensors, delving into specific applications and use cases where different sensor types are employed to capture and analyze data in real time. By examining these scenarios, we can gain insights into the practical implementation of sensors in diverse domains, shedding light on their potential and uncovering new opportunities for innovation and advancement in the field of CPS. The key factors to consider when selecting sensors and auxiliary components in CPS are shown in Fig. 3.3.

3.1.3 Typical Scenarios for Data Collection

In the rapidly evolving realm of CPS, the possibilities for data collection using sensors are virtually limitless. Sensors serve as the primary means of capturing real-time data from the physical world, enabling valuable insights that can revolutionize

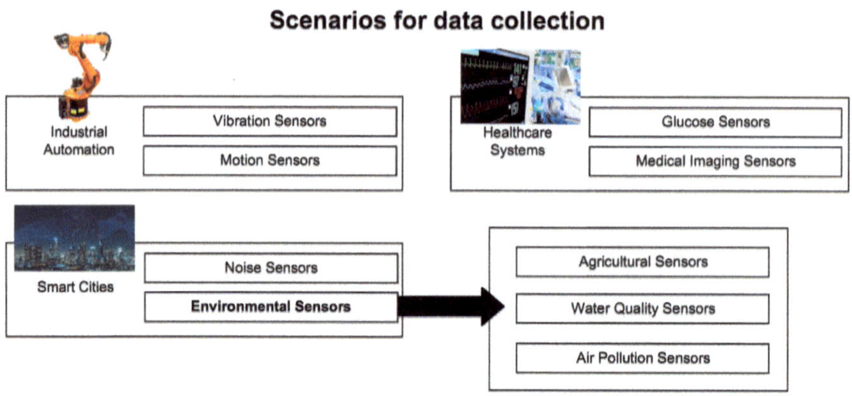

Fig. 3.3 Key factors to consider when selecting sensors and auxiliary components in a CPS include accuracy, response time, reliability, power consumption, and cost

system operation and control. By carefully selecting sensors and devices, designers can optimize the functionality, reliability, and efficiency of CPS solutions. Factors such as measurement range, accuracy, power consumption, compatibility, cost, and environmental conditions must be taken into account during the selection process. Additionally, the evaluation of application scenarios is of paramount importance in identifying the precise sensors and devices necessary for each specific situation. By thoughtfully integrating sensors and devices tailored to the practical needs of the application, CPS can seamlessly operate and unlock a multitude of benefits, including enhanced functionality, improved efficiency, and optimized decision-making processes. Ultimately, by leveraging the capabilities of sensors, CPS can transcend traditional boundaries and shape the future of industries and sectors.

Industrial automation

A typical scenario for data collection using sensors in CPS is industrial automation [32], where sensors play an important role in monitoring and controlling various processes. Temperature sensors, for instance, allow for precise monitoring of equipment and environmental temperatures, ensuring optimal operating conditions and preventing equipment failure that could lead to costly downtime. These sensors not only provide real time data but can also trigger automated actions, such as adjusting cooling systems or issuing alerts when temperature thresholds are exceeded [33]. Furthermore, pressure sensors can provide critical information about fluid pressure levels in industrial environments. By continuously monitoring pressure, these sensors enable accurate process control, ensuring that systems operate within safe and efficient parameters. They can also help detecting potential leaks or anomalies, triggering preventive maintenance measures and minimizing the risk of costly damages.

Motion sensors, another common component of industrial automation, detect movement within the industrial environment. By capturing data on the presence or absence of workers or materials, these sensors facilitate the automation of material handling processes, optimize production workflows, and enhance overall worker safety [34]. The collected data can be analyzed and used to identify bottlenecks or areas for improvement, leading to increased productivity and cost savings.

Smart cities

Sensors are often used to collect relevant data in cities to effectively manage the urban environment and improve the quality of life for residents [35]. Environmental sensors, for example, are deployed strategically throughout the city to monitor pollution levels and air quality. They continuously measure parameters such as particulate matter, nitrogen dioxide, and ozone, providing valuable data that can be used for traffic management, public health initiatives, and environmental policies. This real time information helps authorities identifying pollution hotspots, implementing targeted interventions, and promoting sustainable urban development. Noise sensors, another key component of smart city CPS, measure noise levels in different areas. By capturing data on noise pollution, city planners can assess the impact of various sources of noise, such as traffic or construction, and take measures to mitigate their effects. This can include implementing noise barriers [36], optimizing traffic flow, or regulating construction activities to create quieter and more livable urban spaces. Moreover, sensors integrated into parking spaces can monitor occupancy, allowing for efficient management of parking availability and reducing traffic congestion [37]. By providing real time data on parking occupancy, drivers can be directed to available spaces, minimizing the time spent searching for parking and reducing traffic congestion and emissions. Additionally, sensors integrated into streetlights can optimize energy consumption based on real time data, dimming or brightening lights as needed. This not only reduces energy waste but also enhances safety by ensuring well-lit streets in high-traffic or low-light conditions. By leveraging sensor data, smart city CPS solutions enable effective urban planning, resource management, and responsive services for citizens, fostering sustainable and livable urban environments.

Healthcare systems

Healthcare systems greatly benefit from the use of sensors for data collection and monitoring. Wearable sensors, for instance, have revolutionized the way health data are captured and analyzed [38]. These sensors, integrated into devices such as smartwatches or fitness trackers, enable continuous monitoring of vital signs, including heart rate, blood pressure, and sleep patterns. The data collected by these sensors provide a comprehensive view of an individual's health status, allowing for early detection of anomalies or changes that may require medical attention. Furthermore, wearable sensors can also monitor physical activity levels, providing insights into exercise habits and promoting healthy lifestyles. Additionally, glucose sensors are important for diabetic patients, as they allow for continuous monitoring of blood sugar levels [39]. By collecting real time data on glucose levels, these sensors provide valuable information that can be used to adjust medication dosages, optimize diet plans, and improve overall disease management. In the context of medical devices, sensors integrated into imaging equipment, such as MRI or ultrasound machines, enable high-resolution imaging and diagnostics. These sensors capture detailed data on body structures and functions, facilitating accurate diagnoses and targeted treatment plans. By leveraging the data collected by sensors in healthcare CPS, medical professionals can deliver proactive and personalized care, enhance patient outcomes, and reduce healthcare costs, ultimately improving the overall quality of healthcare services.

Environmental monitoring

Environmental monitoring is another area where sensors play a important role in data collection [40]. For example, in the agricultural sector, temperature and humidity sensors are used to monitor climatic conditions and optimize irrigation. By capturing data on temperature and humidity levels in real time, these sensors enable precise and efficient control of irrigation systems, ensuring optimal crop growth and water conservation. Additionally, water quality sensors detect parameters such as pH, dissolved oxygen, and pollutants, aiding in water resource management and maintaining ecological balance. By continuously monitoring water quality, these sensors provide early detection of water contamination or changes in aquatic ecosystems, enabling timely interventions to mitigate potential risks. Moreover, sensors that monitor air pollution levels play a vital role in assessing and controlling environmental impact. These sensors measure pollutants such as particulate matter, carbon monoxide, and sulfur dioxide, providing valuable data that can be used to implement effective pollution control measures and protect public health. By leveraging sensor data, environmental CPS solutions contribute to sustainable resource management, ecological balance, and public health, ensuring the preservation of our natural environment for future generations.

Smart homes

In the context of smart homes, sensors can create a connected and intelligent living environment. Illuminance sensors, for instance, detect ambient light levels and adjust indoor lighting accordingly [41]. By automatically adjusting lighting based on natural light availability, these sensors optimize energy consumption and create a comfortable ambience for residents. Motion sensors are another component of smart homes, detecting human presence within a designated area. By capturing data on movement or occupancy, these sensors trigger actions such as turning on lights or adjusting temperature settings, enhancing convenience and energy efficiency. For instance, when a person enters a room, the motion sensor can detect their presence and turn on the lights, and when they leave, it can automatically switch off the lights to conserve energy [21]. Humidity sensors are utilized to monitor moisture levels in indoor environments, such as bathrooms or basements. By continuously measuring humidity, these sensors can activate or adjust air conditioning and ventilation systems to maintain optimal indoor air quality, preventing issues such as mould growth or discomfort due to excessive humidity. By integrating sensors into smart home CPS solutions, homeowners can enjoy the benefits of a connected and automated living environment, improving energy efficiency, enhancing security, and providing greater comfort.

The uses of sensors in CPS are numerous and diverse, making them an indispensable component in gathering real time data which informs decision-making processes and system control. For example, sensors can be used to monitor traffic patterns and optimize traffic flow, resulting in improved efficiency and better resource management. They can also be used to monitor and track inventory in a warehouse, providing real-time information on stock levels and helping to prevent stockouts. However, while sensors are undoubtedly powerful tools, there are also numerous considerations and challenges associated with their use. In the next, we will delve into these issues in more detail, exploring specific techniques for data integration, analysis, and security in CPS scenarios. We will also discuss strategies for ensuring the reliability and accuracy of sensor data in various applications. Moreover, we will explore the benefits and limitations of sensor data collection in different applications. For instance, we will examine how sensors can be used to optimize energy consumption in buildings, leading to improved system performance and energy savings. We will also discuss cases of smart homes and wearables, among others, that utilize sensors in a variety of environments to enhance the overall user experience. By leveraging the power of sensors in these ways, CPS solutions can achieve enhanced functionality and improved resource management, leading to significant benefits for users and organizations alike.

Sensors in CPS application scenarios collect real-time data for informed decision-making and system control. The potential applications of sensor data collection are virtually limitless, encompassing a wide spectrum of industries and domains, from industrial automation to smart cities, healthcare, environmental monitoring, and smart homes. By harnessing the power of sensors, CPS solutions can achieve improved efficiency, enhanced functionality, and optimized resource management, leading to increased productivity, cost savings, and sustainability. However, data collection in CPS scenarios presents specific challenges, such as data integration, analysis, and security techniques, that must be addressed to ensure sensor data's reliability and accuracy. Moreover, while sensor data collection provides numerous benefits, it also has limitations that should be considered, such as privacy concerns and cost-effectiveness. Despite these challenges, sensor data collection remains an important component of CPS solutions, contributing to sustainable resource management, ecological balance, and public health, and paving the way for a connected and intelligent future.

3.2 Types of Data

CPS generate a vast amount of data, which can overwhelm organizations that process it. Big data are an umbrella term for any technology used to process such data and is characterized by the "4Vs": value, volume, velocity, and variety [42]. Volume refers to the amount of data generated, which can be challenging to manage and store [43]. Velocity is related to the speed of data generation, which can be extremely fast and continuous, increasing the need for real-time processing and analysis [44]. Variety pertains to the different types of data generated, which can come from various sources and have different formats [45].

CPS collect various types of data, such as one dimensional, image, video, spatial, audio, and textual data. One dimensional data such as temperature and pressure readings, and can provide valuable insights when analyzed properly. Image and video data are collected from cameras, drones, or any device that captures visual information and requires specialized processing and analysis techniques. Technologies such as machine learning help extracting meaningful insights from visualized data. Spatial data are location-based data collected from global positioning system (GPS) sensors or other location-based systems and can be used in tracking vehicles or monitoring traffic flow. Audio data are collected from microphones and includes speech and other sounds. Textual data includes any form of text and can be analyzed using natural language processing techniques to extract insights about customer sentiment or identify emerging trends.

Processing and analyzing these types of data can be challenging but can provide significant benefits for organizations. By gaining a deeper understanding of their data, organizations can make better-informed decisions and improve their overall performance. From predictive maintenance to personalized marketing, leveraging the power of big data in CPS can have endless possibilities. The types of data and data characteristics collected in CPS are shown in Fig. 3.4.

3.2.1 One Dimensional Data

In the world of CPS, one dimensional data collection is an important and widely used method that refers to the gathering of data along a single axis or dimension, as shown in Fig. 3.5. This is a relatively simple and straightforward method to collect, process, and analyze data. Despite its simplicity, one dimensional data collection is still a valuable tool in various scientific and engineering fields, where it is often used to describe physical phenomena such as temperature, pressure, and velocity. The applications of one dimensional data collection are broad and diverse. In environmental monitoring, one dimensional data can be collected to track changes in atmospheric temperature, humidity, and pressure. Similarly, in medical research, one dimensional data can be used to monitor the health of patients by tracking vital signs such as heart rate, blood pressure, and oxygen saturation levels. In the field of industrial quality

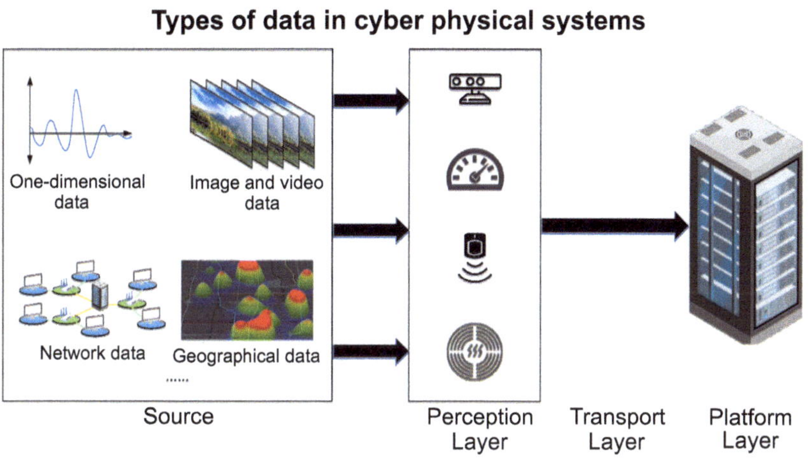

Fig. 3.4 Types of data collected in CPS and data characteristics. One dimensional data, image and video data and other typers of data are collected in the perception layer and uploaded to the platform layer for further processing

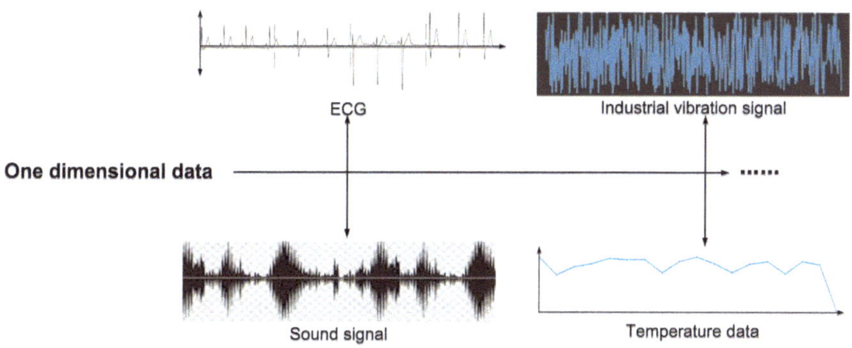

Fig. 3.5 Types of one dimensional signals

control, one dimensional data can be used to monitor the performance of machinery by tracking parameters such as speed, vibration, and temperature. Overall, one dimensional data collection is a valuable tool for data analysis, and its applications are vast and varied. Its simplicity makes it a popular choice for many fields, and its ability to provide valuable insights into physical phenomena continues to make it an important field of study in the scientific and engineering communities [46].

For instance, one dimensional data are an incredibly versatile tool that finds use in a wide range of fields. For instance, in civil engineering, engineers use one dimensional data to monitor the structural integrity of buildings and bridges, looking for any signs of distress or abnormal behavior [47]. Similarly, in the field of medicine, one dimensional data can be used to monitor a patient's vital signs (e.g., heart rate, blood pressure, and temperature) [48]. By monitoring one dimensional data, doctors and nurses can detect any abnormalities that may signal a potential health issue and take appropriate action to ensure the patient's safety. In the field of environmental science, one dimensional data are used to monitor various environmental factors, such as the levels of pollutants in the air or water [49]. These one dimensional data facilitate the study of the impact of pollution on the environment and the identification of any necessary measures to mitigate the effects of pollution.

Furthermore, the versatility of one dimensional data allows it to be combined with other types of data, such as two dimensional or three dimensional data, to provide a more comprehensive understanding of a system [50]. For example, in a weather monitoring application, one dimensional data on wind speed can be combined with two-dimensional data on wind direction to provide a more detailed picture of the weather conditions [51]. The combination of data can lead to new insights and discoveries, and help researchers and engineers better understanding complex systems, making one dimensional data an essential tool in their arsenal.

The main advantage of one dimensional data is that it is relatively simple to collect, process and analyze. Nevertheless, this simplistic nature can double as a limitation, possibly overlooking the intricate aspects of a system or process. As an illustration, one dimensional data might provide insights on temperature, yet fail to encapsulate the entire gamut of contributing factors such as wind velocity or ambient humidity [52]. Despite its inherent restrictions, one dimensional data persist as significant assets in the CPS domain. With inherent user-friendliness and straightforward interpretation, such data proves particularly fitting for an array of applications. Consequently, it remains an invaluable asset for researchers and engineers striving to decode the complexities of our environment. With the development of new technology and sensors, one dimensional data are becoming more powerful and useful, allowing for real time analysis of dynamic systems and processes.

One dimensional data are a simple yet powerful tool for collecting and analyzing data in the field of CPS. Although they may not capture the whole complexity of a system or process, it still has many advantages compared to other methods. For example, one dimensional data can be easily collected in real time and requires less computational power, making it an ideal choice for many applications. Furthermore, the simple interpretation of the data can provide valuable insights that may be hidden in more complex data sets. Since one dimensional data are widely used in different fields, they are important for scientists and engineers to understand its limitations and potential applications. As technology advances, new methods and tools will emerge, but one dimensional data remain irreplaceable in CPS.

3.2.2 Image and Video Data

Image and video data hold a substantial position in the process of recording and evaluating the environment within the context of CPS. These systems rely on various sensors and cameras to collect image and video data, which is then utilized for a wide range of applications. In the field of transportation, for example, image data

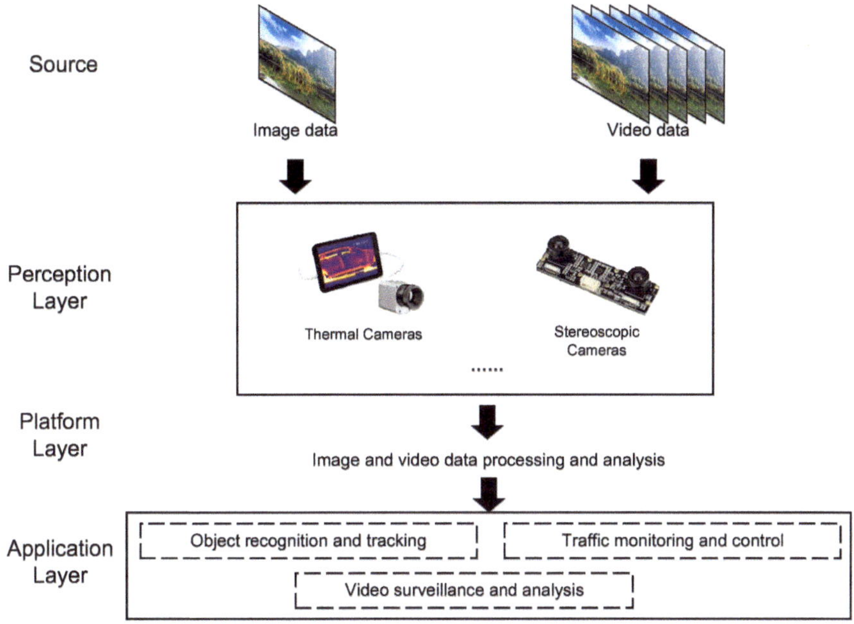

Fig. 3.6 Illustration of the role of image and video data in CPS. The data collected from various cameras enable object recognition, traffic control, and video surveillance to improve the safety and efficiency of intelligent systems

are of utmost importance in CPS. Image and video data enable object recognition, tracking, and classification based on predetermined characteristics. Accurately identifying objects and their movements allows for effective traffic control and the safe operation of autonomous vehicles, ultimately preventing accidents and improving transportation efficiency [53]. The role of image and video data in CPS is shown in Fig. 3.6.

The use of image and video data is not limited to transportation only, as it has a wider range of applications in other industries. For instance, image and video data can be utilized in monitoring and improving manufacturing processes, identifying and detecting defects, and optimizing various parameters to enhance productivity. By utilizing image data, CPS can efficiently monitor and analyze manufacturing operations, which can lead to a significant improvement in quality control and optimized production processes. Additionally, the use of video data can also enable real-time monitoring of various processes, allowing for quick identification and resolution of issues. As such, the use of image and video data has proven to be invaluable in various industries, with its applications extending beyond just transportation [54].

Video data provides an even more detailed understanding of the environment and is particularly useful for surveillance and monitoring purposes. When analyzing video data, it is possible to identify and predict not only objects and human behavior, but also changes in the environment over time. Video data are especially useful in the context of transportation, where video surveillance can effectively monitor traffic conditions, detect accidents, and ensure overall road safety. Moreover, the analysis of video data plays a important role in detecting criminal activities and tracking suspicious behavior, thereby enhancing public safety. With advances in machine learning and computer vision technologies, we can now process video data more efficiently, accurately and at scale. Video data allow us to extract even more insights from the data, such as identifying patterns and anomalies in behavior, predicting future events, and generating actionable insights. The use of video data is a powerful tool for understanding and improving our environment. By leveraging this technology effectively, we can enhance safety, optimize operations, and create a more sustainable future for all [55].

However, collecting and processing image and video data in a CPS requires significant computing resources and storage capacity. This can be attributed to the fact that the data itself is often large and complex, thus presenting several challenges in terms of management and processing [56]. However, with the advent of advanced algorithms and techniques, it becomes possible to transform the collected data into valuable insights that can significantly improve safety, security, and efficiency within CPS. Moreover, the use of advanced technologies such as machine learning can further enhance the analysis and interpretation of the data, allowing for the identification of patterns and anomalies that might not be apparent to the human eye. Thus, the integration of such technologies can potentially revolutionize the way in which CPS operates, paving the way for new applications and advancements in the field.

By incorporating image and video data, CPS can offer a multitude of benefits. The use of image data enables the identification of potential hazards, preventing accidents and ensuring the safety of individuals. Simultaneously, the analysis of video data facilitates the detection of criminal activities and the tracking of suspicious behaviour, thereby bolstering public safety efforts. Moreover, CPS that incorporates image and video data can be leveraged to enhance energy efficiency, reduce waste, and optimize resource allocation, benefiting both the environment and overall system performance. Therefore, image and video data are vital components of CPS, supporting the collection, processing, and analysis necessary to ensure the safety, security, and efficiency of CPS applications. The continued development of efficient algorithms and techniques is important for effectively managing and processing this data while exploring new avenues to leverage its potential for societal benefits. As CPS becomes increasingly prevalent, the importance of image and video data will continue to grow, shaping the future of these systems.

3.2.3 Other Types of Data

In addition to one dimensional data and image/video data, CPS also collect other vital information to ensure efficient and reliable operation. These include three-dimensional data, location data, and network data. These diverse data types enhance CPS capabilities by providing a comprehensive understanding of the physical environment, enabling precise localization and tracking, and facilitating seamless communication and coordination between interconnected components.

In CPS, three dimensional data plays a vital role in capturing and representing the physical world. This data are collected through various sensors and devices, such as depth cameras, light detection and ranging (LiDAR) scanners, or stereo vision systems [57]. These sensors capture information about the physical environment in three dimensions, allowing for the creation of detailed spatial representations.

In robotics, three dimensional data are used for object recognition, scene understanding, and mapping. For example, a robot equipped with a three dimensional sensor can create a point cloud representation of its surroundings, enabling it to detect and navigate around obstacles [58]. In manufacturing processes, three dimensional data are utilized for quality control and inspection, ensuring accurate measurements and precise alignments. In virtual reality and augmented reality applications, three dimensional data are essential for creating immersive and realistic virtual environments.

The analysis of three dimensional data involves a range of techniques that are employed to extract meaningful information. These techniques include point cloud processing, surface reconstruction, and object recognition. Point cloud processing is the process of converting a set of unordered points into a structured representation. Surface reconstruction, on the other hand, is the process of constructing a surface from a set of points. Object recognition is the process of identifying objects in a scene and classifying them based on their features. By employing these techniques, CPS can make informed decisions and interact effectively with the physical world, thereby enhancing their capabilities and enabling them to perform a range of complex tasks.

Location data in CPS refers to information that describes the geographical position of objects, assets, or individuals within the system. It involves collecting data from various positioning technologies, including GPS, Wi-Fi triangulation, Bluetooth beacons, radio frequency identification (RFID), or even indoor positioning systems [59].

In transportation and logistics, location data enables real time tracking of vehicles, shipments, or personnel [60]. It helps optimizing routes, monitoring delivery progress, and enhancing overall supply chain efficiency. Emergency services also rely on location data to quickly locate and respond to incidents. In smart cities, location data are utilized for managing public transportation, analyzing traffic patterns, and providing location-based services to citizens.

Location data are often combined with other data sources, such as sensor data or weather information, to derive additional insights. Geospatial analysis techniques are used to process and visualize location data, enabling effective decision-making and resource allocation in CPS [61].

Network data refers to the information related to communication and connectivity in CPS. In a CPS, numerous interconnected devices and components exchange data and coordinate actions through communication networks. This network data encompasses various aspects, including network topology, traffic patterns, latency, bandwidth utilization, and packet-level details [62]. Analyzing network data in CPS involves techniques such as network traffic analysis, anomaly detection, and performance monitoring. These analyses enable proactive management and optimization of the network infrastructure, ensuring efficient resource utilization, enhancing network reliability, and facilitating effective communication and coordination among the interconnected devices and components. By leveraging network data analysis, CPS can gain valuable insights into network behavior, identify potential issues or threats, and make informed decisions to optimize the performance and functionality of the system. The analysis of network data is essential for the smooth operation and successful functioning of CPS.

Monitoring and analyzing network data in real-time are important for maintaining the reliability, security, and performance of the overall system [63]. By monitoring network traffic, it becomes possible to identify abnormal behavior, detect potential security threats, and prevent network congestion. Furthermore, network data analysis helps optimizing network resources, ensuring efficient data transfer, minimizing latency, and improving overall user experience. Additionally, network data analysis can provide insights into usage patterns, allowing for more informed decision-making when it comes to network infrastructure planning and expansion. It also enables network administrators to identify potential bottlenecks and proactively address them before they become a problem. Therefore, it is essential to conduct regular network data analysis to ensure the reliable and secure operation of the network infrastructure.

In industrial settings, network data plays a important role in managing various distributed control systems [64]. These control systems allow for the seamless and efficient operation of complex industrial processes and machinery. Additionally, network data are also used for supervisory control and data acquisition (SCADA) systems, which are responsible for gathering and analyzing data from various sensors and control systems in real-time [65]. By doing so, SCADA systems provide operators with valuable insights that help them making informed decisions and keeping industrial operations running smoothly. In the context of smart grids, network data are equally important. With the rise of renewable energy sources and the growing demand for energy, it is important to balance the distribution of energy across the grid. Network data plays a critical role in this process by providing real-time information on power generation and consumption. Additionally, network data enables operators to respond quickly to fluctuations in energy demand and adjust energy distribution accordingly. This ensures that the grid remains stable and reliable, even during times of peak demand or unexpected events.

CPS can collect different types of data that require specialized tools and techniques to process and analyze. Three-dimensional data, location data, and network data are becoming increasingly important in CPS applications. As CPS continue to evolve, it is likely that new types of data will emerge, and new tools and techniques will be required to handle them. By leveraging the insights provided by these different types of data, CPS can help improving efficiency, reducing costs, and enhancing overall performance.

3.3 Real Time and Latency

Achieving real time responsiveness and minimizing latency in CPS requires a comprehensive and multidimensional approach. This approach must consider the interplay of various factors such as edge computing, predictive analytics, data compression, prioritized communication, hardware acceleration, energy-efficient designs, fault-tolerant mechanisms, appropriate wireless technologies, and secure protocols. To further elaborate, edge computing involves processing data closer to the source, thereby reducing the amount of data that needs to be transmitted over long distances. Predictive analytics allows for the identification of patterns and trends in data, which can be used to optimize system performance. Data compression techniques can reduce the size of data packets, thereby reducing latency. Prioritized communication can improve the speed and reliability of data transmission by prioritizing critical data. Hardware acceleration involves the use of specialized hardware to enhance processing speed and reduce latency. Energy-efficient designs can optimize energy consumption, enabling devices to operate for longer periods without the need for frequent charging. Fault-tolerant mechanisms can ensure that the system continues to function even in the presence of hardware or software failures. Appropriate wireless technologies and security protocols can ensure that data transmission is secure and reliable.

In addition to the above strategies, evaluating the performance of real time operations and latency in CPS requires a comprehensive assessment of various factors. These include response time, predictability, meeting deadlines, the dynamic nature of the system, the impact of network communication, safety requirements, and fault tolerance. Response time is the time it takes for the system to generate feedback to inputs. Predictability involves the ability to accurately predict system behavior. Meeting deadlines involves ensuring that critical tasks are completed within a specified time frame. The dynamic nature of the system refers to the fact that CPS is often deployed in environments where conditions can change rapidly. The impact of network communication refers to the effect that network latency can have on system performance [66]. Safety requirements are critical in CPS as they ensure that the system operates safely and does not pose a risk to users. Fault tolerance ensures that the system can continue to function even in the presence of hardware or software failures.

To effectively evaluate CPS performance, appropriate evaluation techniques, and methodologies must be employed. Simulation-based evaluation involves the use of computer simulations to test system performance under various conditions. Real-world testing involves testing the system in real-world scenarios to identify potential issues and areas for improvement. Ongoing monitoring and optimization involve continually monitoring the system and making adjustments to optimize system performance. By employing appropriate evaluation techniques and methodologies, CPS can be continually improved and successfully applied in diverse real time scenarios.

3.3.1 Techniques for Reducing Latency

In order to achieve real time response and improve systematicity, latency in CPS needs to be reduced. This discussion will explore techniques that can effectively mitigate latency in CPS, ensuring timely and accurate data processing and decision-making. The key techniques used to reduce CPS latency are shown in Fig. 3.7.

Edge computing is a common technique for reducing latency in CPS [67]. It enables real time data processing, analysis, and decision-making at the network edge, eliminating the need to transmit data to a centralized server. By deploying edge nodes or edge servers in close proximity to the data-generating devices or sensors, latency can be significantly reduced. This approach enhances the responsiveness of CPS, enabling swift reactions to critical events and reducing reliance on centralized cloud-based architectures. It is important to note that edge computing encompasses two distinct architectures: mobile edge computing (MEC) [68] and multi-access edge computing (MEC) [69]. MEC involves deploying edge nodes or servers in close proximity to end-user devices such as smartphones or tablets. It enables real time data processing and analysis at the edge, reducing the need for data transmission to a

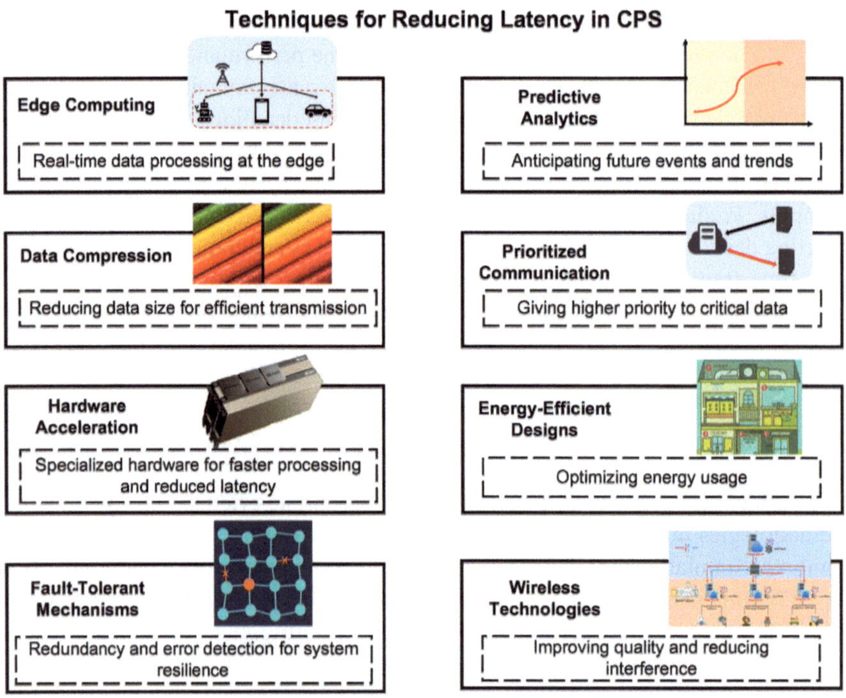

Fig. 3.7 Schematic of the key technologies used to reduce CPS latency. These techniques include edge computing, data compression, hardware acceleration, fault-tolerant mechanisms, predictive analytics, prioritized communication, energy-efficient designs, and wireless technologies

centralized server. On the other hand, multi-access edge computing (MEC) involves deploying edge nodes or servers in proximity to network access points such as base stations or routers. It enables real time data processing and analysis at the network edge, minimizing the need for data transmission to a centralized server. Both MEC and multi-access edge computing (MEC) significantly contribute to reducing latency in CPS, enabling real time responsiveness and enhancing overall system performance.

Another technique for reducing latency in CPS is the use of predictive analytics [70]. Predictive analytics can anticipate future events and trends by leveraging historical data, statistical models, and machine learning algorithms. This proactive approach allows for pre-computed decisions to be made based on predictions, reducing the reliance on real time data processing and consequent latency. By preemptively making decisions based on these predictions, CPS can avoid the latency associated with waiting for real time data. Predictive analytics can optimize system performance and reduce the time required for decision-making, ensuring timely responses in dynamic environments. Moreover, the use of predictive analytics in CPS can be further enhanced by combining it with other techniques such as anomaly detection and fault prediction. Anomaly detection can identify abnormal behavior or events in the system, enabling proactive measures to be taken before they escalate into issues. Fault prediction can anticipate potential system failures or malfunctions, enabling preemptive maintenance to be carried out before they occur. By combining these techniques with predictive analytics, CPS can operate more efficiently in dynamic environments, reducing latency and enhancing system resilience.

Data compression is a technique that reduces CPS latency. In these systems, a substantial amount of data are generated and transmitted over the network [71]. By compressing data before transmission, the amount of information that needs to be transferred can be minimized. This reduction in data size reduces the time required for data transmission and reception, thereby reducing overall latency. Advanced compression algorithms, such as delta encoding or Huffman coding [72], can effectively reduce the size of data packets without compromising their integrity. Employing data compression techniques in CPS leads to improved real time performance and reduced network congestion, allowing for faster and more efficient data processing. Additionally, data compression techniques can reduce the cost of data transmission, making CPS more cost-effective and scalable.

Moreover, in CPS, not all data and communication require the same level of priority [73]. By implementing prioritized communication, where important and time-sensitive data are given higher priority, latency can be minimized. This can be achieved by employing Quality of Service (QoS) mechanisms that prioritize real time data packets over non-critical traffic [74]. By ensuring that critical information reaches its destination without delay, CPS can operate with minimal latency, enabling swift and accurate responses. Prioritized communication can be achieved through techniques such as traffic shaping, packet scheduling, or the use of different network channels dedicated to real time data. By effectively managing the flow of data and giving priority to time-critical information, latency in CPS can be significantly reduced. Furthermore, prioritized communication can improve the overall reliability and security of CPS.

Furthermore, hardware acceleration is a technique that involves using specialized hardware components to perform specific computational tasks more efficiently than general-purpose processors [75]. In CPS, certain computational workloads can be highly demanding and time-sensitive, leading to increased latency if processed solely by traditional processors. Wireless latency can be significantly reduced by offloading these computationally intensive tasks to dedicated hardware accelerators. Hardware accelerators, such as graphics processing units (GPUs) or field-programmable gate arrays (FPGAs), are designed to excel in specific types of computations. By utilizing these accelerators, CPS can benefit from faster processing speeds, enabling real time data analysis and decision-making with minimal latency. Hardware acceleration not only reduces latency but also enhances the overall performance and energy efficiency of the system.

To ensure optimal real time responsiveness and minimal latency in CPS, it is important to consider the trade-offs between latency and energy consumption. Energy-efficient designs and techniques can help minimizing the latency introduced by power constraints. For example, the use of low-power processors, energy-efficient algorithms, and power management strategies can significantly reduce the energy consumption of CPS while maintaining real time responsiveness. By carefully optimizing the energy usage of the system, organizations can strike a balance between performance and energy efficiency, resulting in reduced latency and improved overall system operation.

Additionally, the design of fault-tolerant mechanisms and redundancy can contribute to latency reduction in CPS. Redundancy can involve duplicating critical components or data paths to ensure that if one fails, a backup is immediately available. This redundancy helps preventing latency caused by system failures, ensuring uninterrupted operation. Additionally, fault-tolerant mechanisms, such as error detection and recovery, can quickly identify and rectify errors, minimizing the time required to restore normal operation. By incorporating these fault-tolerant techniques, CPS can effectively mitigate latency due to failures or errors, maintaining real time responsiveness and system integrity.

Moreover, in the context of wireless communication in CPS, the choice of appropriate wireless technologies and protocols can significantly impact latency. For instance, leveraging advanced wireless communication standards like 5G can provide lower latency and higher data transmission rates compared to previous generations. By adopting technologies that offer reduced latency, such as Ultra-Reliable Low-Latency Communication (URLLC) in 5G [76], CPS can achieve real time responsiveness with minimal delay. Additionally, using techniques like beamforming and antenna diversity can enhance wireless communication reliability and reduce latency by improving signal quality and reducing interference.

Lastly, the security aspects of CPS need to be taken into account at the same time as reducing latency. Implementing robust security measures, such as encryption, authentication, and access control, is vital to protect sensitive data and prevent cyberattacks. However, these security measures should be designed in a way that minimizes additional latency. Employing efficient encryption algorithms, lightweight cryptographic protocols, and hardware acceleration for security tasks can help reducing the computational overhead and latency associated with securing data in CPS.

Achieving real-time response and minimizing latency in CPS requires an integrated, multidimensional approach. By employing strategies such as edge computing, predictive analytics, data compression, prioritized communication, hardware acceleration, energy-efficient designs, fault-tolerant mechanisms, appropriate wireless technologies, and secure protocols, organizations can effectively minimize latency and enhance system performance. These techniques not only reduce latency but also improve the overall efficiency and reliability of CPS, enabling them to operate seamlessly in dynamic and time-critical environments. By staying up-to-date with the latest developments in CPS and leveraging these techniques to their advantage, organizations can create new business models, improve customer experience, and gain a competitive edge in the market.

3.3.2 Key Considerations of Real Time and Latency

Real time responsiveness and low latency are important considerations in the design and implementation of CPS. We'll explore five key considerations that are essential for achieving real time performance and minimizing latency in CPS. These considerations encompass various aspects of system architecture, communication protocols, data processing, and decision-making, all aimed at ensuring the timely and accurate operation of CPS in dynamic environments. Key considerations for real-time response and low latency are shown in Fig. 3.8.

System architecture

System architecture affects CPS real time responsiveness and latency. A well-designed architecture should prioritize local processing and decision-making to minimize the need for data transmission to centralized servers. This means that the system should be designed with multiple processing points distributed across the network, enabling data processing at various locations instead of relying on a central point. A distributed architecture, such as edge computing, can enable the processing and analysis of data at the edge of the network, closer to the data sources [77]. This reduces the reliance on long-distance data transmission, resulting in lower latency and improved real time performance. Furthermore, the architecture should be scalable and flexible to accommodate the growing number of connected devices and changing system requirements without compromising real time responsiveness. This means that the system should be designed so that it can easily adapt to changes in the number of devices connected to it and to changes in the requirements of those devices.

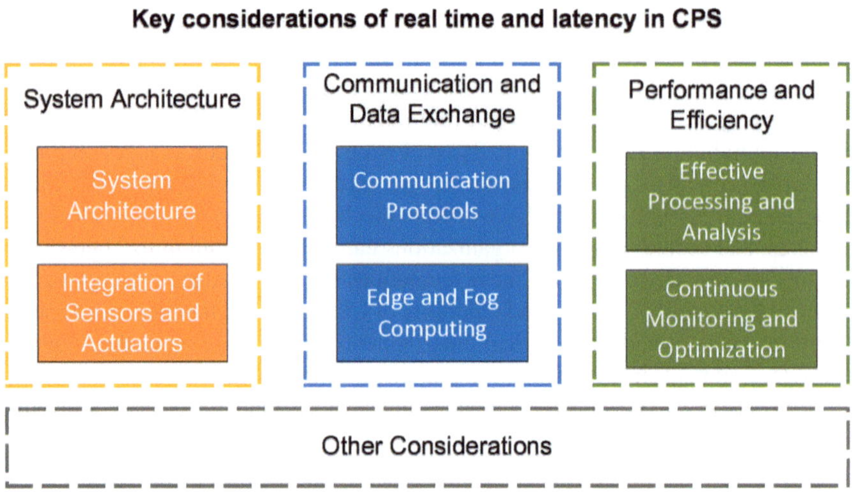

Fig. 3.8 Key considerations for real-time and latency issues. These include system architecture, communication and data exchange, and performance and efficiency

Communication protocols

Efficient, low-latency communication protocols ensure timely data exchange among devices, sensors, and actuators in CPS. Prioritizing protocols that enable real time data transmissions, such as the real time Publish-Subscribe (RTPS) protocol or the Message Queuing Telemetry Transport (MQTT) protocol with QoS mechanisms, is important for minimizing latency [78]. These protocols facilitate fast and reliable communication, minimizing latency and ensuring that critical information reaches its intended destination without delay. Additionally, the use of lightweight protocols and optimized network configurations further contributes to reducing latency in CPS by conserving bandwidth and minimizing data transfer time.

Effective processing and analysis

Effective processing and analysis of data play a substantial role in facilitating real-time responsiveness in CPS. Selecting appropriate data processing algorithms and techniques capable of handling large volumes of data in a time-sensitive manner is important. Techniques such as stream processing and complex event processing (CEP) enable real time data analysis and decision-making by continuously processing and aggregating data as it arrives [79]. Moreover, leveraging predictive analytics and machine learning algorithms provides valuable insights and enables proactive decision-making, reducing the reliance on real time data processing and minimizing latency. These techniques should be designed to handle large data volumes quickly and accurately, ensuring real time responsiveness.

Integration of sensors and actuators

The integration of sensors and actuators in CPS is an important aspect of achieving real time responsiveness [80]. The selection of sensors and actuators should prioritize low-latency communication and rapid response times. Considering the physical characteristics and capabilities of the devices is important to ensure they can operate within the required time constraints. Optimizing the placement and deployment of sensors and actuators minimizes signal propagation delays, enabling timely data collection and control actions. Therefore, the chosen sensors and actuators should facilitate quick and accurate communication with the rest of the system, and their placement should minimize data travel time within the system.

Continuous monitoring and optimization

Continuous monitoring and optimization facilitates real-time identification and resolution of latency issues in CPS [81]. Monitoring system performance metrics such as response times, data transmission rates, and processing delays provides valuable insights into areas that require improvement. Proactive optimization measures, including load balancing, resource allocation, and adaptive control strategies, can be employed to mitigate latency and enhance overall system performance. Implementing feedback mechanisms and closed-loop control strategies allows for dynamic adaptation of system behavior based on real time conditions and requirements, ensuring optimal real time responsiveness with minimal latency. This continuous monitoring and optimization approach ensures that the system operates as efficiently as possible.

Edge computing and fog computing

In recent years, advances in edge computing and fog computing have offered new opportunities for reducing latency in CPS [82]. Edge computing involves processing data closer to the edge of the network, while fog computing extends this concept by enabling data processing and analysis at multiple points within the network infrastructure. By leveraging edge and fog computing paradigms, latency can be further reduced by minimizing the distance data needs to travel for processing and decision-making. This distributed computing approach enhances real time responsiveness and reduces reliance on centralized servers or cloud-based architectures. However, it is important to carefully design and manage the distribution of computing resources in edge and fog computing environments to ensure optimal performance and minimize latency.

Other considerations

Furthermore, with ongoing technological advancements, it's essential to consider potential factors affecting real-time responsiveness and latency in CPS. For example, the rise of 5G networks presents new challenges and opportunities, warranting further research and exploration. 5G networks offer significantly lower latency and higher data transmission rates compared to previous generations. Leveraging the capabilities of 5G can further enhance the real time performance of CPS by enabling faster and more reliable communication. However, it is essential to carefully evaluate the deployment and integration of 5G networks within CPS to ensure seamless compatibility and optimal performance. Additionally, as CPS becomes increasingly interconnected with other emerging technologies such as the IoT and artificial intelligence (AI), the potential synergies and challenges in achieving real time responsiveness and minimizing latency should also be considered. Integrating CPS with IoT devices and AI algorithms can introduce additional complexity and potential latency sources, requiring careful system design and optimization to maintain real time operation.

Achieving real time responsiveness and minimizing latency in CPS require a holistic approach that considers various aspects of system design, communication protocols, data processing techniques, sensor and actuator integration, continuous monitoring, and optimization. By addressing these considerations and staying abreast of emerging technologies and trends, organizations can unlock the full potential of CPS and harness its benefits in diverse applications. With the ability to process, analyze, and respond to data in real time with minimal latency, CPS can revolutionize industries, improve efficiency, and enhance the overall human experience. One example of the potential benefits of CPS is its impact on the transportation industry, where it can enhance road safety, optimize traffic flow, and reduce congestion.

3.3.3 Evaluating the Performance

The real time operations of the CPS ensure the responsiveness and efficiency of the system. These operations refer to the ability of the CPS to respond to events and process data within specified time constraints. Evaluating the performance of real time operations in CPS involves assessing various factors such as response time, predictability, the dynamic nature of the system, the impact of network communication, safety requirements, and fault tolerance, as shown in Fig. 3.9. However, conducting a comprehensive evaluation of CPS's real time performance is a complex task that requires a thorough understanding of these factors and the application of appropriate evaluation techniques and methodologies.

Response time evaluation is an important aspect of evaluating CPS's real time operations [83]. It involves assessing the time taken by the system to react and provide meaningful outputs in response to an input or event. The process entails analyzing the different stages of data processing, communication, and decision-making

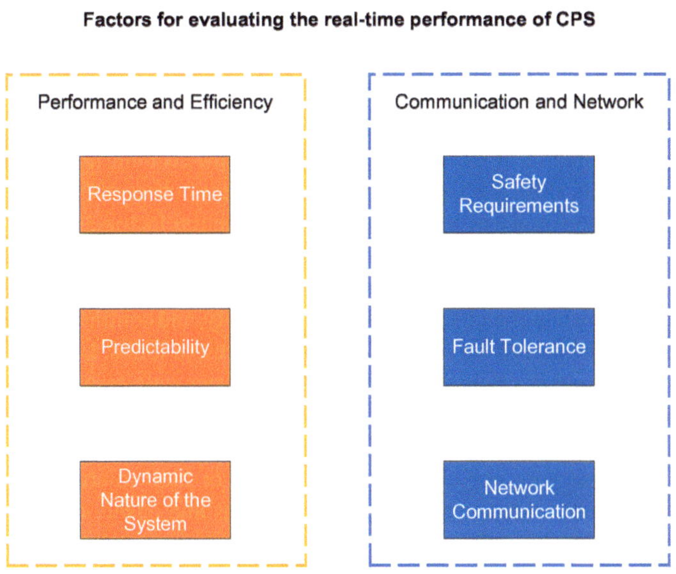

Fig. 3.9 Factors for evaluating the real-time performance of CPS such as response time, predictability, dynamic nature of the system, impact of network communication, security requirements and fault tolerance

within the system. Researchers and engineers can identify potential bottlenecks and optimize the system's performance by measuring the time required for data acquisition, transmission, processing, and generation of appropriate responses. Response time evaluation helps in improving CPS's real time performance by ensuring that the system can meet timing requirements and deliver timely outputs.

Predictability evaluation is another consideration in CPS's real time operations [84]. It refers to the ability to estimate or determine the system's behavior and performance with a high degree of accuracy. The evaluation entails analyzing the system's response under different conditions, such as varying workloads, network congestion, or component failures. Predictability evaluation enables the identification of potential sources of uncertainty or variability that may affect the system's real time performance. Researchers and engineers can design robust CPS architectures, algorithms, and control strategies by quantifying predictability to ensure reliable and predictable real time operations. By understanding and quantifying the system's behavior under different scenarios, they can make informed decisions to optimize performance and mitigate risks.

Deadline needs to be met for CPS running in real-time scenarios [85]. Analyzing the system's ability to accomplish tasks or deliver outputs within predetermined time limits is part of the evaluation process. The process involves assessing the timing constraints imposed by the application domain, the specific requirements of the system, and the available computing and communication resources. Evaluating deadline compliance helps researchers and engineers ensuring that CPS performs within the required temporal bounds, enabling timely decision-making and effective control in time-critical environments. It involves studying the scheduling algorithms, task allocation strategies, and resource management techniques to ensure that tasks are prioritized and executed in a manner that meets the specified deadlines.

The dynamic nature of CPS needs to be taken into account in the performance evaluation process [86]. CPS operate in environments characterized by changing conditions such as variations in workload, environmental factors, or system configurations. Evaluating the system's real time performance under dynamic conditions involves analyzing its adaptability, responsiveness, and robustness. It requires assessing how the system handles changes, maintains stability, and continues to meet timing requirements in the face of varying inputs, disturbances, or faults. Understanding and quantifying the system's behavior in dynamic scenarios help researchers and engineers designing mechanisms that enable CPS to maintain reliable real time operations despite changing circumstances. This may involve developing adaptive control strategies, dynamic resource allocation techniques, or fault tolerance mechanisms that can dynamically respond to changing conditions and ensure continued system performance.

The impact of network communication on real time operations is another important factor in CPS evaluation [87]. CPS often rely on communication networks to exchange data and synchronize their operations. Evaluating the impact of network communication involves analyzing factors such as network latency, jitter, bandwidth, and reliability. It requires assessing how network characteristics affect the system's ability to meet real time requirements and deliver timely responses. By considering the impact of network communication, researchers and engineers can design communication protocols, network architectures, and quality-of-service mechanisms that optimize the system's real time performance. This may involve techniques like network traffic analysis, communication scheduling, or redundancy mechanisms to mitigate the effects of network delays or failures.

Safety requirements and fault tolerance are significant aspects to evaluate in CPS's real time operations [88]. CPS often operate in safety-critical domains where failures can have severe consequences. Evaluating safety and fault tolerance involves assessing the system's ability to detect, tolerate, and recover from faults or errors that may occur during real time operations. It requires analyzing fault detection mechanisms, fault recovery strategies, and the system's overall resilience. Evaluating safety and fault tolerance helps ensuring that CPS can maintain reliable and secure real time operations, even in the presence of potential failures or disruptions. This may involve techniques such as fault injection, fault tree analysis, or redundancy mechanisms to enhance system reliability and safety.

Evaluating the performance of real time operations in CPS is a complex task that requires a comprehensive assessment of factors such as response time, predictability, meeting deadlines, the dynamic nature of the system, the impact of network communication, safety requirements, and fault tolerance. By employing appropriate evaluation techniques and methodologies, researchers and engineers can gain insights into the system's behavior, optimize performance, and ensure the reliable and efficient operation of CPS in various domains. The combination of simulation-based evaluation, real-world testing, and ongoing monitoring and optimization contributes to the continual improvement of CPS performance and its successful application in diverse real time scenarios.

3.4 Data Quality and Reliability Issues

Data quality and reliability represent elements within the CPS framework, wherein the smooth interfacing between physical and computational components is highly beneficial. However, CPS face challenges in ensuring the accuracy, consistency, and dependability of the data collected from various sensors and devices. These challenges arise due to factors such as noise, outliers, missing values, inconsistencies, data integration issues, and the increasing complexity of CPS architectures. Therefore, these challenges must be addressed to ensure the proper operation and performance of the CPS.

CPS generate vast amounts of data from numerous sensors and devices, making data quality a primary concern. Various factors contribute to data quality challenges in CPS. Firstly, noise and outliers present in the collected data can significantly impact the accuracy and reliability of the overall system. Erroneous or inconsistent readings can lead to incorrect interpretations and decisions. Secondly, missing values within the data pose challenges in data analysis and system operation. Incomplete data can hinder the extraction of meaningful insights and compromise the reliability of the system's responses. Additionally, inconsistencies and integration issues arise when merging data from heterogeneous sources, such as different sensor types or communication protocols. Therefore, consistency and accuracy of data are required for CPS to maintain proper operation.

In CPS, data reliability encompasses aspects such as fault tolerance, availability, and data consistency. Maintaining data reliability is challenging due to the distributed and dynamic nature of CPS. Sensor failures, communication errors, and network disruptions can lead to the loss or corruption of data, impacting system reliability. Fault tolerance mechanisms, such as redundancy and error detection/correction techniques, must be employed to mitigate these challenges. Moreover, ensuring the availability of data in real-time is important for timely decision-making and system operation. Unreliable data availability can lead to delays or incorrect actions, jeopardizing the performance and reliability of CPS. Furthermore, ensuring data consistency among distributed components and resolving synchronization issues are important for maintaining system integrity and reliability.

Address data quality and reliability to ensure optimal CPS operation and performance. Reliable and accurate data are vital for informed decision-making, system control, and overall system performance. By employing effective data preprocessing techniques such as data cleaning, integration, transformation, and reduction, CPS can improve data quality and enhance its reliability. Techniques like data validation and verification, statistical analysis, redundancy checks, and error correction codes can help ensuring the accuracy and consistency of the collected data. Furthermore, data management strategies, including data storage, retrieval, and synchronization, play an important role in maintaining data quality and reliability. By addressing these issues, CPS can optimize their performance, enhance their resilience to failures, and improve their overall functionality and efficiency.

Data quality and reliability issues pose significant challenges in the operation and performance of CPS. Ensuring accurate, consistent, and reliable data are vital for informed decision-making, system control, and overall system performance. By addressing data quality and reliability challenges through effective data preprocessing techniques, validation mechanisms, and data management strategies, CPS can enhance their reliability, efficiency, and resilience. Ongoing advancements and research in data quality and reliability will continue to play a important role in further improving the accuracy and dependability of CPS, leading to more robust and trustworthy systems.

3.4.1 Data Preprocessing Techniques

Data quality and reliability are important to the proper functioning of CPS. However, CPS often face challenges in ensuring the accuracy and dependability of the data collected from various sensors and devices [89]. These challenges arise due to factors such as noise, outliers, missing values, inconsistencies, and data integration issues. To overcome these challenges, data preprocessing techniques play a pivotal role. This essay delves into the significance of data preprocessing techniques in addressing data quality and reliability issues in CPS, emphasizing their benefits in enhancing data accuracy, reducing noise, handling missing values, improving data integration, and overall system performance. By employing effective data preprocessing techniques, CPS can enhance their reliability, and efficiency, leading to more robust and dependable CPS. The role of data preprocessing techniques in data collection is shown in Fig. 3.10.

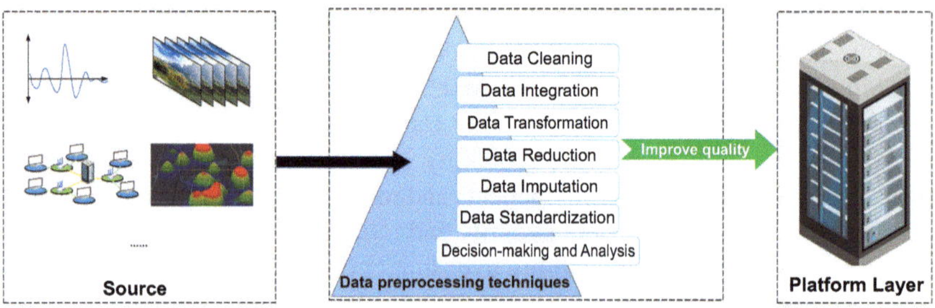

Fig. 3.10 Collected data requires data preprocessing. The data preprocessing techniques include data cleaning, data integration, data transformation, data reduction, data imputation, data standardization, decision-making and analysis. completed. Data preprocessing is completed before further transmission

Data cleaning

Data cleaning is aimed at identifying and rectifying errors or inconsistencies within gathered data. In CPS, where large volumes of data are generated by numerous sensors, the chances of data corruption, noise, and outliers are high. Data cleaning techniques such as outlier detection, noise removal, and error correction algorithms help eliminating inaccurate and unreliable data points [90]. By identifying and rectifying these issues, data cleaning ensures the reliability and accuracy of the data used in CPS, enabling better decision-making and improved system performance. Additionally, data cleaning techniques aid in identifying and resolving data integration issues by standardizing data formats, removing duplicates, and resolving inconsistencies across different data sources. This integration of data cleaning and integration processes ensures the overall quality and reliability of data in CPS.

Moreover, data cleaning can help in various ways. For instance, it can also help maintaining consistency in data sets, which is important for long-term analysis. Additionally, data cleaning can help identifying hidden patterns, trends, and correlations within the data, which can provide valuable insights for decision-makers. It can also help in identifying and handling missing data points, which can otherwise hinder the analysis and interpretation of the data [91]. Data cleaning is thus an important preprocessing step that can not only improve data quality but also enhance the overall reliability and usability of data in CPS.

Data integration

Data integration focuses primarily on improving data quality and reliability, especially in the context of CPS. In these systems, data are collected from multiple sources, which may have different data formats, structures, and levels of accuracy. Data integration techniques help in combining heterogeneous data from diverse sources into a unified and consistent format. These techniques involve data transformation, schema mapping, and semantic reconciliation to ensure that data from different sources can be effectively integrated [92]. By resolving semantic conflicts, handling data inconsistencies, and ensuring data compatibility, data integration enhances the overall data quality and reliability in CPS. It enables seamless data exchange, facilitates data analysis, and supports reliable decision-making processes. Moreover, data integration techniques assist in managing data timeliness, ensuring that data from various sources is synchronized and up-to-date.

Data transformation

Data transformation technology improves data quality and reliability in CPS. These techniques involve transforming raw data into a more suitable format for analysis and interpretation. Data transformation includes processes such as normalization, aggregation, and feature scaling [93]. By normalizing data values, aggregating data from different sources, and scaling features to a common range, data transformation ensures consistent and reliable data representation. This preprocessing step is important for improving data quality, enabling efficient data analysis, and enhancing the accuracy of results obtained from CPS. Moreover, data transformation techniques help in handling missing values, reducing noise, and addressing data integration challenges by aligning data attributes and ensuring data consistency across different sources. Through these transformations, data can be standardized and prepared for further analysis, enhancing its reliability and usability in CPS.

Data reduction

Data reduction techniques aim to reduce the complexity and size of the data while preserving its essential information. In CPS, where large amounts of data are generated in real time, data reduction techniques are necessary to handle the volume and velocity challenges. These techniques include sampling, dimensionality reduction, and feature selection [94]. By reducing the data size and eliminating redundant or irrelevant features, data reduction improves data quality by focusing on the most relevant information. It not only enhances the efficiency of data storage and processing but also improves the overall reliability and performance of CPS by reducing the computational burden. Data reduction techniques also facilitate faster data analysis and enable real-time decision-making, important aspects of reliable operation in CPS. Furthermore, data reduction techniques contribute to addressing privacy concerns by anonymizing or obfuscating sensitive information, ensuring the confidentiality and integrity of data in CPS.

Data imputation

Data imputation is a preprocessing technique used to handle missing values in CPS. In real-world scenarios, data collected from sensors and devices may have missing entries due to various reasons such as sensor failures, communication errors, or data transmission issues. Data imputation techniques help in estimating and filling in the missing values to ensure the completeness and reliability of the dataset [95]. These techniques can be based on statistical methods, machine learning algorithms, or domain knowledge. By imputing missing values, data imputation enhances the overall data quality and reliability in CPS, allowing for more accurate analysis and decision-making.

Data standardization

Data standardization is a preprocessing technique that aims to transform data into a common format or scale, ensuring consistency and comparability. In CPS, data may come from different sensors or devices with varying measurement units, scales, or ranges. Data standardization techniques such as z-score normalization or min-max scaling bring the data to a common scale, making it easier to interpret and analyze [96]. Standardization also helps in reducing the impact of outliers and improving the accuracy and reliability of data in CPS. By standardizing data, this preprocessing technique promotes interoperability, facilitates data integration, and ensures consistent and reliable analysis across different components of the CPS.

Data preprocessing techniques significantly contribute to mitigating data quality and reliability concerns in CPS. The techniques discussed above, including data cleaning, data integration, data transformation, data reduction, data imputation, and data standardization, are essential for ensuring the accuracy, consistency, and reliability of data used in CPS. By applying these techniques, CPS can improve decision-making processes, enhance system performance, and enable reliable operation in various domains, such as smart grids, transportation systems, and industrial automation. Data preprocessing is a fundamental step towards harnessing the full potential of CPS and ensuring their effective and reliable functioning in the digital age. As CPS continue to evolve and become more complex, the development and application of advanced data preprocessing techniques, including addressing data timeliness, privacy concerns, missing values, and standardization, will be critical to maintaining data quality and reliability in these systems.

3.4.2 Impact of Data Redundancy on Reliability

Ensuring data quality and reliability in CPS, where physical and computational components are integrated. One of the most significant challenges in achieving data quality and reliability is data redundancy. Our research aims to explore the effects of data redundancy on CPS reliability by analyzing its implications, benefits, and challenges. By comprehending the relationship between data redundancy and reliability, researchers and practitioners can develop effective strategies to improve the overall performance and resilience of CPS.

Data redundancy refers to duplicate or replicated data within a system. In the context of CPS, redundancy can be intentional or unintentional, occurring at various levels, including sensors, communication channels, and data storage. Redundancy is a fundamental technique employed to enhance the reliability of data, providing additional copies or versions of the same data [97]. The presence of redundant data serves as a mechanism to detect and correct errors, mitigate data loss, and enhance fault tolerance in CPS.

One significant advantage of data redundancy is error detection and correction. Multiple copies of data allow for the identification of inconsistencies and discrepancies, enabling the system to rectify errors and maintain data integrity. This improved data accuracy and consistency reduce the likelihood of erroneous or inconsistent information being propagated throughout the system. Furthermore, redundancy enhances fault tolerance by providing alternative data sources when primary sources fail or provide erroneous readings [98]. Redundant data can facilitate seamless system operation and reduce the impact of sensor failures on data reliability. Additionally, redundancy enhances the system's resilience against network and communication failures. With redundant communication channels or multiple paths for data transmission, the system can continue to function even if some connections are disrupted, ensuring data availability and reducing vulnerability to single points of failure.

However, data redundancy also introduces several challenges that must be addressed to ensure optimal reliability in CPS. One of the significant challenges is the efficient management of redundant data, as it increases storage and computational requirements. Storing and processing multiple copies of data demands additional resources and may impact the overall system performance. Moreover, the synchronization and consistency of redundant data across distributed CPS components can be complex, particularly in large-scale systems [99]. Ensuring the timely updates and coherence of redundant data while maintaining system efficiency poses considerable challenges for designers and operators.

To maximize the benefits of data redundancy and mitigate associated challenges, several strategies can be employed in CPS. Firstly, careful design and selection of redundancy mechanisms are essential [100]. This involves identifying critical data elements that require redundancy, determining appropriate redundancy levels, and implementing efficient redundancy schemes. Moreover, effective data management techniques, such as data compression, deduplication, and intelligent data placement, can optimize storage and computational overhead while maintaining reliability. Additionally, advanced synchronization protocols and consistency models can be employed to ensure the coherence of redundant data across distributed components. These strategies help striking a balance between data redundancy and resource utilization, ensuring optimal reliability and performance in CPS.

Data redundancy plays an important role in enhancing the reliability of data in CPS. By providing error detection and correction, fault tolerance, improved data accuracy, and resilience to failures, redundancy contributes to the overall reliability of CPS. However, the challenges associated with managing redundant data, such as increased storage requirements and synchronization complexities, must be carefully addressed. By implementing appropriate strategies for optimizing data redundancy, CPS designers and operators can ensure reliable data processing and resilient system performance in the face of uncertainties and failures.

3.4.3 Data Validation Techniques

Data validation improves the efficiency and reliability of the data processed in CPS, given that these systems are predicated on the effective integration of physical and computational components. To achieve this, various data validation techniques are employed, including statistical analysis, redundancy and consistency checks, error correction codes, data fusion, model-based validation, real-time monitoring, trustworthiness and security validation and continuous improvement and adaptation.

Statistical analysis

Statistical analysis is a fundamental technique for data validation in CPS [101]. It involves the application of statistical methods to identify and handle anomalies or outliers in the collected data. By analyzing the distribution, variance, and trends of the data, statistical techniques can identify discrepancies that may indicate errors or malicious activities. Statistical analysis also enables the detection of data drifts and facilitates the recalibration of system parameters to maintain accuracy [102]. Advanced statistical techniques, such as machine learning algorithms, can be utilized to uncover complex patterns and relationships in the data for more accurate validation.

Redundancy

Redundancy and consistency checks do serve as vital techniques that are commonly utilized to validate data integrity in CPS, further ensuring their reliable functioning. Redundancy involves the replication of data across multiple sources or sensors to verify their consistency. By comparing the data from different sources, inconsistencies or discrepancies can be identified, indicating potential errors. Additionally, consistency checks involve verifying the coherence and logical relationships between different data elements, ensuring that they align with predefined rules or models. If any inconsistencies are detected, appropriate actions can be taken, such as discarding or revalidating the data. Redundancy and consistency checks can be enhanced by utilizing fault-tolerant architectures and distributed consensus algorithms to ensure accurate and consistent data validation in CPS [103].

Error correction code

Error correction codes are widely used in CPS to detect and correct errors introduced during data transmission or storage [104]. These codes add redundancy to the data by incorporating additional bits or symbols that enable the identification and correction of errors. Popular error correction codes, such as Hamming codes or Reed-Solomon codes, can detect and correct a certain number of errors, thus ensuring the accuracy and reliability of the data. Error correction codes are particularly beneficial in environments where data transmission or storage is prone to noise, interference, or other sources of error. Advanced error correction techniques, such as low-density parity-check codes [105] or turbo codes [106], can be employed for more robust error detection and correction in CPS.

Data fusion

Data fusion is another powerful technique employed in CPS to enhance accuracy and reliability [107]. It involves the integration of data from multiple sources or sensors to derive more accurate and robust information. By combining data from different sensors, CPS can compensate for individual sensor limitations, mitigate noise or outliers, and provide a more comprehensive understanding of the system's state. Data fusion algorithms, such as Kalman filters [108], particle filters [109], or Bayesian networks [110], merge and analyze the data, reducing uncertainties and enhancing the system's accuracy. Data fusion enables CPS to make more informed decisions, improve situational awareness, and enhance system performance in various domains, including transportation, healthcare, and industrial automation. Advanced techniques, such as distributed data fusion and adaptive fusion algorithms, can be utilized to handle large-scale CPS with heterogeneous data sources and dynamic conditions [111].

Model-based validation

Model-based validation techniques involve the comparison of collected data with mathematical models or simulations that represent the expected behavior of the CPS [112]. By comparing the observed data with the model's predictions, discrepancies can be identified, which may indicate potential errors in the collected data. Model-based validation helps validating the correctness of the system, detecting deviations from expected performance, and enabling early detection of anomalies. This technique is particularly useful in CPS where accurate models can be developed based on the system's characteristics, behaviors, and environmental factors. By continuously comparing the collected data with the model, ongoing validation ensures the accuracy of the CPS over time, even in dynamic and evolving environments.

Real-time monitoring

Real-time monitoring and feedback mechanisms can be considered significant elements in the process of data validation in CPS, providing invaluable insights and prompt adjustments [113]. These techniques involve continuously monitoring the data streams in real-time and providing immediate feedback on the accuracy and quality of the collected data. Real-time monitoring allows for the detection of anomalies, errors, or deviations as they occur, enabling prompt corrective actions. Feedback mechanisms can range from simple alerts or notifications to automated control actions that adjust system parameters or trigger error recovery mechanisms. By integrating real-time monitoring and feedback into the data validation process, CPS can quickly respond to data inaccuracies and maintain the integrity and reliability of the system.

Trustworthiness and security

In addition to ensuring accuracy, data validation in CPS must also address trustworthiness and security concerns [114]. Trustworthiness validation involves verifying the authenticity, integrity, and provenance of the collected data. Techniques such as digital signatures [115], cryptographic hashes, and blockchain-based solutions can be employed to establish and validate the trustworthiness of data sources [85]. Security validation techniques aim to detect and prevent malicious activities that can compromise the accuracy and reliability of the data. This includes techniques such as anomaly detection, intrusion detection systems, and secure communication protocols. By incorporating trustworthiness and security validation into the data validation process, CPS can ensure that only trustworthy and secure data are used for decision-making and system operation.

Data validation techniques are critical for ensuring the accuracy and reliability of data in CPS. By employing a range of techniques such as statistical analysis, redundancy and consistency checks, error correction codes, data fusion, model-based validation, real-time monitoring, trustworthiness, and security validation, and continuous improvement and adaptation, CPS can establish robust mechanisms to validate data and support the effective operation of the system. These techniques enable CPS to make reliable decisions, maintain data integrity, and mitigate risks. The ongoing advancement and refinement of data validation techniques will continue to play a vital role in enhancing the accuracy and reliability of CPS in various domains.

3.5 Summary

This section emphasizes the importance of sensors and devices in CPS and the challenges of data-driven in CPS. The discussion also addresses the selection of suitable sensors and devices for CPS, as well as the importance of maintaining data quality and reliability. Achieving real-time responsiveness and minimizing latency in CPS requires a multidimensional approach, including factors such as edge computing, predictive analytics, data compression, prioritized communication, hardware acceleration, energy-efficient designs, fault-tolerant mechanisms, appropriate wireless technologies, and secure protocols. CPS generate vast amounts of data and processing and analyzing it can provide significant benefits for organizations. Appropriate evaluation techniques and methodologies must be employed to continually improve the performance of CPS in real-time scenarios.

Competing Interests

1. The research is supported by the National Key Research and Development Project under Grant 2023YFB3307300, the National Natural Science Foundation of China under Grants 62373014, 62422301, 62125301, and 92467205.
2. The authors have no conflicts of interest to declare that are relevant to the content of this chapter.

References

1. F.-J. Wu, Y.-F. Kao, and Y.-C. Tseng, "From wireless sensor networks towards cyber physical systems," *Pervasive and Mobile Computing*, vol. 7, pp. 397–413, Aug. 2011.
2. C. Bi, J. Huang, G. Xing, L. Jiang, X. Liu, and M. Chen, "SafeWatch: A Wearable Hand Motion Tracking System for Improving Driving Safety," *ACM Transactions on Cyber-Physical Systems*, vol. 4, pp. 1–21, Jan. 2020.
3. E. Balestrieri, P. Daponte, L. De Vito, and F. Lamonaca, "Sensors and Measurements for Unmanned Systems: An Overview," *Sensors*, vol. 21, p. 1518, Feb. 2021.
4. Ling Li, Shancang Li, and Shanshan Zhao, "QoS-Aware Scheduling of Services-Oriented Internet of Things," *IEEE Transactions on Industrial Informatics*, vol. 10, pp. 1497–1505, May 2014.
5. B. K. Sinha and M. S. Patel, "Recent developments in high precision quartz and langasite pressure sensors for high temperature and high pressure applications," in *2016 IEEE International Frequency Control Symposium (IFCS)*, pp. 1–13, IEEE, 2016.
6. A. S. Uluagac, V. Subramanian, and R. Beyah, "Sensory channel threats to cyber physical systems: A wake-up call," in *2014 IEEE Conference on Communications and Network Security*, pp. 301–309, IEEE, 2014.
7. C. Ho, A. Robinson, D. Miller, and M. Davis, "Overview of Sensors and Needs for Environmental Monitoring," *Sensors*, vol. 5, pp. 4–37, Feb. 2005.
8. C. Labrado, H. Thapliyal, S. Prowell, and T. Kuruganti, "Use of Thermistor Temperature Sensors for Cyber-Physical System Security," *Sensors*, vol. 19, p. 3905, Sept. 2019.
9. K. Huang, C. Zhou, Y.-C. Tian, S. Yang, and Y. Qin, "Assessing the Physical Impact of Cyberattacks on Industrial Cyber-Physical Systems," *IEEE Transactions on Industrial Electronics*, vol. 65, pp. 8153–8162, Oct. 2018.
10. O. Younis and N. Moayeri, "Employing Cyber-Physical Systems: Dynamic Traffic Light Control at Road Intersections," *IEEE Internet of Things Journal*, vol. 4, pp. 2286–2296, Dec. 2017.
11. G. Horváth and G. Erdős, "Gesture control of cyber physical systems," *Procedia Cirp*, vol. 63, pp. 184–188, 2017.
12. M. Leccadito, T. Bakker, R. Klenke, and C. Elks, "A survey on securing UAS cyber physical systems," *IEEE Aerospace and Electronic Systems Magazine*, vol. 33, pp. 22–32, Oct. 2018.
13. F. Hu, Q. Hao, Q. Sun, X. Cao, R. Ma, T. Zhang, Y. Patil, and J. Lu, "Cyberphysical System With Virtual Reality for Intelligent Motion Recognition and Training," *IEEE Transactions on Systems, Man, and Cybernetics: Systems*, pp. 1–17, 2016.
14. E. Yeniaras, J. Lamaury, Zhigang Deng, and N. V. Tsekos, "Towards a new cyber-physical system for MRI-guided and robot-assisted cardiac procedures," in *Proceedings of the 10th IEEE International Conference on Information Technology and Applications in Biomedicine*, (Corfu, Greece), pp. 1–5, IEEE, Nov. 2010.
15. D. Cogliati, M. Falchetto, D. Pau, M. Roveri, and G. Viscardi, "Intelligent cyber-physical systems for industry 4.0," in *2018 First International Conference on Artificial Intelligence for Industries (AI4I)*, pp. 19–22, IEEE, 2018.
16. V. Gunes, S. Peter, T. Givargis, and F. Vahid, "A survey on concepts, applications, and challenges in cyber-physical systems.," *KSII Trans. Internet Inf. Syst.*, vol. 8, no. 12, pp. 4242–4268, 2014.
17. P. Krishnamurthy, F. Khorrami, R. Karri, D. Paul-Pena, and H. Salehghaffari, "Process-Aware Covert Channels Using Physical Instrumentation in Cyber-Physical Systems," *IEEE Transactions on Information Forensics and Security*, vol. 13, pp. 2761–2771, Nov. 2018.
18. B. Bordel, D. S. De Rivera, and R. Alcarria, "Plug-and-play transducers in cyber-physical systems for device-driven applications," in *2016 10th International Conference on Innovative Mobile and Internet Services in Ubiquitous Computing (IMIS)*, pp. 316–321, IEEE, 2016.
19. S. Han, M. Xie, H.-H. Chen, and Y. Ling, "Intrusion detection in cyber-physical systems: Techniques and challenges," *IEEE systems journal*, vol. 8, no. 4, pp. 1052–1062, 2014.

20. T. Bures, D. Weyns, C. Berger, S. Biffl, M. Daun, T. Gabor, D. Garlan, I. Gerostathopoulos, C. Julien, F. Krikava, *et al.*, "Software engineering for smart cyber-physical systems–towards a research agenda: report on the first international workshop on software engineering for smart cps," *ACM SIGSOFT Software Engineering Notes*, vol. 40, no. 6, pp. 28–32, 2015.
21. Y. Fang, Y. Lim, S. E. Ooi, C. Zhou, and Y. Tan, "Study of Human Thermal Comfort for Cyber–Physical Human Centric System in Smart Homes," *Sensors*, vol. 20, p. 372, Jan. 2020.
22. F. Bocklisch, G. Paczkowski, S. Zimmermann, and T. Lampke, "Integrating human cognition in cyber-physical systems: A multidimensional fuzzy pattern model with application to thermal spraying," *Journal of Manufacturing Systems*, vol. 63, pp. 162–176, 2022.
23. Y. Liu, Y. Peng, B. Wang, S. Yao, and Z. Liu, "Review on cyber-physical systems," *IEEE/CAA Journal of Automatica Sinica*, vol. 4, pp. 27–40, Jan. 2017.
24. X. Yu and Y. Xue, "Smart grids: A cyber–physical systems perspective," *Proceedings of the IEEE*, vol. 104, no. 5, pp. 1058–1070, 2016.
25. S. K. Gupta, T. Mukherjee, G. Varsamopoulos, and A. Banerjee, "Research directions in energy-sustainable cyber–physical systems," *Sustainable Computing: Informatics and Systems*, vol. 1, pp. 57–74, Mar. 2011.
26. M. Andronie, G. Lăzăroiu, M. Iatagan, I. Hurloiu, and I. Dijmărescu, "Sustainable Cyber-Physical Production Systems in Big Data-Driven Smart Urban Economy: A Systematic Literature Review," *Sustainability*, vol. 13, p. 751, Jan. 2021.
27. S. Ali, S. Qaisar, H. Saeed, M. Khan, M. Naeem, and A. Anpalagan, "Network Challenges for Cyber Physical Systems with Tiny Wireless Devices: A Case Study on Reliable Pipeline Condition Monitoring," *Sensors*, vol. 15, pp. 7172–7205, Mar. 2015.
28. Y. Lu and M. R. Asghar, "Semantic communications between distributed cyber-physical systems towards collaborative automation for smart manufacturing," *Journal of manufacturing systems*, vol. 55, pp. 348–359, 2020.
29. Z. Wang, H. Song, D. W. Watkins, K. G. Ong, P. Xue, Q. Yang, and X. Shi, "Cyber-physical systems for water sustainability: challenges and opportunities," *IEEE Communications Magazine*, vol. 53, no. 5, pp. 216–222, 2015.
30. K. Sampigethaya and R. Poovendran, "Aviation cyber–physical systems: Foundations for future aircraft and air transport," *Proceedings of the IEEE*, vol. 101, no. 8, pp. 1834–1855, 2013.
31. J. Wurm, Y. Jin, Y. Liu, S. Hu, K. Heffner, F. Rahman, and M. Tehranipoor, "Introduction to cyber-physical system security: A cross-layer perspective," *IEEE Transactions on Multi-Scale Computing Systems*, vol. 3, no. 3, pp. 215–227, 2016.
32. N. Jazdi, "Cyber physical systems in the context of industry 4.0," in *2014 IEEE international conference on automation, quality and testing, robotics*, pp. 1–4, IEEE, 2014.
33. A. Garro, A. Tundis, D. Bouskela, A. Jardin, N. Thuy, M. Otter, L. Buffoni, P. Fritzson, M. Sjölund, W. Schamai, and H. Olsson, "On formal cyber physical system properties modeling: A new temporal logic language and a Modelica-based solution," in *2016 IEEE International Symposium on Systems Engineering (ISSE)*, pp. 1–8, Oct. 2016.
34. M. Lewandowski, M. Gath, D. Werthmann, and M. Lawo, "Agent-based Control for Material Handling Systems in In-House Logistics—Towards Cyber-Physical Systems in In-House-Logistics Utilizing Real Size," in *Smart SysTech 2013; European Conference on Smart Objects, Systems and Technologies*, pp. 1–5, June 2013.
35. A. A. Ghaemi, "A cyber-physical system approach to smart city development," in *2017 IEEE International Conference on Smart Grid and Smart Cities (ICSGSC)*, (Singapore, Singapore), pp. 257–262, IEEE, July 2017.
36. T. Gurriet, P. Nilsson, A. Singletary, and A. D. Ames, "Realizable Set Invariance Conditions for Cyber-Physical Systems," in *2019 American Control Conference (ACC)*, (Philadelphia, PA, USA), pp. 3642–3649, IEEE, July 2019.
37. C. Chen, X. Liu, T. Qiu, and A. K. Sangaiah, "A short-term traffic prediction model in the vehicular cyber–physical systems," *Future Generation Computer Systems*, vol. 105, pp. 894–903, Apr. 2020.

38. L. K. Ramasamy, F. Khan, M. Shah, B. V. V. S. Prasad, C. Iwendi, and C. Biamba, "Secure Smart Wearable Computing through Artificial Intelligence-Enabled Internet of Things and Cyber-Physical Systems for Health Monitoring," *Sensors*, vol. 22, p. 1076, Jan. 2022.
39. J. Pawlick and Q. Zhu, "Strategic Trust in Cloud-Enabled Cyber-Physical Systems With an Application to Glucose Control," *IEEE Transactions on Information Forensics and Security*, vol. 12, pp. 2906–2919, Dec. 2017.
40. C. Stary, "Digital Twin Generation: Re-Conceptualizing Agent Systems for Behavior-Centered Cyber-Physical System Development," *Sensors*, vol. 21, p. 1096, Feb. 2021.
41. J. Criado, J. Asensio, N. Padilla, and L. Iribarne, "Integrating Cyber-Physical Systems in a Component-Based Approach for Smart Homes," *Sensors*, vol. 18, p. 2156, July 2018.
42. R. Ranchal, P. Bastide, X. Wang, A. Gkoulalas-Divanis, M. Mehra, S. Bakthavachalam, H. Lei, and A. Mohindra, "Disrupting Healthcare Silos: Addressing Data Volume, Velocity and Variety With a Cloud-Native Healthcare Data Ingestion Service," *IEEE Journal of Biomedical and Health Informatics*, vol. 24, pp. 3182–3188, Nov. 2020.
43. R. Birke, M. Bjoerkqvist, L. Y. Chen, E. Smirni, and T. Engbersen, "({Big) Data} in a virtualized world: Volume, velocity, and variety in cloud datacenters," in *12th USENIX Conference on File and Storage Technologies (FAST 14)*, pp. 177–189, 2014.
44. N. B. C. E. . Jamil, I. B. Ishak, F. Sidi, L. S. Affendey, and A. Mamat, "A Systematic Review on the Profiling of Digital News Portal for Big Data Veracity," *Procedia Computer Science*, vol. 72, pp. 390–397, 2015.
45. S. Sagiroglu and D. Sinanc, "Big data: A review," in *2013 International Conference on Collaboration Technologies and Systems (CTS)*, (San Diego, CA, USA), pp. 42–47, IEEE, May 2013.
46. D. W. McKee, S. J. Clement, J. Almutairi, and J. Xu, "Survey of advances and challenges in intelligent autonomy for distributed cyber-physical systems," *CAAI Transactions on Intelligence Technology*, vol. 3, no. 2, pp. 75–82, 2018.
47. W.-H. Zhou and L.-S. Zhao, "One-Dimensional Consolidation of Unsaturated Soil Subjected to Time-Dependent Loading with Various Initial and Boundary Conditions," *International Journal of Geomechanics*, vol. 14, pp. 291–301, Apr. 2014.
48. I. Lee and O. Sokolsky, "Medical cyber physical systems," in *Proceedings of the 47th design automation conference*, pp. 743–748, 2010.
49. S. J. Oks, A. Fritzsche, and K. M. Möslein, "An application map for industrial cyber-physical systems," *Industrial internet of things: Cybermanufacturing systems*, pp. 21–46, 2017.
50. B. Chen, S.-Y. Tsai, and J.-Y. J. Chang, "Development of intelligent coil leveling machine for smart manufacturing," in *2016 23rd International Conference on Mechatronics and Machine Vision in Practice (M2VIP)*, pp. 1–6, Jan. 2016.
51. S. Harbola and V. Coors, "One dimensional convolutional neural network architectures for wind prediction," *Energy Conversion and Management*, vol. 195, pp. 70–75, Sept. 2019.
52. R. Liu, M. Li, J. Guzmán, and F. Rodríguez, "A fast and practical one-dimensional transient model for greenhouse temperature and humidity," *Computers and Electronics in Agriculture*, vol. 186, p. 106186, July 2021.
53. Q. Hao and L. Qin, "The Design of Intelligent Transportation Video Processing System in Big Data Environment," *IEEE Access*, vol. 8, pp. 13769–13780, 2020.
54. A. Sezer and A. Altan, "Detection of solder paste defects with an optimization-based deep learning model using image processing techniques," *Soldering and Surface Mount Technology*, vol. ahead-of-print, Aug. 2021.
55. C. Amrutha, C. Jyotsna, and J. Amudha, "Deep Learning Approach for Suspicious Activity Detection from Surveillance Video," in *2020 2nd International Conference on Innovative Mechanisms for Industry Applications (ICIMIA)*, pp. 335–339, Mar. 2020.
56. C. Morikawa, M. Kobayashi, M. Satoh, Y. Kuroda, T. Inomata, H. Matsuo, T. Miura, and M. Hilaga, "Image and video processing on mobile devices: A survey," *The Visual Computer*, vol. 37, pp. 2931–2949, Dec. 2021.
57. F. Villa, F. Severini, F. Madonini, and F. Zappa, "SPADs and SiPMs Arrays for Long-Range High-Speed Light Detection and Ranging (LiDAR)," *Sensors*, vol. 21, p. 3839, June 2021.

58. R. C. Luo and C.-W. Kuo, "Intelligent Seven-DoF Robot With Dynamic Obstacle Avoidance and 3-D Object Recognition for Industrial Cyber–Physical Systems in Manufacturing Automation," *Proceedings of the IEEE*, vol. 104, pp. 1102–1113, May 2016.
59. L. Kuang, L. Yu, L. Huang, Y. Wang, P. Ma, C. Li, and Y. Zhu, "A Personalized QoS Prediction Approach for CPS Service Recommendation Based on Reputation and Location-Aware Collaborative Filtering," *Sensors*, vol. 18, p. 1556, May 2018.
60. R. Iqbal, F. Doctor, B. More, S. Mahmud, and U. Yousuf, "Big Data analytics and Computational Intelligence for Cyber–Physical Systems: Recent trends and state of the art applications," *Future Generation Computer Systems*, vol. 105, pp. 766–778, Apr. 2020.
61. S. Kumar, U. Dohare, K. Kumar, D. Prasad Dora, K. Naseer Qureshi, and R. Kharel, "Cybersecurity Measures for Geocasting in Vehicular Cyber Physical System Environments," *IEEE Internet of Things Journal*, vol. 6, pp. 5916–5926, Aug. 2019.
62. T. Benson, A. Akella, and D. A. Maltz, "Network traffic characteristics of data centers in the wild," in *Proceedings of the 10th ACM SIGCOMM Conference on Internet Measurement*, (Melbourne Australia), pp. 267–280, ACM, Nov. 2010.
63. F. Chen, Y. Tang, C. Wang, J. Huang, C. Huang, D. Xie, T. Wang, and C. Zhao, "Medical Cyber–Physical Systems: A Solution to Smart Health and the State of the Art," *IEEE Transactions on Computational Social Systems*, vol. 9, pp. 1359–1386, Oct. 2022.
64. Feng-Li Lian, J. Moyne, and D. Tilbury, "Network design consideration for distributed control systems," *IEEE Transactions on Control Systems Technology*, vol. 10, pp. 297–307, Mar. 2002.
65. H. Hayashi, Y. Takabayashi, H. Tsuji, and M. Oka, "Rapidly increasing application of Intranet technologies for SCADA (supervisory control and data acquisition system)," in *IEEE/PES Transmission and Distribution Conference and Exhibition*, vol. 1, (Yokohama, Japan), pp. 22–25, IEEE, 2002.
66. I. Lee, O. Sokolsky, S. Chen, J. Hatcliff, E. Jee, B. Kim, A. King, M. Mullen-Fortino, S. Park, A. Roederer, *et al.*, "Challenges and research directions in medical cyber–physical systems," *Proceedings of the IEEE*, vol. 100, no. 1, pp. 75–90, 2011.
67. Q. Qi, D. Zhao, T. Liao, and F. Tao, *Modeling of Cyber-Physical Systems and Digital Twin Based on Edge Computing, Fog Computing and Cloud Computing Towards Smart Manufacturing*. American Society of Mechanical Engineers, June 2018.
68. J. Zhou, H.-N. Dai, and H. Wang, "Lightweight Convolution Neural Networks for Mobile Edge Computing in Transportation Cyber Physical Systems," *ACM Transactions on Intelligent Systems and Technology*, vol. 10, pp. 1–20, Nov. 2019.
69. D. Borsatti, G. Davoli, W. Cerroni, and C. Raffaelli, "Enabling Industrial IoT as a Service with Multi-Access Edge Computing," *IEEE Communications Magazine*, vol. 59, pp. 21–27, Aug. 2021.
70. J. Kim and J. Y. Lee, "Server-Edge dualized closed-loop data analytics system for cyber-physical system application," *Robotics and Computer-Integrated Manufacturing*, vol. 67, p. 102040, Feb. 2021.
71. R. Atat, L. Liu, J. Wu, G. Li, C. Ye, and Y. Yang, "Big Data Meet Cyber-Physical Systems: A Panoramic Survey," *IEEE Access*, vol. 6, pp. 73603–73636, 2018.
72. U. U. Tariq, H. Ali, L. Liu, J. Panneerselvam, and X. Zhai, "Energy-efficient Static Task Scheduling on VFI-based NoC-HMPSoCs for Intelligent Edge Devices in Cyber-physical Systems," *ACM Transactions on Intelligent Systems and Technology*, vol. 10, pp. 1–22, Nov. 2019.
73. T. Dillon, V. Potdar, J. Singh, and A. Talevski, "Cyber-physical systems: Providing quality of service (qos) in a heterogeneous systems-of-systems environment," in *5th IEEE International Conference on Digital Ecosystems and Technologies (IEEE DEST 2011)*, pp. 330–335, IEEE, 2011.
74. K.-J. Park, J. Kim, H. Lim, and Y. Eun, "Robust Path Diversity for Network Quality of Service in Cyber-Physical Systems," *IEEE Transactions on Industrial Informatics*, vol. 10, pp. 2204–2215, Nov. 2014.
75. T. Grimm, B. Janßen, O. Navarro, and M. Hübner, "The value of FPGAs as reconfigurable hardware enabling Cyber-Physical Systems," in *2015 IEEE 20th Conference on Emerging Technologies & Factory Automation (ETFA)*, pp. 1–8, Sept. 2015.

76. A. Vora and K.-D. Kang, "Effective 5G Wireless Downlink Scheduling and Resource Allocation in Cyber-Physical Systems," *Technologies*, vol. 6, p. 105, Nov. 2018.
77. E. Lee, "The Past, Present and Future of Cyber-Physical Systems: A Focus on Models," *Sensors*, vol. 15, pp. 4837–4869, Feb. 2015.
78. S. Yun, J.-H. Park, and W.-T. Kim, "Data-centric middleware based digital twin platform for dependable cyber-physical systems," in *2017 Ninth International Conference on Ubiquitous and Future Networks (ICUFN)*, pp. 922–926, July 2017.
79. L. Vegh and L. Miclea, "Secure and efficient communication in cyber-physical systems through cryptography and complex event processing," in *2016 International Conference on Communications (COMM)*, pp. 273–276, June 2016.
80. X. Huang and J. Dong, "Reliable Control Policy of Cyber-Physical Systems Against a Class of Frequency-Constrained Sensor and Actuator Attacks," *IEEE Transactions on Cybernetics*, vol. 48, pp. 3432–3439, Feb. 2018.
81. B. Bonakdarpour, J. V. Deshmukh, and M. Pajic, "Opportunities and Challenges in Monitoring Cyber-Physical Systems Security," in *Leveraging Applications of Formal Methods, Verification and Validation. Industrial Practice* (T. Margaria and B. Steffen, eds.), vol. 11247, pp. 9–18, Cham: Springer International Publishing, 2018.
82. O. Bouachir, M. Aloqaily, L. Tseng, and A. Boukerche, "Blockchain and Fog Computing for Cyberphysical Systems: The Case of Smart Industry," *Computer*, vol. 53, pp. 36–45, Sept. 2020.
83. R. Alguliyev, Y. Imamverdiyev, and L. Sukhostat, "Cyber-physical systems and their security issues," *Computers in Industry*, vol. 100, pp. 212–223, Sept. 2018.
84. Y. Liu, Y. Peng, B. Wang, S. Yao, and Z. Liu, "Review on cyber-physical systems," *IEEE/CAA Journal of Automatica Sinica*, vol. 4, pp. 27–40, Jan. 2017.
85. W. Zhao, C. Jiang, H. Gao, S. Yang, and X. Luo, "Blockchain-Enabled Cyber–Physical Systems: A Review," *IEEE Internet of Things Journal*, vol. 8, pp. 4023–4034, Mar. 2021.
86. T. Bures, V. Matena, R. Mirandola, L. Pagliari, and C. Trubiani, "Performance Modelling of Smart Cyber-Physical Systems," in *Companion of the 2018 ACM/SPEC International Conference on Performance Engineering*, (Berlin Germany), pp. 37–40, ACM, Apr. 2018.
87. C. Lu, A. Saifullah, B. Li, M. Sha, H. Gonzalez, D. Gunatilaka, C. Wu, L. Nie, and Y. Chen, "Real-Time Wireless Sensor-Actuator Networks for Industrial Cyber-Physical Systems," *Proceedings of the IEEE*, vol. 104, pp. 1013–1024, May 2016.
88. Y. Xu, I. Koren, and C. M. Krishna, "AdaFT: A Framework for Adaptive Fault Tolerance for Cyber-Physical Systems," *ACM Transactions on Embedded Computing Systems*, vol. 16, pp. 1–25, Aug. 2017.
89. L. Hu, N. Xie, Z. Kuang, and K. Zhao, "Review of Cyber-Physical System Architecture," in *2012 IEEE 15th International Symposium on Object/Component/Service-Oriented Real-Time Distributed Computing Workshops*, pp. 25–30, Apr. 2012.
90. J. Li, Y. Yang, J. S. Sun, K. Tomsovic, and H. Qi, "ConAML: Constrained Adversarial Machine Learning for Cyber-Physical Systems," in *Proceedings of the 2021 ACM Asia Conference on Computer and Communications Security*, (Virtual Event Hong Kong), pp. 52–66, ACM, May 2021.
91. P. M. Seeger, Z. Yahouni, and G. Alpan, "Literature review on using data mining in production planning and scheduling within the context of cyber physical systems," *Journal of Industrial Information Integration*, vol. 28, p. 100371, July 2022.
92. C. Hildebrandt, A. Köcher, C. Küstner, C.-M. López-Enríquez, A. W. Müller, B. Caesar, C. S. Gundlach, and A. Fay, "Ontology Building for Cyber–Physical Systems: Application in the Manufacturing Domain," *IEEE Transactions on Automation Science and Engineering*, vol. 17, pp. 1266–1282, July 2020.
93. D. Lavrova, M. Poltavtseva, and A. Shtyrkina, "Security analysis of cyber-physical systems network infrastructure," in *2018 IEEE Industrial Cyber-Physical Systems (ICPS)*, pp. 818–823, May 2018.
94. B. Chen, D. W. C. Ho, W.-A. Zhang, and L. Yu, "Distributed Dimensionality Reduction Fusion Estimation for Cyber-Physical Systems Under DoS Attacks," *IEEE Transactions on Systems, Man, and Cybernetics: Systems*, vol. 49, pp. 455–468, Feb. 2019.

95. R. Razavi-Far, E. Hallaji, M. Farajzadeh-Zanjani, R. Aljoudi, and M. Saif, "A Critical Study on the Impact of Missing Data Imputation for Classifying Intrusions in Cyber-Physical Water Systems," in *IECON 2021—47th Annual Conference of the IEEE Industrial Electronics Society*, pp. 1–6, Oct. 2021.
96. S. Krishna Narayanan, S. Dhanasekaran, and V. Vasudevan, "An effective parameter tuned deep belief network for detecting anomalous behavior in sensor-based cyber-physical systems," *Theoretical Computer Science*, vol. 931, pp. 142–151, Sept. 2022.
97. C. Machado and A. A. M. Fröhlich, "Iot data integrity verification for cyber-physical systems using blockchain," in *2018 IEEE 21st international symposium on real-time distributed computing (ISORC)*, pp. 83–90, IEEE, 2018.
98. M. S. Chong, H. Sandberg, and A. M. Teixeira, "A Tutorial Introduction to Security and Privacy for Cyber-Physical Systems," in *2019 18th European Control Conference (ECC)*, pp. 968–978, June 2019.
99. Z. Shu, J. Wan, D. Zhang, and D. Li, "Cloud-Integrated Cyber-Physical Systems for Complex Industrial Applications," *Mobile Networks and Applications*, vol. 21, Oct. 2016.
100. U. Bodkhe, D. Mehta, S. Tanwar, P. Bhattacharya, P. K. Singh, and W.-C. Hong, "A Survey on Decentralized Consensus Mechanisms for Cyber Physical Systems," *IEEE Access*, vol. 8, pp. 54371–54401, 2020.
101. M. Al-Mhiqani, R. Ahmad, Z. Zainal Abidin, W. Mohamed, A. Hassan, and N. Clarke, "A NEW TAXONOMY OF INSIDER THREATS; AN INITIAL STEP IN UNDERSTANDING AUTHORIZED ATTACK," *International Journal of Information Systems and Management*, vol. 1, June 2018.
102. M. Krishna, S. M. B. Chowdary, P. Nancy, and V. Arulkumar, "A Survey on Multimedia Analytics in Security Systems of Cyber Physical Systems and IoT," in *2021 2nd International Conference on Smart Electronics and Communication (ICOSEC)*, pp. 1–7, Oct. 2021.
103. D. Levshun, A. Chechulin, I. Kotenko, and Y. Chevalier, "Design and Verification Methodology for Secure and Distributed Cyber-Physical Systems," in *2019 10th IFIP International Conference on New Technologies, Mobility and Security (NTMS)*, pp. 1–5, June 2019.
104. A. Behfarnia and A. Eslami, "Error Correction Coding Meets Cyber-Physical Systems: Message-Passing Analysis of Self-Healing Interdependent Networks," *IEEE Transactions on Communications*, vol. 65, pp. 2753–2768, July 2017.
105. M. Kuc, W. Sułek, and D. Kania, "FPGA-Oriented LDPC Decoder for Cyber-Physical Systems," *Mathematics*, vol. 8, p. 723, May 2020.
106. M. Zhan, J. Wu, H. Wen, and P. Zhang, "A Novel Error Correction Mechanism for Energy-Efficient Cyber-Physical Systems in Smart Building," *IEEE Access*, vol. 6, pp. 39037–39045, 2018.
107. A. Miloslavov and M. Veeraraghavan, "Sensor Data Fusion Algorithms for Vehicular Cyber-Physical Systems," *IEEE Transactions on Parallel and Distributed Systems*, vol. 23, pp. 1762–1774, Sept. 2012.
108. M. H. Basiri, J. G. Thistle, J. W. Simpson-Porco, and S. Fischmeister, "Kalman Filter Based Secure State Estimation and Individual Attacked Sensor Detection in Cyber-Physical Systems," in *2019 American Control Conference (ACC)*, pp. 3841–3848, July 2019.
109. H. Li, X. He, Y. Zhang, and W. Guan, "Attack Detection in Cyber-Physical Systems Using Particle Filter: An Illustration on Three-Tank System," in *2018 IEEE 8th Annual International Conference on CYBER Technology in Automation, Control, and Intelligent Systems (CYBER)*, pp. 504–509, July 2018.
110. X. Lyu, Y. Ding, and S.-H. Yang, "Bayesian Network Based C2P Risk Assessment for Cyber-Physical Systems," *IEEE Access*, vol. 8, pp. 88506–88517, 2020.
111. E. Blasch, I. Kadar, L. L. Grewe, R. Brooks, W. Yu, A. Kwasinski, S. Thomopoulos, J. Salerno, and H. Qi, "Panel summary of cyber-physical systems (cps) and internet of things (iot) opportunities with information fusion," in *Signal Processing, Sensor/Information Fusion, and Target Recognition XXVI*, vol. 10200, pp. 171–188, SPIE, 2017.
112. L. C. Silva, H. O. Almeida, A. Perkusich, and M. Perkusich, "A Model-Based Approach to Support Validation of Medical Cyber-Physical Systems," *Sensors*, vol. 15, pp. 27625–27670, Nov. 2015.

113. X. Guan, B. Yang, C. Chen, W. Dai, and Y. Wang, "A comprehensive overview of cyber-physical systems: From perspective of feedback system," *IEEE/CAA Journal of Automatica Sinica*, vol. 3, pp. 1–14, Jan. 2016.
114. H. Tao, M. Z. A. Bhuiyan, M. A. Rahman, T. Wang, J. Wu, S. Q. Salih, Y. Li, and T. Hayajneh, "TrustData: Trustworthy and Secured Data Collection for Event Detection in Industrial Cyber-Physical System," *IEEE Transactions on Industrial Informatics*, vol. 16, pp. 3311–3321, May 2020.
115. S. Vattaparambil Sudarsan, O. Schelén, and U. Bodin, "A Model for Signatories in Cyber-Physical Systems," in *2020 25th IEEE International Conference on Emerging Technologies and Factory Automation (ETFA)*, vol. 1, pp. 15–21, Sept. 2020.

Chapter 4
Data Storage and Management in Cyber Physical Systems

Abstract Data-driven cyber physical systems (DDCPS) represent a complex network paradigm that enables the seamless integration of physical systems and information systems. Critical to the operation of these systems is the data storage aspect, which can be classified through different perspectives, including storage medium, location, scale, and performance. Effective data storage solutions are essential to ensure the real-time responsiveness, robustness, and efficiency of DDCPS, especially given the need to process large amounts of data ranging from sensor readings to machine learning models. Starting from the data storage of CPS, this chapter discusses the types of data storage solutions suitable for CPS. The classification of data storage types is further discussed, and an example of data storage application in DDCPS is provided. By exploring the above, this chapter in a comprehensive understanding of the role of data storage in DDCPS, thereby facilitating the development of these systems in various fields.

This chapter focuses on data storage types, data management and processing techniques for DDCPS, and big data processing techniques for DDCPS. The specific framework is shown in Fig. 4.1.

4.1 Types of Data Storage for DDCPS

DDCPS are complex network systems composed of CPS, which can realize the seamless integration of physical systems and information systems. The data storage solutions of such systems can be classified according to different perspectives, such as storage medium, storage location, storage scale and storage performance. Section 4.1 focuses on types of data storage solutions for CPS. Section 4.1.1 begins with an introduction to data storage, respectively in CPS, data storage type, an instance of the data storage.

F. Li et al., *Data-Driven Cyber Physical Systems*,
https://doi.org/10.1007/978-981-96-8709-1_4

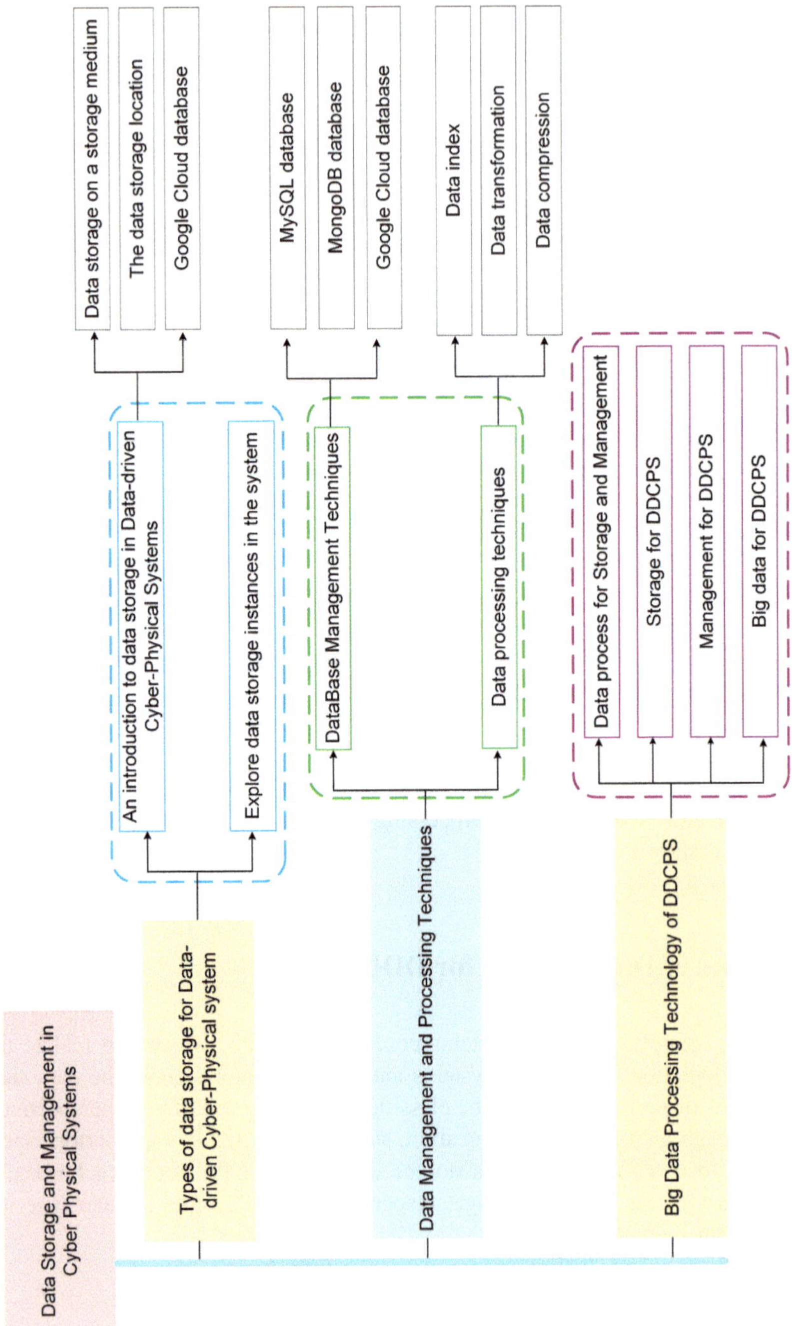

Fig. 4.1 Overall system diagram. The main components of this chapter are presented, namely types of data storage for DDCPS, data management and processing techniques and big data processing technology

4.1.1 An Introduction to Data Storage in DDCPS

Data storage is a cornerstone of DDCPS, complex integrations of computational and physical components that are now central to various sectors such as autonomous vehicles, smart homes [1], and industrial automation. The importance of data storage in DDCPS is amplified by the need for real-time interaction with the physical environment, as the system's ability to respond dynamically to environmental changes is dependent on its capacity to efficiently store [2], process, and utilize vast quantities of data, ranging from sensor readings and control signals to complex machine learning models and high-definition video feeds. Moreover, the efficiency of data management, which includes storage, retrieval, integrity maintenance [3], security, and availability across system components, is essential to the reliability, safety, and performance of DDCPS. Additionally, the decentralization common in many DDCPS introduces a layer of complexity to data storage, as data may need to be stored across multiple locations, each with unique constraints related to storage capacity, network connectivity, and computational powercite [4].

4.1.1.1 Data Storage on a Storage Medium

Data storage, a critical element in DDCPS, leverages various technologies, each offering unique advantages, as presented in Fig. 4.2. Magnetic storage, including hard disk drives, uses magnetic coatings on platters for data storage [5], with areal density being a critical factor affecting its performance and capacity. Optical storage devices, such as CDs and DVDs, utilize laser technology to write and read data, offering substantial yet inexpensive storage capacity [6]. Solid-state drives, with their flash memory chips and absence of mechanical parts, provide fast, reliable, and energy-efficient data storage. Storage on tape is slow, but offers a cost-effective option for storing large amounts of data, often for backup or archiving [7]. Lastly, cloud storage, the newest form in the spectrum, provides data storage on remote servers accessible via the internet, offering scalability and accessibility advantages. These diverse storage technologies collectively enhance the capabilities of DDCPS. The five storage options are explained below [8].

Magnetic storage. Among the oldest forms of data storage, the hard disk drive (HDD) stores data on platters coated with magnetic coatings, with read-write heads traversing the surface of the platters to read and write data [9]. The principle of magnetic recording technology underpins these storage mechanisms, which comprises magnetic storage media, read/write heads, recording channel electronics for data encoding and clock logic, and the technology that manages the gap between the head and the media. In such systems, the fundamental parameter is area density, which is defined as the linear bit density per inch (BPI) along the track and the number of tracks per inch (TPI) [10]. This parameter is integral as it influences both the

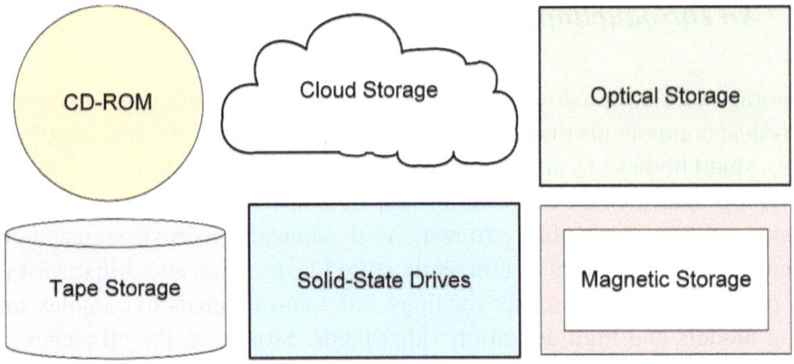

Fig. 4.2 Data storage platform. Data storage platform. Storage platforms include CD-ROM, coud storage, optical storage, tape storage, solid-state drives, and magnetic storage. The details of these storage methods are as follows

capacity and the performance of the storage system and ultimately dictates the cost per megabyte of storage [11].

The evolution of storage technology has predominantly centered around the enhancement of areal density as a means to boost the capacity of tape and disk devices [12]. Various techniques, such as the utilization of thinner particle coatings with superior magnetic characteristics, the employment of improved head materials, and the implementation of advanced manufacturing techniques facilitating smaller recording head gap lengths, contribute to this increase in areal density [13]. Furthermore, the reduction in the gap between the head and the magnetic surface, paired with more accurate head positioning, is instrumental in achieving capacity improvements [14].

However, the value of increased track density is less pronounced for half-inch tape devices due to constraints from legacy technology and compatibility requirements [15]. The growth in track density of moving head disks is an essential aspect in the overall improvement of areal density, but it is not as significant as linear density. The lateral resolution and positioning accuracy of the read/write heads primarily limit track density [16].

On the one hand, the performance of storage devices hinges on the time taken to access the start of a data block and the time to transfer it to the main storage. For large sequential record processing, the effective data rate is the primary concern. On the other hand, for interactive systems, the rate of access achievable within a reasonable response time is a critical measure [17]. Sequential processing performance has seen considerable advancements, primarily attributed to increases in the intrinsic data rate of tapes and disks and the development of multi-head access structures. This structure permits a conceptual cylindrical data track at each access location, facilitating an almost continuous data flow via electronic track switching [18]. These advancements in data storage have led to significant enhancements in DDCPS.

Optical storage. Optical storage, characterized by its use of light to read and write data, incorporates technologies such as CDs, DVDs, and Blu-ray discs. An inherent advantage of optical storage is its durability, given its resistance to environmental threats that may degrade other storage forms [19]. Initially, optical storage gained traction for its role in distributing media content, thanks to its high-capacity offerings in comparison to early magnetic storage options and the simplicity of mass production [20]. These properties made it a popular choice for music, movie, and software dissemination [21].

Despite these challenges, optical storage maintains relevance in specific scenarios where its attributes, such as longevity and portability, are desired [22]. Optical storage can provide a stable, long-term data storage solution's ability to help it in support of the DDCPS today continue to play a role in portfolios of data storage [23].

Solid-state drives. Flash drives, more commonly recognized as USB drives or thumb drives, exemplify portability and convenience in data storage mediums. They utilize the same core technology as SSDs, flash memory, but are designed to be small and highly portable [24].

At the crux of flash drives is their ability to store data on an integrated circuit [25]. This non-volatile memory type, akin to SSDs, can hold data without requiring power. Flash drives are notable for their plug-and-play functionality, enabling users to easily transfer data between different computers or devices with USB ports [26]. Although they lack the high storage capacities of larger SSDs, HDDs, or cloud storage solutions, flash drives offer substantial utility in their portability and simplicity [27]. They are an ideal solution for physically transporting data or quickly sharing files between machines. The affordability of flash drives also makes them accessible for personal use and business applications alike [28].

In the context of DDCPS, flash drives have limited direct application due to their relatively lower storage capacity and lack of network connectivity [29]. However, a temporary storage or quick data transfer are applications where they are useful. As with any technology, flash drives are not without their constraints [30]. They have limited write/erase cycles, after which they may begin to fail. Nonetheless, for many users, the convenience and portability of flash drives outweigh these limitations. Flash drives hold a unique place in the spectrum of data storage technologies, even though they may not form the backbone of a DDCPS solution [31].

Tape storage. Tape storage represents one of the oldest forms of digital storage and remains relevant in modern data management primarily due to its extensive capacity and cost-effectiveness [32]. As with HDDs, magnetic tape is used to record data, similar to how data is written to them. In the early days of computing, tape storage was commonly used for both primary and backup storage. However, with the advent of faster, more convenient storage technologies like HDDs and SSDs, the role of tape storage has been significantly refined [33]. Data archiving and backup are the most common techniques.

Tape storage systems, given their sequential access nature, are not designed for rapid data retrieval. Unlike the random-access capabilities of HDDs, SSDs, or even optical storage, data on tapes must be read sequentially, meaning it can take longer to locate specific data points [34]. Therefore, tape storage isn't ideal for applications demanding instant data access. However, for long-term storage, particularly in backup scenarios where large quantities of data are written and stored sequentially, tape storage excels [35]. With DDCPS generating vast amounts of data, there's an escalating demand for efficient, reliable, and cost-effective long-term data storage solutions [36].

In spite of the rise of modern storage technology, tape storage remains a crucial part of the data storage ecosystem, especially for large-scale, long-term data storage needs [37]. By providing an economical solution for mass storage, tape has a place in comprehensive data storage strategies, including those deployed in DDCPS.

Cloud Storage. Cloud storage marks the latest advancement in the evolution of data storage, bringing with it a shift from localized hardware-based solutions to remote, network-based storage infrastructure. Data is stored across arrays of servers in data centers operated by third parties [38].

At the core of cloud storage is the principle of virtualization. Data is distributed across a network of virtual servers instead of being stored on a specific physical device. The servers may be located in different locations, but they can be accessed and operated as a single entity. Data storage solutions provided by cloud storage are scalable, flexible, and can be adapted to meet the needs of different users [39].

Compared to traditional storage mediums, cloud storage provides several unique advantages. The most significant of these is accessibility. A remote storage facility allows users to access data from anywhere with an internet connection. By enabling real-time collaboration, remote work, and seamless data sharing, businesses and individuals can completely transform how they manage their digital assets [40].

The scalability of cloud storage is another critical advantage, as HDDs and SSDs are traditional storage media with fixed capacities. A user will need to invest in additional storage once that limit is reached [41]. Conversely, cloud storage can easily be scaled up or down to match the user's needs, often via a simple adjustment in the service subscription. Businesses with fluctuating data storage requirements benefit from its flexibility and cost-effectiveness [42]. However, cloud storage does come with a few challenges. Data security and privacy are primary concerns, as sensitive information is stored off-site on servers operated by third-party providers [43]. Hence, robust encryption and security protocols are essential to protect data.

The cloud storage platform is mainly composed of storage devices network connected to the data centre as shown in Fig. 4.3, in order to ensure high reliability and availability of data, the cloud storage platform usually adopts redundant storage and backup strategies, and also needs to be equipped with monitoring systems and alarm mechanisms, as well as a perfect fault handling process [44]. In short, a cloud storage platform is an ecosystem of multiple components working in tandem, and each part is very important [45].

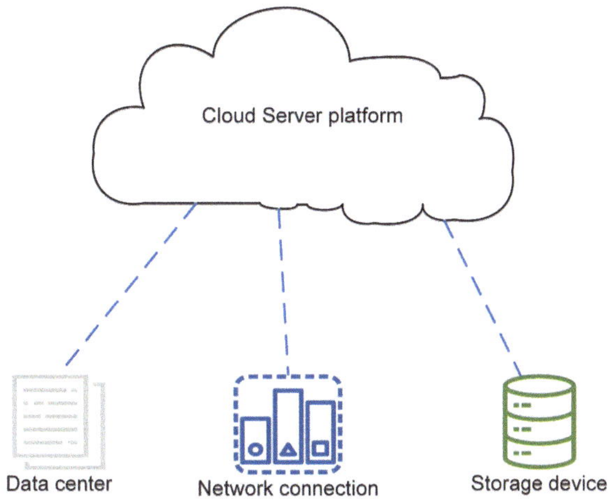

Fig. 4.3 Introduction to cloud storage platforms. The main components are storage devices, network connectivity and data centres

4.1.1.2 The Data Storage Location

The location of data storage can affect the speed of data access. Data stored near users can be accessed faster due to lower network latency. Data storage facilities should be reliable and stable to prevent data loss [46]. In some countries and regions, certain laws and regulations dictate where data must be stored, for example, cloud service providers often store backups of data in multiple locations around the world [47]. Data must be stored within the European Union or appropriately protected locations in compliance with the General Data Protection Regulation (GDPR) [48]. Laws and regulations may limit the choices of data storage locations. Choosing the right location for data storage can also impact data security [49]. For example, storing data on your own servers might require investment in hardware and human resources, while using cloud storage services may involve service fees [50]. Imperfect security systems make it easy for data to be accessed or stolen, and different data storage locations may bring different costs [51]. In general, choosing a data storage location is a decision that requires balancing multiple factors, including performance, availability, reliability, compliance, security, and cost [52].

There are five kinds of data storage locations: file storage, relational database, non-relational database, data warehouse and cloud storage [53]. File Storage presents a straightforward method to store data in various file formats like comma separated values (CSV), Java Script object notation (JSON), and extensible markup language (XML), offering universal compatibility at the expense of handling complexity [54]. Relational Databases provide a structured approach using tables, facilitating data queries and manipulations, yet may grapple with scalability and unstructured data. On the other hand, Non-relational Databases prioritize flexibility, scalability, and performance, making them apt for big data and real-time applications [55]. Data Warehouses cater to the storage and analysis of vast, diverse data sources, promoting data-driven decision-making, though their batch processing nature makes them less suitable for transactional operations [56]. Lastly, Cloud Storage allows remote data storage and access over the internet, cost-effective solution accessible from anywhere. The choice of storage method depends on the specific needs of the DDCPS.

File storage. In Sect. 4.1.1.2, file storage, a basic method of storing information on a computer, will be introduced. We will discuss the advantages and disadvantages of different file formats such as text files, CSV, JSON, XML, and so on.

In DDCPS, file storage is necessary in many situations, especially when dealing with tasks that require real-time feedback or have strict latency requirements [57]. For example, many critical decisions in autonomous vehicles need to be computed and stored locally in the vehicle, as relying on remote storage may cause unacceptable delays [58].

CSV (Comma separated values) files provide a more structured way to store data. They are text files that use commas to separate values, often representing data in a tabular form [59]. CSV files are straightforward and widely compatible with various applications, making them an excellent choice for simple datasets. Hierarchical and complex data structures cannot be stored in these formats [60].

JSON (JavaScript bbject notation) and XML (extensible markup language) files are designed to handle complex data structures. In both cases, data can be represented hierarchically using a human-readable, text-based format. JSON, being lighter and easier to read and write, is commonly used in web applications. XML, although more verbose, provides a robust schema and namespaces, making it suitable for complex, large-scale applications [61].

Each of these file formats serves specific needs, and their utilization depends on the type of data, the scale of the data, and the specific requirements of the data storage system [62]. In the realm of DDCPS, choosing the appropriate file storage format is critical to ensure efficient data management, accurate data representation, and effective data utilization.

Relational databases. Relational database is the mainstream database system at present, including MySQL, PostgreSQL, Oracle and SQL Server, and they all have their own advantages and disadvantages [63]. MySQL, an open-source relational database management system, employs a specific mechanism for data storage. The data structure within MySQL comprises tables organized into related groups [64]. An entry is stored in each row and column of these tables. Nevertheless, MySQL's storage engine system, particularly InnoDB, dictates the way data is stored and retrieved and is the crux of the data storage system [65]. Upon creating a new database, MySQL sets up a corresponding directory within the server's data directory. Each table in this database correlates with one or more files housed within this directory.

For tables managed with the InnoDB storage engine, the data is encapsulated in ibd files for individual tables or in ibdata files for a shared tablespace. Within these files, the data and the indexes of a table are organized into pages, serving as the basic unit for storage and retrieval operations. Tables are structured in a way that allows MySQL to efficiently access and manage data [66].

Furthermore, to ensure data integrity in the event of a system crash, InnoDB uses redo log files. Keeping track of every modification to the data helps in data recovery after a system failure [67]. MySQL also creates binary log files, logging all data alterations for replication and recovery purposes. The execution flow of MySQL is shown in Fig. 4.4 and can be summarised in the following steps: the client sends a query request and establishes a connection to the server through a half-duplex communication protocol. The connector parses and validates the query request and forwards it to the execution engine. If the query hits the cache, the cached results are returned directly; otherwise, it is parsed to generate a parse tree. The parse tree is syntactically and semantically analysed by the parser and transformed into an executable internal representation. The optimiser optimises the parse tree and selects the best execution plan. The execution plan is used by the executor to perform specific query operations, including data reading, sorting and filtering. During execution, the analyser records performance metrics such as execution time and resource consumption. For queries involving multiple tables or databases, data integration operations are performed and results are returned. The storage engine reads the data according to the selected engine type. Eventually, the execution results are returned to the client. This whole process, the parser and the performance metrics recorded by the parser are very helpful for subsequent performance analysis and optimisation work.

PostgreSQL stores data using a mechanism called Multiversion Concurrency Control (MVCC) [68]. The system can handle a large number of concurrent transactions without needing to lock excessively. In the process of storing data, an identification number is assigned to each transaction. The transaction ID is used to indicate when a new version of a row (tuple) became valid in the database. Older versions of rows are eventually removed by a process known as vacuuming. Moreover, PostgreSQL uses a page-based storage model where the unit of storage is a disk block, typically 8 KB in size. Indexes and table data are stored in heap files. Row data is stored in a heap file, and its location is tracked by index files, providing efficient random access to rows [69].

Oracle uses a storage structure known as tablespaces for storing the database's data [70]. A tablespace in Oracle is a logical container for physical data files, which in turn store the data for database tables and indexes. Each tablespace contains one or more data files, each of which is physically a large file in the file system of the machine hosting the Oracle instance [71].

When data is stored, Oracle uses a mechanism similar to PostgreSQL's MVCC for transaction handling and concurrency control [72]. However, in Oracle, the previous versions of a row are stored in a separate area known as undo segments, rather than alongside the current data. The undo data is used to provide a consistent view of data for other transactions and for rolling back transactions when necessary [73].

SQL Server organizes databases into multiple files of three types: primary data files, secondary data files, and log files. The primary data file hosts the system level data and any other data not included in secondary files [74]. Secondary data files hold user data, indexes, and large objects such as images or comments. Log files, as the name suggests, store all the log information used to recover the database.

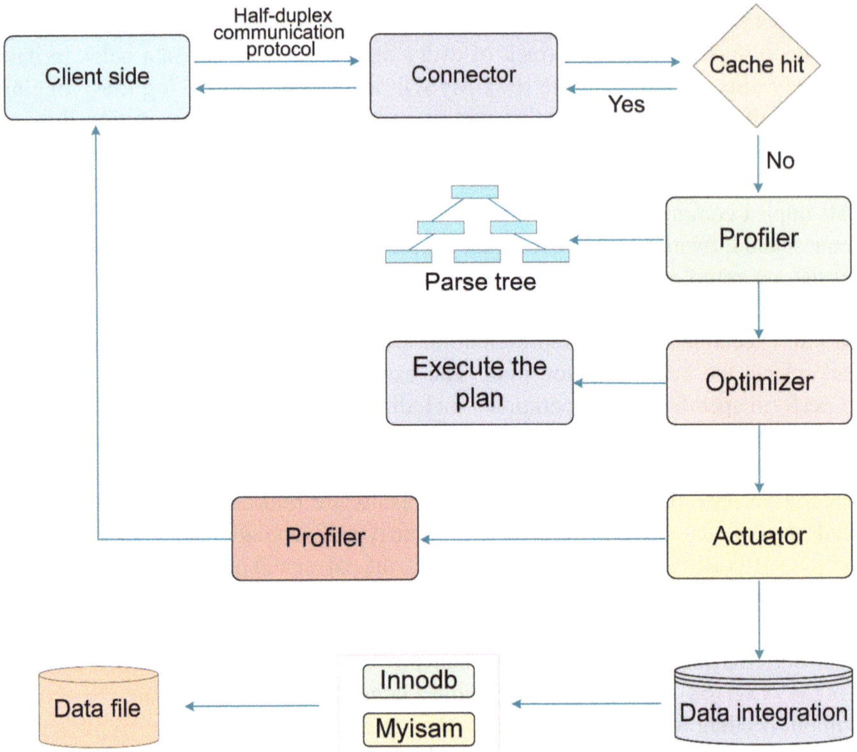

Fig. 4.4 Flow chart of MySQL execution process. The client sends a query request and establishes a connection with the server through a half-duplex communication protocol. The connector parses and validates the query request and forwards it to the execution engine

To ensure consistency, SQL Server uses a pessimistic concurrency control mechanism by default, unlike PostgreSQL and Oracle. However, SQL Server also supports optimistic concurrency control similar to MVCC with its Snapshot Isolation and Read Committed Snapshot Isolation levels [75].

Non-relational databases. Taking a closer look at non-relational databases will help us learning more about databases without fixed table structures. MongoDB, Cassandra, Redis, and other popular non-relational databases will be discussed in detail [76].

MongoDB is a NoSQL document-oriented database that stores data in a binary JSON format (BSON), allowing for flexible and dynamic schemas. MongoDB organizes data into collections of documents, each with a different set of fields, offering a great deal of flexibility [77]. Documents have an id field that functions as a primary key. The data is stored in a directory named dbpath when MongoDB is set up during installation. MongoDB uses a memory-mapped files mechanism to map the entire data file to byte arrays in memory, allowing the system's virtual memory to handle the caching of data and indexes [78].

Apache Cassandra utilizes a decentralized system, distributing data across all nodes in a cluster, with each node responsible for a designated range of data determined by a partition key. A distributed environment with this distribution mechanism provides high scalability across multiple servers, supporting efficient data management [79].

The process of data storage in Cassandra begins with data being recorded in a commit log for durability, followed by its placement in a memtable, an in-memory data structure. Sorted String Table (SSTable) is an immutable data file created when the memtable is filled. Data written at a given point is immutable, which ensures consistency. Rather than modifying the SSTables, Cassandra updates or deletes the row, keeping a history of changes [80].

To optimize storage and improve read performance, Cassandra employs a mechanism called compaction. SSTables are merged periodically and old or deleted data is discarded, thereby reducing storage overhead. Overall, Cassandra's data storage is marked by its unique partitioning strategy, commit logs for durability, memtables for in-memory storage, SSTables for disk storage, and compaction for storage optimization [81].

REmote DIctionary Server (Redis) is an in-memory data structure store, used as a database, cache, and message broker. Data in Redis is stored as key-value pairs, with support for various types of values (such as strings, lists, sets, and more). Redis periodically persists data to disk using either point-in-time snapshots or by appending each command to a write-ahead log (the append-only file or AOF). AOF provides better durability than snapshots when backing up and restoring [82].

Data warehouses. Data warehouses are systems for analyzing and reporting on large quantities of data. A discussion of data warehousing solutions such as Amazon Redshift, Google BigQuery, and Snowflake will be provided [83].

Amazon Redshift is a fully managed, petabyte-scale data warehouse service by Amazon Web Services (AWS). The system uses columnar storage to compress data and perform high-speed queries. Each node in Redshift has its own CPU, memory, and hard disk space, and data is distributed across the nodes. Various encoding schemes are used to compress data and reduce I/O operations for read queries [84].

Google BigQuery is a web service from Google that is used for handling and analyzing big data. The architecture is based on Google's Dremel tech, and columnar storage is used as well. However, unlike Redshift, BigQuery abstracts the underlying hardware, storage, and resources. The storage and computing layers are separated, allowing them to scale independently. Data is stored in Capacitor, BigQuery's proprietary columnar storage format, enabling fast SQL queries and high compression ratios [85].

Architecture: Redshift uses a classic massively parallel processing (MPP) architecture [86]. BigQuery is built on Dremel and abstracts the hardware and resource layer. Snowflake separates storage and computation, but also organizes data in micro-partitions. Scaling: In Redshift, you scale by adding more nodes to a cluster. BigQuery and Snowflake both allow for independent scaling of storage and compute resources, but BigQuery handles this automatically.

Data Storage: All three use columnar storage, but they handle data partitioning and compression differently. Redshift distributes data across nodes, BigQuery uses its Capacitor format, and Snowflake automatically divides data into micro-partitions [87].

Cloud storage. Cloud storage is a way of using cloud computing platform to store and manage data. Popular cloud storage services such as Amazon S3, Google Cloud Storage, and Microsoft Azure Blob Storage, we will look at how they do data storage. Cloud storage is a model of data storage where digital data is stored in logical pools across a network of physical servers. Servers are managed and maintained by cloud storage service providers, which keep the data available and accessible, and protect it from system failures, physical damage, and data loss [88].

Data is stored in buckets within the Amazon S3 service. A file and optional metadata are the components of an object. Uploading a file to a bucket is how you store an object in Amazon S3. Permissions and metadata can be set when you upload a file. Amazon S3 creates a unique version ID and assigns it to the object. If you enable versioning for the bucket, Amazon S3 applies the versioning configuration to all objects in the bucket [89]. The data is stored redundantly across multiple facilities and is designed for durability.

Google Cloud Storage stores data as blobs or objects within a hierarchical namespace, analogous to files within a directory structure. Each object is identified by a unique identifier within this namespace. Google Cloud Storage employs a distributed system for data storage and retrieval, providing strong data consistency. The data is automatically replicated across various Google data centers for high durability and high availability [90]. Google Cloud Storage also offers different storage classes that can be selected based on how frequently the data is accessed and how quickly it needs to be retrieved.

Azure Blob Storage stores large amounts of unstructured object data, such as text or binary data. Data is stored as blobs (binary large objects) within a container in the storage account [91]. Each blob consists of a set of blocks that hold the data, a block list that specifies the block order, and system properties such as size and time of last modification. Blob Storage offers three types of blobs: block blobs, append blobs, and page blobs. Each type is useful for different types of applications. For example, block blobs are ideal for storing text and binary data in the cloud, append blobs are ideal for append operations like logging data from virtual machines, and page blobs are designed for frequent read/write operations and provide the ability to write to a range of bytes in a blob. Azure automatically replicates data across data centers for high availability and durability [92].

4.1.1.3 Data Storage at the Storage Scale

The scale of data is growing at an astonishing rate. From personal smartphones and wearable devices, to industrial sensor networks, and large datasets in scientific research, we are producing and consuming vast amounts of data [93]. In a standalone or small-scale network environment, local storage or network-attached storage (NAS) can meet most requirements. Data storage at the storage scale is becoming increasingly important as data sizes grow and data needs to be stored across multiple devices or locations. Data storage, management, access, and persistence pose specific challenges to DDCPS, which typically involves large volumes of data streams that need to be processed and analyzed in real-time or near real-time. Therefore, choosing the right storage technologies and using them effectively is crucial for supporting the operation of DDCPS and optimizing its performance [94].

Data storage techniques at storage scale include, local storage, distributed storage and large scale data storage techniques as shown in Fig. 4.5. The following describes how local and distributed storage are used for data storage in DDCPS [95]. Local storage in DDCPS refers to data storage directly on the devices used in these systems. The data may be stored on the memory of a sensor, the hard drive of a server, or the flash memory of a user's device. Local storage is often fast and reliable, allowing quick access to data. However, the volume of data that can be stored is limited by the device's storage capacity. Data could also be lost if the device fails or is damaged because local storage does not provide redundancy [96].

Distributed storage systems in DDCPS split and store data across multiple devices, often located in different geographical locations. Using this approach enhances device reliability by enabling more efficient data storage as well as adding redundancy [97]. Distributed storage systems use techniques such as data replication and sharding to distribute data. Replication involves storing copies of the same data on multiple devices, while sharding splits a dataset into smaller parts that are stored on different devices. Examples of distributed storage systems include Hadoop Distributed File System (HDFS) and Amazon's DynamoDB [98]. For the largest scales of data, special storage technologies are needed. Cassandra and HBase are distributed databases. Amazon S3 is an object storage system. Google Cloud Storage is a distributed file system [99]. Data can be stored and managed across thousands of storage devices, storing petabytes of data or more. Hardware failures, network issues, and other challenges related to operating at this scale are handled through them. DDCPS also benefits from powerful tools and APIs to query and analyze the data [100].

4.1.2 Explore Data Storage Instances in the System

Defining data storage instances. Instances of data storage, typically databases, are single instances of data storage that enable data to be stored, retrieved, and manipulated. A database instance may exist as a relational database, a NoSQL database, a data lake, or a data warehouse, depending on the system's overall

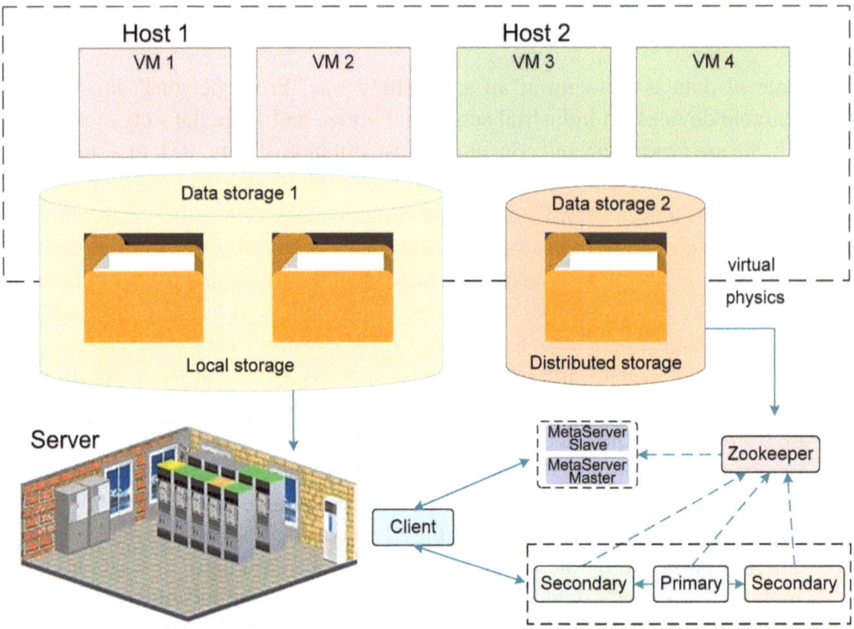

Fig. 4.5 Data storage at a storage location, mainly through the host and virtual machine with the storage, storage is mainly divided into local storage and distributed storage, local storage directly into the server to save, distributed storage is a cyclic process

requirement [101]. A data storage instance is an entity in a storage system that can store, retrieve, and manipulate data. It could be a database, a dataset, a data file, or some other forms of data container. The main goal of a data storage instance is to provide the system with a reliable and efficient way to process data. Data storage instances can be classified into different types based on the type and structure of data they store, as well as the operations and functions they support. For example, relational database instances store structured data and support SQL queries. NoSQL database instances can store structured, semi-structured, or unstructured data and support flexible and scalable data operations [102]. Data lake instances store raw and unprocessed data and support big data processing and analytics.Data storage instances can also be classified into different types based on their storage schema and architecture. For example, a stand-alone database instance runs on a single server, the distributed database instance runs on multiple servers and is connected via network. Cloud database instances run in a cloud environment and provide services over the Internet.Selecting and designing a data storage instance is a critical system design decision that affects system performance, reliability, scalability, and cost. Therefore, understanding the concept and types of data storage instances and how to select and design them according to the requirements and constraints of the system is an important part of the design and implementation of DDCPS [103].

Role of data storage instances in DDCPS. In DDCPS, the data storage instance plays a key role. Data generated by system components as well as data that drives decision-making are stored in them. The data can range from sensor readings to predictive analytics outputs [104].

Characteristics of effective data storage instances. Not all data storage instances are created equal. Certain characteristics, such as scalability, reliability, and performance, can greatly impact the effectiveness of a data storage instance within a DDCPS. To better understand the role of data storage instances in DDCPS, we'll examine a real-world case study. The purpose of this study is to illuminate how data storage instances function within a fully operational DDCPS, as well as the insights gained from this data-driven approach. Data storage instances can include distributed file systems, relational databases, and no-SQL databases [105].

Distributed file systems: When dealing with large volumes of unstructured or semi-structured data, distributed file systems like Hadoop Distributed File System (HDFS) shine. Data is broken down into smaller blocks and distributed across multiple nodes, allowing for parallel processing and exceptional scalability [106]. In a DDCPS, distributed file systems contribute to handling high-volume, high-velocity data, making them instrumental in big data analytics, machine learning, and other data-intensive tasks.

Relational databases: When data integrity and consistency are paramount, relational databases (RDBMS) like MySQL, PostgreSQL, or Oracle database become valuable. Table-based structures and robust transaction mechanisms ensure high data accuracy and consistency [107]. In a DDCPS context, an RDBMS can play a vital role in areas requiring strict transactional control or complex queries involving multiple tables, for example, user account management or financial transactions.

NoSQL databases: These databases, such as MongoDB (a document store), Cassandra (a column store), or Redis (a key-value store), provide a high degree of flexibility and scalability, ideal for diverse data structures and high-volume transactions [108]. In a DDCPS, NoSQL databases can contribute significantly to real-time processing, personalization features, or in scenarios where data does not conform to a strict schema.

To illustrate this in a real-world scenario, consider a DDCPS for a smart city application. Sensor data from across the city can be stored in a distributed file system like HDFS for processing and trend analysis. At the same time, relational databases could handle the structured data related to city administration, like resident records or public service data. Simultaneously, a NoSQL database could be used to manage diverse data types from different sources like social media feeds, traffic data, etc., enabling real-time insights and decisions.

Distributed file system: In a distributed file system, data is stored and distributed on multiple servers or nodes, which enables the system to process and store large amounts of data. A data storage instance in this case may be a specific data set or file, stored in a part of the system [109]. For example, Hadoop Distributed File System (HDFS) is a widely used distributed file system that can store large datasets and support large-scale data analysis applications. Such data storage instances can provide high data throughput and fault tolerance, making them advantageous when dealing with large-scale data.

In the context of DDCPS, the contribution of data storage instances in a distributed file system like Hadoop Distributed File System (HDFS) is vital. HDFS, specifically designed for storing and processing large datasets across clusters of computers, can seamlessly handle data generated from various cyber-physical system components.

For example, consider a DDCPS for a smart city, with numerous IoT sensors scattered across urban infrastructure for monitoring traffic, air quality, energy consumption, etc. Each sensor generates continuous data streams which need to be stored, processed, and analyzed in real-time. Here, each sensor's output can be viewed as an instance of data storage in the distributed file system [110]. The data from each sensor is distributed across multiple nodes in the HDFS, ensuring high fault tolerance and accessibility.The distributed nature of HDFS allows for efficient parallel processing of these data streams. This is crucial in real-time or near real-time applications, where timely processing and analysis of data directly contribute to DDCPS functionality. For instance, in our smart city example, real-time analysis of traffic sensor data can provide instant feedback to traffic control systems, optimizing traffic flow and reducing congestion.Furthermore, the inherent redundancy in HDFS ensures data persistence and fault tolerance. This characteristic is essential for DDCPS, ensuring the reliability and robustness of the system. By ensuring efficient storage and processing of large amounts of data, facilitating real-time analysis, and enhancing fault tolerance and reliability, HDFS instances contribute significantly to DDCPS' overall functionality.

Relational database: A relational database is a database based on the relational model, where data is stored in the form of a table. Each table has multiple columns (representing data fields) and multiple rows (representing data records). A data storage instance in this case could be a database, a table, or a specific record. For example, MySQL and PostgreSQL are both widely used relational database management systems. Structured queries (such as SQL) and data integrity guarantees ensure the integrity of structured data [111].

NoSQL databases: NoSQL databases are non-relational databases such as Fig. 4.6, which do not rely on the traditional relational model but use various data models such as key-value pairs, documents, column families and graphs. A data storage instance in this case could be a database, a collection (in the case of a document database), or a specific data item. For example, MongoDB is a document-oriented NoSQL database and Cassandra is a column-oriented NoSQL database. Such data storage instances can provide a high degree of scalability and flexibility, making it advantageous when dealing with large-scale, diverse, and rapidly changing data [112].

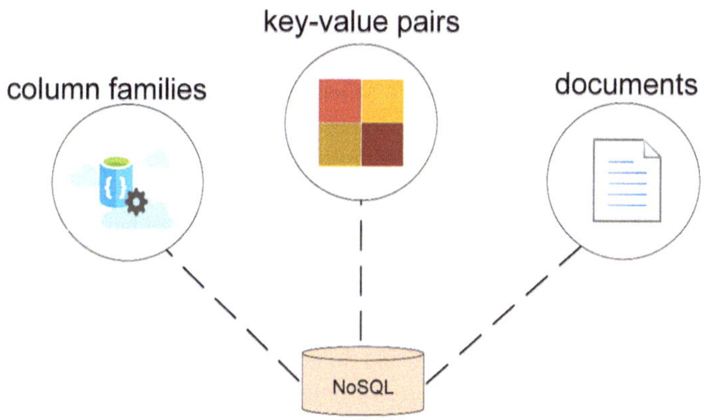

Fig. 4.6 NoSQL database is a new type of database for big data processing, which does not use the traditional tabular-relational model, but a more flexible data model, such as key-value, document, graph, or column family, etc., with high scalability, high performance, and high availability

4.2 Data Management and Processing Techniques

In the current era, we are surrounded by massive amounts of data generated from various sources, such as sensor data, device logs, network traffic information, and so on. In DDCPS, data comes from various sources, such as Internet of Things devices, industrial control systems, traffic control systems, and so on. Smart manufacturing, energy management, healthcare, etc., can benefit from effective data handling and processing as shown in Fig. 4.7.

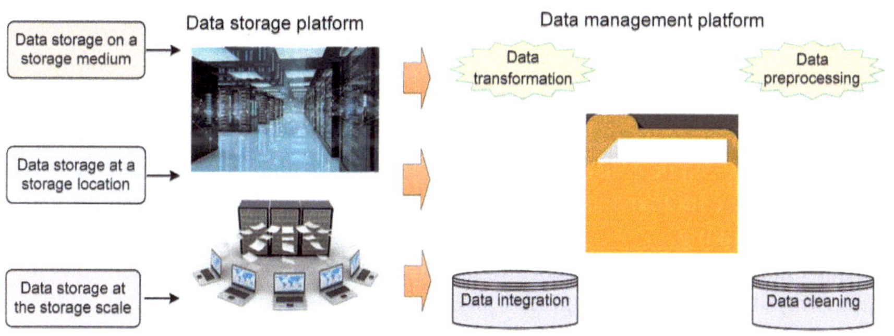

Fig. 4.7 Introduction to data storage platforms and data management platforms. A data storage platform is a software or hardware infrastructure used to store, manage and access data. It includes various components such as file systems, databases, distributed storage systems, and cloud storage services. Data management platform is a software system that provides functionality to manage and control all aspects of the entire lifecycle of data. It includes data integration, data transformation, and data pre-processing

Relationship between data management and data processing. Data management and data processing are two interrelated aspects of dealing with data, each with its unique role, but they usually work together to extract valuable information from the data [113]. Data, database, and database management system are three basic concepts closely related to database technology.

Data. Data is a symbol used to record things. A symbol can be a number, a text, a graphic, an image, an audio or a video. Computers store digitally processed information. In modern CPS systems, the concept of data has been expanded significantly. While early CPS systems primarily dealt with numerical data for scientific computing, including integers, real numbers, and floating-point numbers, today's computers store and process an extensive variety of objects, leading to increasingly complex data representations [114].

Data base. On the surface, a database can be seen as a digital warehouse for storing data. However, instead of a physical location, it resides on a computer storage device and employs a specific format for data storage. Essentially, a database is a computer-based long-term storage that houses a sizable collection of organized and shareable data. The data within a database is structured, described, and stored based on a particular data model. It offers low redundancy, high data independence and scalability, enabling multiple users to access and utilize it.

Data management system. The data management system is a scientific organization and storage of data, efficient access and maintenance of data. Data management is mainly concerned with data acquisition, verification, storage, protection, and availability.Data management encompasses the practices, architectural techniques, and procedures to manage the full data lifecycle needs of an organization. It includes data collection, data validation, storage, protection, and processing. Effective data management ensures that the data is accurate, accessible, consistent, and complete.
Data processing. Processing data involves using and analyzing them to derive meaningful information or knowledge. Data processing involves the conversion of raw data into a more meaningful format. Data cleaning, transformation, and aggregation, along with further analysis and visualization, are typically used to accomplish this. Data processing can be considered a component or a result of data management. The data curated by data management practices is used to provide valuable information for decision-making.

Data management is a critical component of DDCPS. Complex systems that integrate computation, networking, and physical processes generate and rely on large amounts of data. The data is used to monitor system status, make real-time adjustments, and even train machine learning models to improve system performance [115]. In essence, data management sets the foundation for data processing. Without effective data management, data processing would be dealing with data that might be incomplete, inconsistent, or inaccurate, resulting in less reliable outputs. Conversely, effective data management practices enable more efficient and reliable data processing, which can lead to more accurate insights and decision-making.In a broader sense, data management and data processing are parts of a larger system designed to maximize the value of data for an organization. This system includes various other components like data governance, data analytics, data science, and more, all working together to turn raw data into actionable insights.

The relationship between data management and data processing can be seen as the former providing the basis and framework for the latter. Without good data management, data processing cannot be carried out effectively. For example, if the quality of the data is poor (due to errors, missing values, inconsistencies, etc.), the results obtained during the data processing phase can be misleading. On the contrary, with good data management, data processing can be carried out more smoothly and accurately, resulting in more valuable results.In general, data management and data processing go hand in hand, and together they form a core part of data science, supporting the transformation from raw data to useful information [116].

4.2.1 Database Management Techniques

Database management is a layer of data management software that sits between the user and the operating system as shown in Fig. 4.8. The database management system is the basic software of the computer as well as the operating system. It is also a large and complex software system. Its main functions include the following aspects:

(1) Data definition function: The database management system provides data definition language (DDL), through which users can easily define the composition and structure of data objects in the database.
(2) Data organization, storage and management: The database management system needs to classify, organize, store, and manage various data, including data dictionary, user data, data access path, etc. It needs to determine which file structure and access method to organize these data at the storage level, and how to realize the connection between the data. The basic goal of data organization and storage is to improve storage space utilization and facilitate access, and provide multiple access methods (such as index search, hash search, sequential search, etc.) to improve access efficiency.
(3) Data manipulation function: The database management system also provides data manipulation language (DML); users can use it to manipulate data to achieve the basic operation of the database, such as query, insert, delete, and modify.
(4) Transaction management and operation management of the database: The database is managed and controlled by the database management system during the establishment, operation, and maintenance of the database so as to ensure the correct operation of the transaction, ensure the safety and integrity of the data, support the concurrent use of the data by multiple users, and provide the system recovery function after the failure occurs.

DDCPS contain a large amount of data, so they may use various types of databases, including but not limited to relational databases (such as MySQL, PostgreSQL),

Fig. 4.8 An important component of database management technology. They consist of MongoDB database management, MySQL database management and Google Cloud database management

SQL-less databases (such as MongoDB, Cassandra), time series databases (such as InfluxDB) and graph databases (such as Neo4j). The use of those databases for data management will be determined according to the specific needs and application scenarios of CPS. The following is to select several of these databases to introduce how to manage data.

4.2.1.1 MySQL Database Management

MySQL, as a relational database management system, stands as one of the most widely used relational databases today. SQL, a structured query language, is purposefully designed for database interaction, offering a straightforward and efficient means for data retrieval and modification. SQL caters to nearly all database management systems (DBMS), and its statements are formed from a concise set of words. Likewise, MySQL employs standard SQL language, operating on a client-server model, broadly bifurcated into a server layer and a storage engine layer. The server layer envelops the majority of MySQL's core functionalities, such as connectors, optimizers, executors, query cache, and analyzers, along with all inbuilt functions, triggers, views, and stored procedures. The storage engine layer is tasked with the storage and retrieval of data. MySQL supports a range of storage engines, including but not limited to InnoDB and MyISAM, which offer plug-and-play capabilities.

MySQL database for data management method. MySQL, like any other relational database management system (RDBMS), can be used to manage data in CPS, but it's important to note that MySQL isn't specifically designed for CPS. MySQL can be used to store and manage large amounts of data generated by sensors and other devices in CPS. This may include system state data, sensor measurements, user input, etc. MySQL includes functions for querying and analyzing the data stored in it. This can be used to understand the state of CPS, detect anomalies, predict future states, etc. MySQL can be used as an intermediary for communication between different parts of the CPS. For example, one part of the system can write data to a MySQL database and another part of the system can read that data and operate on it. MySQL has data redundancy and backup functions, which is very important in CPS because data loss may cause serious real-world consequences [117].

Using MySQL (or any traditional RDBMS) with CPS also presents some challenges. 1. Real-time or near real-time data processing is usually required in CPS, while traditional RDBMS may encounter difficulties, especially when dealing with large amounts of data. CPS can generate large amounts of data, and extend a traditional RDBMS to handle this data can be difficult. 3, CPS can generate a wide variety of data, including structured, semi-structured and unstructured data. Traditional RDBMSS are best suited for structured data, but may struggle with other types of data. CPSS are often distributed systems, and traditional RDBMSS are not always well suited for such systems. A distributed database or data store may be more suitable [118].

4.2.1.2 MongoDB Database Management

CPS are the integration of physical and networked systems that interact through computing and communication technologies. Databases such as MongoDB, a non-relational database with a flexible data structure, are often used in CPS applications since they process large amounts of real-time data. Below are some MongoDB approaches to managing data in CPS [119].

Oracle database for data management method. The basic unit of data stored in MongoDB is a document, which is made up of key-value pairs, similar to JSON objects. In a traditional relational database, documents are arranged in tables called collections, which are similar to tables in this structure [120]. However, unlike tables, collections do not need to define any schema.MongoDB supports distributing data across multiple servers, which is called sharding. This can provide higher throughput and data redundancy.

MongoDB supports indexes for any field, which can improve query performance. Indexes enhance query performance by speeding up queries on the table's data. An index contains the values of one or more columns and Pointers to the rows in the table associated with these values [121]. When a query is executed, MongoDB can quickly find rows that match the query, if the index is available, without needing to scan the entire table.

The MongoDB aggregation framework allows you to perform complex processes on your data, and it has a variety of security features, including access control, encryption, and auditing.

4.2.1.3 Google Cloud Database Management

Google Cloud Spanner is a globally distributed relational database service from Google that seamlessly scales globally while providing transactional consistency and SQL semantics.

InfluxDB database for data management method. Google Cloud Spanner is a fully managed, mission-critical, relational database service that offers transactional consistency at a global scale, schemas, SQL (ANSI 2011 with extensions), and automatic, synchronous replication for high availability.

Google Cloud Spanner is a powerful, globally consistent relational database service whose data management process is highly organic and coherent. First, when data is written, Spanner utilizes Google's TrueTime API to assign a globally consistent timestamp to each transaction, ensuring external consistency of reads and writes as well as globally consistent reads [122]. After data is written, Spanner automatically replicates data synchronously in multiple regions, ensuring high availability and durability of data. Splits are automatically created by Spanner and distributed among load-balanced servers as needed. At the same time, being a relational database, Spanner supports schema and SQL queries, provides full ACID transactions, and automatically optimizes SQL queries [123]. As a fully managed service, all operational work related to running the database, such as hardware configuration, setup and configuration, capacity planning, backup, and failure recovery, is handled by Google. Finally, to ensure data security, Spanner implements a number of security measures including encryption of data at rest and in transmission, identity and access management roles, and auditing using Cloud Audit Logs. Google Cloud Spanner makes global transactional consistency, high scalability, and high availability possible while keeping data secure and optimizing performance.

4.2.2 Data Processing Techniques

Data processing is the basis of data storage. However, real-world storage is often very difficult. Processing and storing data are closely related and must be done efficiently and precisely [124]. Data processing is divided into data index, data transformation and data compression as in Fig. 4.9.

4.2.2.1 Data Index

Using an index can improve query performance for large-scale data storage and query. According to the query requirements, select the appropriate field to create an index to speed up efficient data retrieval [125].

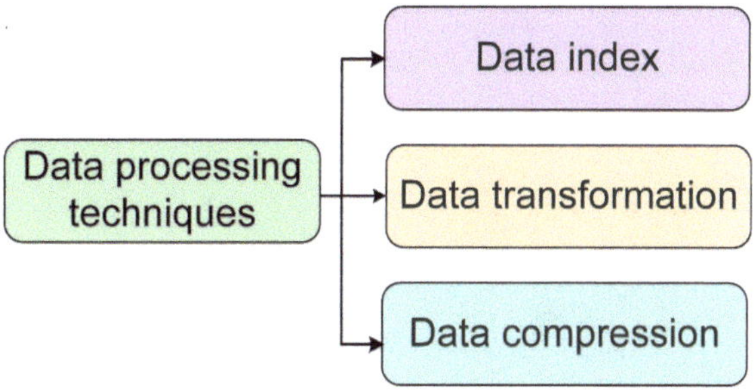

Fig. 4.9 An important component of database management technology. They consist of MongoDB database management, MySQL Database Management and Google Cloud database management

B-tree (B-Tree) index. B-tree is a balanced multi-tree often used in database systems. It improves the query efficiency by storing multiple key values in each node, which is suitable for range query and exact matching query.

B + tree (B + Tree) index. B + tree is an optimized index structure based on the B tree. All the B + tree data is stored on the leaf node, and the non-leaf node only stores the key value information, which can accelerate the speed of range query and sequential access.

Hash index. Hash index uses a hash function to map keywords to index items and directly locates data location by calculating a hash value. Hash index is suitable for equivalent query but does not support range query.

Full-text index. Full-text index is mainly used to search for text-type data, such as articles, blogs, etc. It will segment the text and establish an inverted index to support the fuzzy matching of keywords and full-text search.

4.2.2.2 Data Transformation

In computer programming, it is an important task to deal with the conversion between different data types. Especially in the process of data storage, it is often necessary to convert data from one type to another to ensure correct storage and retrieval. Two concepts are introduced: mandatory type conversion and automatic type conversion [126]. Section 4.2.2.2 discusses the application of these two types of conversion to data storage.

Automatic type conversion. As shown in Fig. 4.10, automatic type conversion occurs when a programming language automatically converts one data type to another under certain circumstances without requiring explicit conversion.The data is usually converted in this way to ensure accuracy when operating on or calculating the expression [127]. A type conversion using automatic data storage is implemented in the following scenarios:

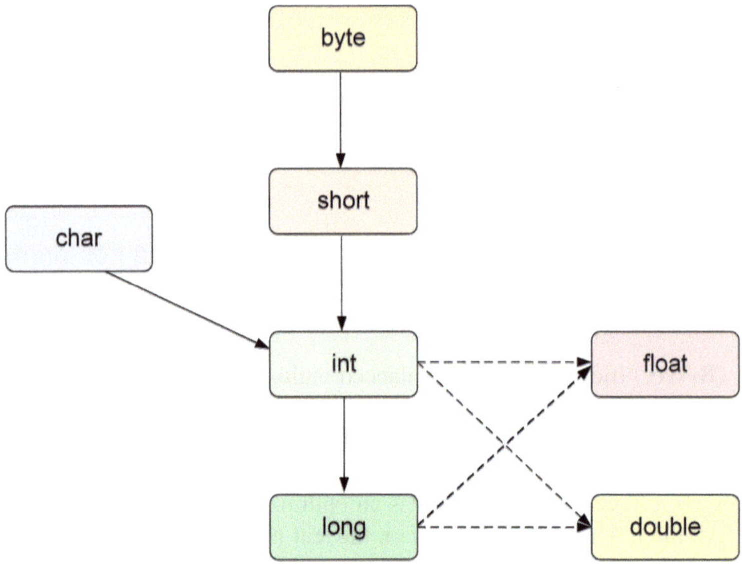

Fig. 4.10 Automatic type conversion. When an expression or operation involves variables of different data types, the system performs type conversions according to an order from low-precision to high-precision to ensure that they are compatible. This order is from byte to short, to int, then long, then float, and finally double

(1) Variable assignment: When assigning a value to a variable, the programming language automatically converts the type according to the context to ensure data compatibility.
(2) Data structure storage: When storing data into arrays, lists, or other data structures, automatic type conversion can ensure the type matching of data to store and process data correctly.
(3) Expression calculation: When performing complex mathematical or logical expression calculations, the programming language will automatically perform type conversion to perform the calculation and obtain accurate results correctly.

Forced type conversion. A forced type conversion or explicit type conversion is an explicit instruction in code that instructs the programming language to convert one data type to another data type as in Fig. 4.11. The programmer must specify the type of conversion to be implemented and use a specific syntax or function to accomplish the conversion. In data storage, forced type conversions are mainly used in the following situations.

Precision control: When we need to make precise conversions between different data types, such as converting floating-point numbers to integers, mandatory type conversion allows us to specify conversion rules and ensure the accuracy of the results.

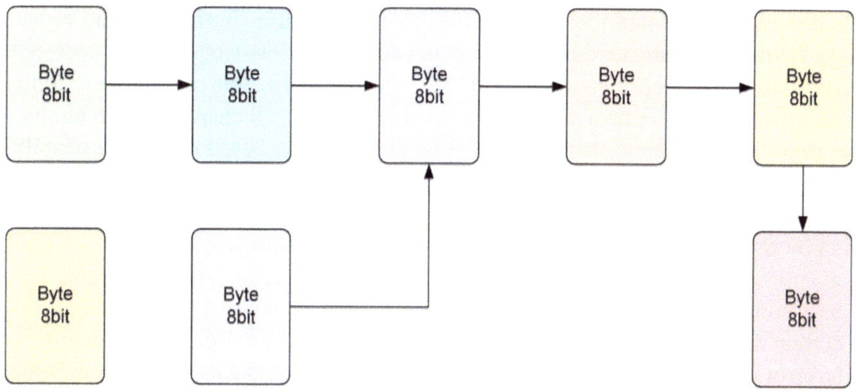

Fig. 4.11 A force transformation diagram shows the conversion of one data type to another. Byte is an 8-bit data type

Data truncation: Sometimes, we may need to truncate the data to a smaller range, such as converting a larger integer into a smaller integer type. Forced type conversion can help us achieving this goal while ensuring that the truncated data remains consistent.

Data serialization and deserialization: In the process of data serialization and deserialization, forced type conversion is used to convert data from one type to a byte stream or other format for storage or transmission. During the deserialization process, the data will be converted back to the original data type in order to recover the original form of the data.

In data storage, forced type conversion and automatic type conversion are common and necessary operations. Automatic type conversion ensures data compatibility and correct processing, while forced type conversion allows us to control the conversion process of data types accurately. Programmers should be aware of the type conversion rules and behaviors of programming languages and use these two types of conversions carefully to ensure the accuracy and consistency of data storage and processing [128].

4.2.2.3 Data Compression

Lossless compression technology: Lossless compression technology means that after the original data is compressed and decompressed, the decompressed data is completely consistent with the information contained in the original data. Among them, according to the different principles of the algorithm, lossless compression technology is divided into compression algorithms based on a statistical model and compression algorithms based on a dictionary model [129].

The core idea of the compression algorithm based on a statistical model is to use the frequency of each character in the data to achieve coding compression. Among these algorithms, Huffman coding, arithmetic coding, and run-length coding are common. The main idea of Huffman coding is to use shorter codes to replace high-frequency characters and longer codes to replace low-frequency characters to achieve the goal of compressing data [130]. The basic principle of arithmetic coding is to encode the entire data to be compressed and encode all characters into numbers between 0 and 1. Input characters must be pre-analyzed based on their probability and encoded accordingly. The higher the accuracy of probability analysis, the better the compression effect [131]. Run-length coding achieves the purpose of compression by representing the same symbol that appears continuously.

Lossy compression technology: Lossy compression technology makes the data have a certain degree of information error in the compression and decompression process. The error has little effect on the interpretation of the original data. Compressed video files, for example, are commonly used in scenes where accuracy is not critical [132]. Although the video compression degree of different clarity is different, the viewing experience is not significantly affected. In addition, the historical data in the industrial field has the characteristics of large storage, strong correlation, and high redundancy. The application of lossy compression can effectively increase the storage space and reduce data storage. Lossy compression technology removes part of the non-core information from the original data, which has a limited impact on the main meaning of the original data. This technology can significantly improve the data compression rate within the allowable accuracy range [133].

4.3 Big Data Processing Technology of DDCPS

Section 4.3 focuses on the data flow of storage and management, the storage of DDCPS, the management of DDCPS, and the big data processing techniques of DDCPS, as shown in Fig. 4.12.

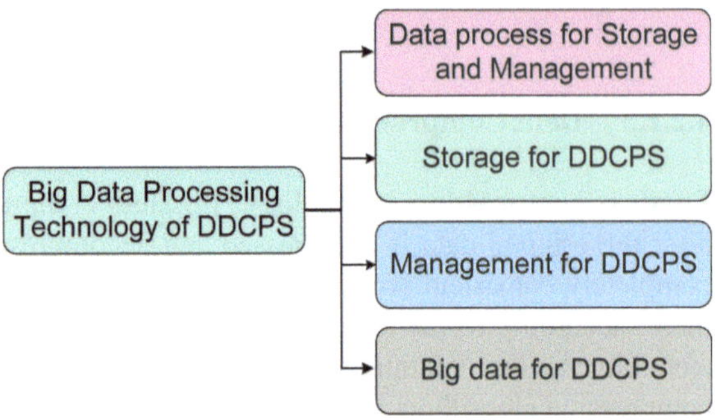

Fig. 4.12 Big data processing technology of DDCPS consists of data process for storage and management, storage for DDCPS, management for DDCPS and big data for DDCPS, which explains how storage management and big data are applied in DDCPS

4.3.1 Data Process for Storage and Management

Data processing plays a key role in improving data storage and management. Through effective data processing, the quality, security and availability of data can be optimized, thus improving the efficiency and performance of data storage and management. Here are some ways to improve data storage and management in DDCPS.

Data compression and archiving: For historical data or less accessed data, data compression and archiving techniques can be applied. Compression can reduce the storage space footprint of data, while archiving can save storage costs by migrating data from online storage to a lower cost storage medium such as tape or cloud storage.

Real-time data processing: Data generated in real time should be processed in a timely manner in DDCPS. Using streaming data processing technology, data can be processed, transmitted, and stored in real time, ensuring that data is accessed and utilized in a timely manner to support real-time decision making and control.

Data consolidation and analysis: In DDCPS, data usually comes from multiple sources and may be heterogeneous. Data integration and analysis are very important to help the system to obtain useful information from multiple data sources. Using appropriate data integration techniques and data analysis models can optimize the efficiency of data utilization.

4.3.2 Storage for DDCPS

In DDCPS, enforcing rules is an important task. These complex systems often involve the collection, processing, and analysis of large amounts of real-time data to support intelligent decision making and autonomous behavior [134]. A large scale and diversity of these big data techniques makes them an ideal candidate for regularization to improve their application and understanding. The regularity of Data storage is a key step, which ensures that data can be collected, stored and managed efficiently, safely and reliably in the whole system. The systems integrate physical entities with computer systems, allowing for intelligent decision-making and autonomous control [135]. The following are the main aspects of regularizing Data stores in DDCPS: DDCPS generate data in real time, so efficient mechanisms for real-time data processing and storage are required. Stream data processing technology is used to ensure that the data can be transmitted, processed and stored in real time after being collected to support real-time decision-making and control [136].

Distributed data storage: Data-driven intelligent physical systems may involve multiple geographical locations and computing resources, so a distributed data storage architecture is required. The risk of single points of failure can be reduced and fault tolerance and availability of the system can be improved [137].

Data security and privacy: Data in intelligent physical systems often contain sensitive information, including device status, user privacy, etc. Therefore, measures such as data encryption, authority control and authentication must be taken to ensure the security and privacy protection of data during transmission and storage [138].

Regular backup and data recovery procedure can prevent data loss due to hardware failure, natural disasters, and unexpected events. Ensure the reliability and integrity of data backups so that data can be restored in a timely manner when needed [139].

Data quality management: Data quality directly affects system decisions and performance. Hence, the need for a data quality management strategy is vital for ensuring data accuracy and consistency [140].

Storage system performance optimization: Aiming at the storage requirements of data-driven intelligent physical systems, this paper optimizes the read/write speed and data access efficiency of storage systems to meet the requirements of real-time data processing and decision-making.

The regularization of data storage is crucial in DDCPS. Our data storage strategies and technical and security measures can guarantee reliable, secure, and efficient data, allowing us to support intelligent decision-making and system optimization, as well as realize the sustainable development of intelligent physical systems [141].

4.3.3 Management for DDCPS

In DDCPS, the regularization of Data management is the process of ensuring that data is collected, stored, processed, and utilized efficiently throughout the system life cycle. Data management plays a crucial role in these complex systems, which involves data organization, integration, cleaning, analysis and application [142]. The following are the main aspects of regularized data management in DDCPS.

Data acquisition and transmission: Standardize the selection and deployment of data acquisition equipment to ensure accurate and reliable data collection. Furthermore, efficient data transmission channels are designed to ensure real-time decision-making and control are supported by timely data transmission [143].

Data quality management: Ensuring the quality of data is essential for the proper functioning of the system. In the process of data management, data quality management strategies should be implemented, including data validation, error correction and deduplication, to ensure the accuracy, integrity and consistency of data [144].

Data storage and backup: Select appropriate data storage technology and equipment to establish efficient and safe data storage system. In addition, data backup and recovery mechanisms should be established in order to prevent unexpected scenarios, such as data loss or hardware failures, from occurring [145].

Data security and privacy protection: Security measures such as data encryption, authority control and identity authentication should be taken to ensure the security and privacy protection of data during transmission and storage [146]. In particular, regulations and privacy policies must be adhered to for data involving personal privacy.

Data lifecycle management: The lifecycle of data needs to be considered. Define data storage duration, processing methods, and archiving strategies to ensure that data is effectively managed and maintained [147].

Data visualization and application: Through data visualization technology, complex data is transformed into easy-to-understand charts and visual images to help users better understand and apply data to support intelligent decision-making and optimization.

In summary, the regularization of data management is crucial in DDCPS. As a result of developing reasonable data management strategies and adopting appropriate technical and security measures, data reliability, security, and efficiency can be ensured, facilitating intelligent decision-making and optimization of the system, as well as the sustainable development of the intelligent physical system [148].

4.3.4 Big Data for DDCPS

DDCPS is a production system based on data-driven, which realizes safe, efficient and intelligent operation by integrating data analysis, distributed intelligent manufacturing and integrated blockchain technology. In DDCPS, big data technology is used to regularize complex systems in order to better respond to user needs and achieve safe, efficient and intelligent operation.

Data collection and standardization: Through big data technology, DDCPS can collect and integrate a large amount of data from different sources, including sensor data, equipment data, production data, etc. Different formats and structures are often used for this kind of data. Standardizing them with big data technology will allow them to be stored and analyzed more consistently [149].

Transformation tools: For example, Apache NiFi is a streaming data integration and transport tool that helps to format and standardize data from different data sources. By defining data flow and rules, NiFi can automatically process data with different formats and structures to make it conform to a unified standard.

Data analysis and mining: DDCPS uses big data technology to conduct in-depth analysis and mining of the collected data. Machine learning, data mining, statistical analysis, and other methods can be used to discover hidden patterns, trends, and correlations in data. This helps to understand the behavior and performance of complex systems, identify potential problems and risks, and provide decision-making basis for normalizing complex systems.

Real-time monitoring and feedback control: DDCPS can realize real-time monitoring and feedback control of complex systems by using big data technology. Through the collection and analysis of real-time data, state changes, anomalies and potential risks of the system can be detected in time, and corresponding control strategies can be adopted. This can ensure the safe operation of the system and realize the standardized management of complex systems [150].

4.4 Summary

There are three main sections: data storage types, data management and processing techniques, and DDCPS big data processing techniques. The first part of this chapter focuses on the many approaches and technologies for storing and organising data, including traditional relational databases, emerging NoSQL databases, columnar databases and distributed file systems. The advantages and disadvantages are explored in detail for each storage type. Data management and processing techniques provides an in-depth look at the techniques and algorithms used for data management and processing. Topics covered include data integration, data cleansing and preprocessing, data transformation, indexing, querying, and data analysis. The final session on big data processing techniques for DDCPS focuses on techniques for processing big data in the context of DDCPS, which involves the integration of physical processes with data analytics and control systems to enable real-time decision making. Techniques and architectures for efficient and scalable big data processing in DDCPS are explored.

This chapter provides real-world case studies, examples, and applications to illustrate the concepts and technologies discussed; and provides readers with a comprehensive understanding of data storage, management, and processing technologies, particularly suited to modern DDCPS.

Competing Interests

1. The research is supported by the National Key Research and Development Project under Grant 2023YFB3307300, the National Natural Science Foundation of China under Grants 62373014, 62422301, 62125301, and 92467205.
2. The authors have no conflicts of interest to declare that are relevant to the content of this chapter.

References

1. J. Guanetti, Y. Kim, and F. Borrelli, "Control of connected and automated vehicles: State of the art and future challenges," *Annual reviews in control*, vol. 45, pp. 18–40, 2018.
2. D. J. Cook and S. K. Das, "How smart are our environments? an updated look at the state of the art," *Pervasive and mobile computing*, vol. 3, no. 2, pp. 53–73, 2007.
3. K. Zheng, Z. Yang, K. Zhang, P. Chatzimisios, K. Yang, and W. Xiang, "Big data-driven optimization for mobile networks toward 5g," *IEEE network*, vol. 30, no. 1, pp. 44–51, 2016.
4. E. A. Lee, "Cyber physical systems: Design challenges," in *2008 11th IEEE international symposium on object and component-oriented real-time distributed computing (ISORC)*, pp. 363–369, IEEE, 2008.
5. B. Sprecher, R. Kleijn, and G. J. Kramer, "Recycling potential of neodymium: the case of computer hard disk drives," *Environmental science & technology*, vol. 48, no. 16, pp. 9506–9513, 2014.
6. J. P. Roth, *Rewritable optical storage technology*. Meckler, 1991.

7. "Solid-state drive-wikipedia," 2023.
8. "Tape drive-wikipedia," 2023.
9. G. Scheunert, O. Heinonen, R. Hardeman, A. Lapicki, M. Gubbins, and R. Bowman, "A review of high magnetic moment thin films for microscale and nanotechnology applications," *Applied Physics Reviews*, vol. 3, no. 1, 2016.
10. N. Sharma, M. Sharma, A. Singhal, R. Vyas, H. Malik, A. Afthanorhan, and M. A. Hossaini, "Recent trends in eeg based motor imagery signal analysis and recognition: A comprehensive review.," *IEEE Access*, 2023.
11. J. H. Park, J. Y. Heo, and J. I. Lee, "Techno-economic study of nuclear integrated liquid air energy storage system," *Energy Conversion and Management*, vol. 251, p. 114937, 2022.
12. R. H. Dee, "Magnetic tape for data storage: an enduring technology," *Proceedings of the IEEE*, vol. 96, no. 11, pp. 1775–1785, 2008.
13. Y. Yan, Y. Jiang, E. L. L. Ng, Y. Zhang, C. Owh, F. Wang, Q. Song, T. Feng, B. Zhang, P. Li, et al., "Progress and opportunities in additive manufacturing of electrically conductive polymer composites," *Materials Today Advances*, vol. 17, p. 100333, 2023.
14. C. Campanale, C. Massarelli, D. Losacco, D. Bisaccia, M. Triozzi, and V. F. Uricchio, "The monitoring of pesticides in water matrices and the analytical criticalities: A review," *TrAC Trends in Analytical Chemistry*, vol. 144, p. 116423, 2021.
15. A. P. Cracknell, "The development of remote sensing in the last 40 years," 2018.
16. G. Cherubini, C. C. Chung, W. C. Messner, and S. R. Moheimani, "Control methods in data-storage systems," *IEEE Transactions on control systems technology*, vol. 20, no. 2, pp. 296–322, 2011.
17. Y. Luo, M. R. Abidian, J.-H. Ahn, D. Akinwande, A. M. Andrews, M. Antonietti, Z. Bao, M. Berggren, C. A. Berkey, C. J. Bettinger, et al., "Technology roadmap for flexible sensors," *ACS nano*, vol. 17, no. 6, pp. 5211–5295, 2023.
18. J. S. Pelz, N. Ku, M. A. Meyers, and L. R. Vargas-Gonzalez, "Additive manufacturing of structural ceramics: a historical perspective," *journal of materials research and technology*, vol. 15, pp. 670–695, 2021.
19. T. Schenk, M. Pešić, S. Slesazeck, U. Schroeder, and T. Mikolajick, "Memory technology—a primer for material scientists," *Reports on Progress in Physics*, vol. 83, no. 8, p. 086501, 2020.
20. K. Fall, "A delay-tolerant network architecture for challenged internets," in *Proceedings of the 2003 conference on Applications, technologies, architectures, and protocols for computer communications*, pp. 27–34, 2003.
21. S. Skoog, *Component and system design of a mild hybrid 48 V powertrain for a light vehicle*. Chalmers Tekniska Hogskola (Sweden), 2020.
22. P. H. Pathak, X. Feng, P. Hu, and P. Mohapatra, "Visible light communication, networking, and sensing: A survey, potential and challenges," *IEEE communications surveys & tutorials*, vol. 17, no. 4, pp. 2047–2077, 2015.
23. L. Cao, *Handbook of Artificial Intelligence and Big Data Applications in Investments*. CFA Institute Research Foundation, 2023.
24. L. Null, *The essentials of computer organization and architecture*. Jones & Bartlett Learning, 2023.
25. M. Haselman and S. Hauck, "The future of integrated circuits: A survey of nanoelectronics," *Proceedings of the IEEE*, vol. 98, no. 1, pp. 11–38, 2009.
26. J. King, R. Bose, H.-I. Yang, S. Pickles, and A. Helal, "Atlas: A service-oriented sensor platform: Hardware and middleware to enable programmable pervasive spaces," in *Proceedings. 2006 31st IEEE Conference on Local Computer Networks*, pp. 630–638, IEEE, 2006.
27. L. A. Barroso, U. Hölzle, and P. Ranganathan, *The datacenter as a computer: Designing warehouse-scale machines*. Springer Nature, 2019.
28. E. E. Schadt, M. D. Linderman, J. Sorenson, L. Lee, and G. P. Nolan, "Computational solutions to large-scale data management and analysis," *Nature reviews genetics*, vol. 11, no. 9, pp. 647–657, 2010.
29. A. Al-Fuqaha, M. Guizani, M. Mohammadi, M. Aledhari, and M. Ayyash, "Internet of things: A survey on enabling technologies, protocols, and applications," *IEEE communications surveys & tutorials*, vol. 17, no. 4, pp. 2347–2376, 2015.

30. F. Chen, T. Luo, and X. Zhang, "{CAFTL}: A {Content-Aware} flash translation layer enhancing the lifespan of flash memory based solid state drives," in *9th USENIX Conference on File and Storage Technologies (FAST 11)*, 2011.
31. B. Rose, "Home networks: a standards perspective," *IEEE Communications magazine*, vol. 39, no. 12, pp. 78–85, 2001.
32. K. Goda and M. Kitsuregawa, "The history of storage systems," *Proceedings of the IEEE*, vol. 100, no. Special Centennial Issue, pp. 1433–1440, 2012.
33. G. A. Gibson and R. Van Meter, "Network attached storage architecture," *Communications of the ACM*, vol. 43, no. 11, pp. 37–45, 2000.
34. R. Appuswamy, G. Graefe, R. Borovica-Gajic, and A. Ailamaki, "The five-minute rule 30 years later and its impact on the storage hierarchy," *Communications of the ACM*, vol. 62, no. 11, pp. 114–120, 2019.
35. P. K. Austin, "Data and language documentation," *Essentials of language documentation*, vol. 87112, 2006.
36. G. Bijoy, R. Rajeev, L. Benny, S. Jose, and A. Varghese, "Enzyme immobilization on biomass-derived carbon materials as a sustainable approach towards environmental applications," *Chemosphere*, p. 135759, 2022.
37. D. Al Kez, A. M. Foley, D. Laverty, D. F. Del Rio, and B. Sovacool, "Exploring the sustainability challenges facing digitalization and internet data centers," *Journal of Cleaner Production*, vol. 371, p. 133633, 2022.
38. P. Huang, B. Copertaro, X. Zhang, J. Shen, I. Löfgren, M. Rönnelid, J. Fahlen, D. Andersson, and M. Svanfeldt, "A review of data centers as prosumers in district energy systems: Renewable energy integration and waste heat reuse for district heating," *Applied energy*, vol. 258, p. 114109, 2020.
39. B. Diène, J. J. Rodrigues, O. Diallo, E. H. M. Ndoye, and V. V. Korotaev, "Data management techniques for internet of things," *Mechanical Systems and Signal Processing*, vol. 138, p. 106564, 2020.
40. I. A. T. Hashem, V. Chang, N. B. Anuar, K. Adewole, I. Yaqoob, A. Gani, E. Ahmed, and H. Chiroma, "The role of big data in smart city," *International Journal of information management*, vol. 36, no. 5, pp. 748–758, 2016.
41. F. T. Hady, A. Foong, B. Veal, and D. Williams, "Platform storage performance with 3d xpoint technology," *Proceedings of the IEEE*, vol. 105, no. 9, pp. 1822–1833, 2017.
42. S. J. Bartley and J. H. Golek, "Evaluating the cost effectiveness of online and face-to-face instruction," *Journal of Educational Technology & Society*, vol. 7, no. 4, pp. 167–175, 2004.
43. D. Chen and H. Zhao, "Data security and privacy protection issues in cloud computing," in *2012 international conference on computer science and electronics engineering*, vol. 1, pp. 647–651, IEEE, 2012.
44. V. Gazis, "A survey of standards for machine-to-machine and the internet of things," *IEEE Communications Surveys & Tutorials*, vol. 19, no. 1, pp. 482–511, 2016.
45. M. Díaz, C. Martín, and B. Rubio, "State-of-the-art, challenges, and open issues in the integration of internet of things and cloud computing," *Journal of Network and Computer applications*, vol. 67, pp. 99–117, 2016.
46. J. Ni, K. Zhang, X. Lin, and X. Shen, "Securing fog computing for internet of things applications: Challenges and solutions," *IEEE Communications Surveys & Tutorials*, vol. 20, no. 1, pp. 601–628, 2017.
47. S. Spiekermann and L. F. Cranor, "Engineering privacy," *IEEE Transactions on software engineering*, vol. 35, no. 1, pp. 67–82, 2008.
48. C. D. Cottrill, "Maas surveillance: Privacy considerations in mobility as a service," *Transportation Research Part A: Policy and Practice*, vol. 131, pp. 50–57, 2020.
49. D. J. Fagnant and K. Kockelman, "Preparing a nation for autonomous vehicles: opportunities, barriers and policy recommendations," *Transportation Research Part A: Policy and Practice*, vol. 77, pp. 167–181, 2015.
50. S. Subashini and V. Kavitha, "A survey on security issues in service delivery models of cloud computing," *Journal of network and computer applications*, vol. 34, no. 1, pp. 1–11, 2011.

51. U. Uludag, S. Pankanti, S. Prabhakar, and A. K. Jain, “Biometric cryptosystems: issues and challenges,” *Proceedings of the IEEE*, vol. 92, no. 6, pp. 948–960, 2004.
52. S. Abolfazli, Z. Sanaei, E. Ahmed, A. Gani, and R. Buyya, “Cloud-based augmentation for mobile devices: motivation, taxonomies, and open challenges,” *IEEE Communications Surveys & Tutorials*, vol. 16, no. 1, pp. 337–368, 2013.
53. C. P. Chen and C.-Y. Zhang, “Data-intensive applications, challenges, techniques and technologies: A survey on big data,” *Information sciences*, vol. 275, pp. 314–347, 2014.
54. A. Hazra, M. Adhikari, T. Amgoth, and S. N. Srirama, “A comprehensive survey on interoperability for iiot: Taxonomy, standards, and future directions,” *ACM Computing Surveys (CSUR)*, vol. 55, no. 1, pp. 1–35, 2021.
55. D. Gaspar and I. Coric, *Bridging relational and NoSQL databases*. IGI Global, 2017.
56. Y. Niu, L. Ying, J. Yang, M. Bao, and C. Sivaparthipan, “Organizational business intelligence and decision making using big data analytics,” *Information Processing & Management*, vol. 58, no. 6, p. 102725, 2021.
57. K.-H. Lin, H.-H. Liu, K.-H. Hu, A. Huang, and H.-Y. Wei, “A survey on drx mechanism: Device power saving from lte and 5g new radio to 6g communication systems,” *IEEE Communications Surveys & Tutorials*, 2022.
58. J. Zhang and K. B. Letaief, “Mobile edge intelligence and computing for the internet of vehicles,” *Proceedings of the IEEE*, vol. 108, no. 2, pp. 246–261, 2019.
59. G. J. van den Burg, A. Nazábal, and C. Sutton, “Wrangling messy csv files by detecting row and type patterns,” *Data Mining and Knowledge Discovery*, vol. 33, no. 6, pp. 1799–1820, 2019.
60. Y. Hu, T.-M. Li, L. Anderson, J. Ragan-Kelley, and F. Durand, “Taichi: a language for high-performance computation on spatially sparse data structures,” *ACM Transactions on Graphics (TOG)*, vol. 38, no. 6, pp. 1–16, 2019.
61. C. Benghi, “Automated verification for collaborative workflows in a digital plan of work,” *Automation in Construction*, vol. 107, p. 102926, 2019.
62. J.-R. Roussel, D. Auty, N. C. Coops, P. Tompalski, T. R. Goodbody, A. S. Meador, J.-F. Bourdon, F. De Boissieu, and A. Achim, “lidr: An r package for analysis of airborne laser scanning (als) data,” *Remote Sensing of Environment*, vol. 251, p. 112061, 2020.
63. X. Zhao, X. Zhou, and G. Li, “Automatic database knob tuning: A survey,” *IEEE Transactions on Knowledge and Data Engineering*, 2023.
64. Y. Zhou, T. Bao, X. Shu, Y. Li, and Y. Li, “Bim and ontology-based knowledge management for dam safety monitoring,” *Automation in Construction*, vol. 145, p. 104649, 2023.
65. A. Beaulieu, *Learning SQL: Generate, manipulate, and retrieve data*. O’Reilly Media, 2020.
66. R. Duke, V. Bhat, and C. Risko, “Data storage architectures to accelerate chemical discovery: data accessibility for individual laboratories and the community,” *Chemical Science*, vol. 13, no. 46, pp. 13646–13656, 2022.
67. D. H. Kang, C. Min, S.-W. Lee, and Y. I. Eom, “Making application-level crash consistency practical on flash storage,” *IEEE Transactions on Parallel and Distributed Systems*, vol. 31, no. 5, pp. 1009–1020, 2019.
68. Z. Zhao, W. Lu, H. Zhao, Z. He, H. Li, A. Pan, and X. Du, “T-sql: A lightweight implementation to enable built-in temporal support in mvcc-based rdbmss,” *IEEE Transactions on Knowledge and Data Engineering*, vol. 35, no. 1, pp. 1028–1042, 2021.
69. A. Moffat and J. Mackenzie, “Efficient immediate-access dynamic indexing,” *Information Processing & Management*, vol. 60, no. 3, p. 103248, 2023.
70. M. Kvet and K. Matiasko, “Data block and tuple identification using master index,” *Sensors*, vol. 20, no. 7, p. 1848, 2020.
71. F. F. X. de Barros, “Advanced oracle monitoring agent for zabbix,” 2022.
72. A. Alhomssi and V. Leis, “Scalable and robust snapshot isolation for high-performance storage engines,” *Proceedings of the VLDB Endowment*, vol. 16, no. 6, pp. 1426–1438, 2023.
73. N. S. Barghouti and G. E. Kaiser, “Concurrency control in advanced database applications,” *ACM Computing Surveys (CSUR)*, vol. 23, no. 3, pp. 269–317, 1991.

74. R. Das and I. Turkoglu, "Creating meaningful data from web logs for improving the impressiveness of a website by using path analysis method," *Expert Systems with Applications*, vol. 36, no. 3, pp. 6635–6644, 2009.
75. P.-Å. Larson, S. Blanas, C. Diaconu, C. Freedman, J. M. Patel, and M. Zwilling, "High-performance concurrency control mechanisms for main-memory databases," arXiv:1201.0228, 2011.
76. J. Han, E. Haihong, G. Le, and J. Du, "Survey on nosql database," in *2011 6th international conference on pervasive computing and applications*, pp. 363–366, IEEE, 2011.
77. G. Baruffa, M. Femminella, M. Pergolesi, and G. Reali, "Comparison of mongodb and cassandra databases for spectrum monitoring as-a-service," *IEEE Transactions on Network and Service Management*, vol. 17, no. 1, pp. 346–360, 2019.
78. H. Zhang, G. Chen, B. C. Ooi, K.-L. Tan, and M. Zhang, "In-memory big data management and processing: A survey," *IEEE Transactions on Knowledge and Data Engineering*, vol. 27, no. 7, pp. 1920–1948, 2015.
79. D. Agrawal, A. El Abbadi, S. Antony, and S. Das, "Data management challenges in cloud computing infrastructures," in *International Workshop on Databases in Networked Information Systems*, pp. 1–10, Springer, 2010.
80. J. Lu and I. Holubová, "Multi-model databases: a new journey to handle the variety of data," *ACM Computing Surveys (CSUR)*, vol. 52, no. 3, pp. 1–38, 2019.
81. A. Davoudian, L. Chen, and M. Liu, "A survey on nosql stores," *ACM Computing Surveys (CSUR)*, vol. 51, no. 2, pp. 1–43, 2018.
82. J. Gray, P. McJones, M. Blasgen, B. Lindsay, R. Lorie, T. Price, F. Putzolu, and I. Traiger, "The recovery manager of the system r database manager," *ACM Computing Surveys (CSUR)*, vol. 13, no. 2, pp. 223–242, 1981.
83. B. Dageville, T. Cruanes, M. Zukowski, V. Antonov, A. Avanes, J. Bock, J. Claybaugh, D. Engovatov, M. Hentschel, J. Huang, *et al.*, "The snowflake elastic data warehouse," in *Proceedings of the 2016 International Conference on Management of Data*, pp. 215–226, 2016.
84. A. J. Muller-Molina and T. Shinohara, "Efficient similarity search by reducing i/o with compressed sketches," in *2009 Second International Workshop on Similarity Search and Applications*, pp. 30–38, IEEE, 2009.
85. S. Melnik, A. Gubarev, J. J. Long, G. Romer, S. Shivakumar, M. Tolton, T. Vassilakis, H. Ahmadi, D. Delorey, S. Min, et al., "Dremel: A decade of interactive sql analysis at web scale," *Proceedings of the VLDB Endowment*, vol. 13, no. 12, pp. 3461–3472, 2020.
86. G. Cerar, B. Bertalanič, A. Pirnat, A. Čampa, and C. Fortuna, "On designing data models for energy feature stores," arXiv:2205.04267, 2022.
87. P. Gupta, A. Mhedhbi, and S. Salihoglu, "Columnar storage and list-based processing for graph database management systems," arXiv:2103.02284, 2021.
88. Z. Xiao and Y. Xiao, "Security and privacy in cloud computing," *IEEE communications surveys & tutorials*, vol. 15, no. 2, pp. 843–859, 2012.
89. M. Wittig and A. Wittig, *Amazon web services in action*. Simon and Schuster, 2018.
90. M. Armbrust, A. Fox, R. Griffith, A. D. Joseph, R. Katz, A. Konwinski, G. Lee, D. Patterson, A. Rabkin, I. Stoica, et al., "A view of cloud computing," *Communications of the ACM*, vol. 53, no. 4, pp. 50–58, 2010.
91. S. Tang, B. He, C. Yu, Y. Li, and K. Li, "A survey on spark ecosystem: Big data processing infrastructure, machine learning, and applications," *IEEE Transactions on Knowledge and Data Engineering*, vol. 34, no. 1, pp. 71–91, 2020.
92. G. ZOLLINO, C. SARTORI, and E. A. BULGARELLI, "Big data for the real-time analysis of the cherenkov telescope array observatory," 2019.
93. R. Y. Zhong, S. T. Newman, G. Q. Huang, and S. Lan, "Big data for supply chain management in the service and manufacturing sectors: Challenges, opportunities, and future perspectives," *Computers & Industrial Engineering*, vol. 101, pp. 572–591, 2016.
94. X. Wang, H.-M. Kim, Y. Xiao, and Y.-K. Sun, "Nanostructured metal phosphide-based materials for electrochemical energy storage," *Journal of Materials Chemistry A*, vol. 4, no. 39, pp. 14915–14931, 2016.

95. P. Bellavista, A. Corradi, and A. Reale, "Quality of service in wide scale publish—subscribe systems," *IEEE Communications Surveys & Tutorials*, vol. 16, no. 3, pp. 1591–1616, 2014.
96. C.-Y. Chong and S. P. Kumar, "Sensor networks: evolution, opportunities, and challenges," *Proceedings of the IEEE*, vol. 91, no. 8, pp. 1247–1256, 2003.
97. M. Wirthlin, "High-reliability fpga-based systems: Space, high-energy physics, and beyond," *Proceedings of the IEEE*, vol. 103, no. 3, pp. 379–389, 2015.
98. J. Verbraeken, M. Wolting, J. Katzy, J. Kloppenburg, T. Verbelen, and J. S. Rellermeyer, "A survey on distributed machine learning," *Acm computing surveys (csur)*, vol. 53, no. 2, pp. 1–33, 2020.
99. R. Zafar, E. Yafi, M. F. Zuhairi, and H. Dao, "Big data: the nosql and rdbms review," in *2016 International Conference on Information and Communication Technology (ICICTM)*, pp. 120–126, IEEE, 2016.
100. T. Ojha, S. Misra, and N. S. Raghuwanshi, "Internet of things for agricultural applications: The state of the art," *IEEE Internet of Things Journal*, vol. 8, no. 14, pp. 10973–10997, 2021.
101. C. Diamantini, P. Lo Giudice, D. Potena, E. Storti, and D. Ursino, "An approach to extracting topic-guided views from the sources of a data lake," *Information Systems Frontiers*, vol. 23, pp. 243–262, 2021.
102. K. Sahatqija, J. Ajdari, X. Zenuni, B. Raufi, and F. Ismaili, "Comparison between relational and nosql databases," in *2018 41st international convention on information and communication technology, electronics and microelectronics (MIPRO)*, pp. 0216–0221, IEEE, 2018.
103. P. Darke, G. Shanks, and M. Broadbent, "Successfully completing case study research: combining rigour, relevance and pragmatism," *Information systems journal*, vol. 8, no. 4, pp. 273–289, 1998.
104. S. Ayvaz and K. Alpay, "Predictive maintenance system for production lines in manufacturing: A machine learning approach using iot data in real-time," *Expert Systems with Applications*, vol. 173, p. 114598, 2021.
105. R. Sahal, J. G. Breslin, and M. I. Ali, "Big data and stream processing platforms for industry 4.0 requirements mapping for a predictive maintenance use case," *Journal of manufacturing systems*, vol. 54, pp. 138–151, 2020.
106. P. R. Amestoy, I. S. Duff, J.-Y. L'Excellent, and J. Koster, "A fully asynchronous multifrontal solver using distributed dynamic scheduling," *SIAM Journal on Matrix Analysis and Applications*, vol. 23, no. 1, pp. 15–41, 2001.
107. G. Yu, Y. Hu, and J. Dai, "Topotag: A robust and scalable topological fiducial marker system," *IEEE transactions on visualization and computer graphics*, vol. 27, no. 9, pp. 3769–3780, 2020.
108. S. Khalifa, Y. Elshater, K. Sundaravarathan, A. Bhat, P. Martin, F. Imam, D. Rope, M. Mcroberts, and C. Statchuk, "The six pillars for building big data analytics ecosystems," *ACM Computing Surveys (CSUR)*, vol. 49, no. 2, pp. 1–36, 2016.
109. M. Stonebraker and G. Kemnitz, "The postgres next generation database management system," *Communications of the ACM*, vol. 34, no. 10, pp. 78–92, 1991.
110. A. K. Agarwal and N. N. Mustafi, "Real-world automotive emissions: Monitoring methodologies, and control measures," *Renewable and Sustainable Energy Reviews*, vol. 137, p. 110624, 2021.
111. L. Ferretti, M. Marchetti, M. Andreolini, and M. Colajanni, "A symmetric cryptographic scheme for data integrity verification in cloud databases," *Information Sciences*, vol. 422, pp. 497–515, 2018.
112. B. Kulis, P. Jain, and K. Grauman, "Fast similarity search for learned metrics," *IEEE Transactions on Pattern Analysis and Machine Intelligence*, vol. 31, no. 12, pp. 2143–2157, 2009.
113. A. Gandomi and M. Haider, "Beyond the hype: Big data concepts, methods, and analytics," *International journal of information management*, vol. 35, no. 2, pp. 137–144, 2015.
114. T. Pavlidis, *Algorithms for graphics and image processing*. Springer Science & Business Media, 2012.

115. X. Zhou, X. Xu, W. Liang, Z. Zeng, and Z. Yan, "Deep-learning-enhanced multitarget detection for end–edge–cloud surveillance in smart iot," *IEEE Internet of Things Journal*, vol. 8, no. 16, pp. 12588–12596, 2021.
116. J. Fan, F. Han, and H. Liu, "Challenges of big data analysis," *National science review*, vol. 1, no. 2, pp. 293–314, 2014.
117. F. Lombardi and R. Di Pietro, "Secure virtualization for cloud computing," *Journal of network and computer applications*, vol. 34, no. 4, pp. 1113–1122, 2011.
118. B. Raza, A. Aslam, A. Sher, A. K. Malik, and M. Faheem, "Autonomic performance prediction framework for data warehouse queries using lazy learning approach," *Applied Soft Computing*, vol. 91, p. 106216, 2020.
119. D. Mourtzis and E. Vlachou, "A cloud-based cyber-physical system for adaptive shop-floor scheduling and condition-based maintenance," *Journal of manufacturing systems*, vol. 47, pp. 179–198, 2018.
120. S. Sakr, A. Liu, D. M. Batista, and M. Alomari, "A survey of large scale data management approaches in cloud environments," *IEEE communications surveys & tutorials*, vol. 13, no. 3, pp. 311–336, 2011.
121. S. Chaudhuri, U. Dayal, and V. Narasayya, "An overview of business intelligence technology," *Communications of the ACM*, vol. 54, no. 8, pp. 88–98, 2011.
122. N. Gorelick, M. Hancher, M. Dixon, S. Ilyushchenko, D. Thau, and R. Moore, "Google earth engine: Planetary-scale geospatial analysis for everyone," *Remote sensing of Environment*, vol. 202, pp. 18–27, 2017.
123. R. Hai, C. Koutras, C. Quix, and M. Jarke, "Data lakes: A survey of functions and systems," *IEEE Transactions on Knowledge and Data Engineering*, 2023.
124. R. Li, H. Wang, H. Dai, J. Hong, G. Tong, and X. Chen, "Accurate state of charge prediction for real-world battery systems using a novel dual-dropout-based neural network," *Energy*, vol. 250, p. 123853, 2022.
125. G. Tolosa and A. Mallia, "Many are better than one: Algorithm selection for faster top-k retrieval," *Information Processing & Management*, vol. 60, no. 4, p. 103359, 2023.
126. R. K. Goswami, K. Agrawal, H. M. Upadhyaya, V. K. Gupta, and P. Verma, "Microalgae conversion to alternative energy, operating environment and economic footprint: An influential approach towards energy conversion, and management," *Energy Conversion and Management*, vol. 269, p. 116118, 2022.
127. J. Tian, R. Xiong, W. Shen, J. Lu, and F. Sun, "Flexible battery state of health and state of charge estimation using partial charging data and deep learning," *Energy Storage Materials*, vol. 51, pp. 372–381, 2022.
128. B. Liskov, A. Snyder, R. Atkinson, and C. Schaffert, "Abstraction mechanisms in clu," *Communications of the ACM*, vol. 20, no. 8, pp. 564–576, 1977.
129. H. Li, J. Zhang, Z. Li, J. Liu, and Y. Wang, "Improvement of min-entropy evaluation based on pruning and quantized deep neural network," *IEEE Transactions on Information Forensics and Security*, vol. 18, pp. 1410–1420, 2023.
130. X. Sun, Z. Chen, L. Wang, and C. He, "A lossless image compression and encryption algorithm combining jpeg-ls, neural network and hyperchaotic system," *Nonlinear Dynamics*, pp. 1–31, 2023.
131. Y. Mao, M. Wang, S. Wang, and S. Kwong, "High efficiency rate control for versatile video coding based on composite cauchy distribution," *IEEE Transactions on Circuits and Systems for Video Technology*, vol. 32, no. 4, pp. 2371–2384, 2021.
132. F. M. Moss, K. Wang, F. Zhang, R. Baddeley, and D. R. Bull, "On the optimal presentation duration for subjective video quality assessment," *IEEE Transactions on Circuits and Systems for Video Technology*, vol. 26, no. 11, pp. 1977–1987, 2015.
133. J. Jia, Y. Wu, A. Li, S. Ma, and Y. Liu, "Subnetwork-lossless robust watermarking for hostile theft attacks in deep transfer learning models," *IEEE transactions on dependable and secure computing*, 2022.
134. R.-X. Ding, I. Palomares, X. Wang, G.-R. Yang, B. Liu, Y. Dong, E. Herrera-Viedma, and F. Herrera, "Large-scale decision-making: Characterization, taxonomy, challenges and future

directions from an artificial intelligence and applications perspective," *Information fusion*, vol. 59, pp. 84–102, 2020.
135. L. Wang and A. Haghighi, "Combined strength of holons, agents and function blocks in cyber-physical systems," *Journal of manufacturing systems*, vol. 40, pp. 25–34, 2016.
136. T. Cheng and J. Teizer, "Real-time resource location data collection and visualization technology for construction safety and activity monitoring applications," *Automation in construction*, vol. 34, pp. 3–15, 2013.
137. W. Bartlett and L. Spainhower, "Commercial fault tolerance: A tale of two systems," *IEEE Transactions on dependable and secure computing*, vol. 1, no. 1, pp. 87–96, 2004.
138. M. U. Hassan, M. H. Rehmani, and J. Chen, "Privacy preservation in blockchain based iot systems: Integration issues, prospects, challenges, and future research directions," *Future Generation Computer Systems*, vol. 97, pp. 512–529, 2019.
139. W. Liang, Y. Fan, K.-C. Li, D. Zhang, and J.-L. Gaudiot, "Secure data storage and recovery in industrial blockchain network environments," *IEEE Transactions on Industrial Informatics*, vol. 16, no. 10, pp. 6543–6552, 2020.
140. D. Han, Y. Zhu, D. Li, W. Liang, A. Souri, and K.-C. Li, "A blockchain-based auditable access control system for private data in service-centric iot environments," *IEEE Transactions on Industrial Informatics*, vol. 18, no. 5, pp. 3530–3540, 2021.
141. X. Yu and Y. Xue, "Smart grids: A cyber–physical systems perspective," *Proceedings of the IEEE*, vol. 104, no. 5, pp. 1058–1070, 2016.
142. I. Bose and R. K. Mahapatra, "Business data mining—a machine learning perspective," *Information & management*, vol. 39, no. 3, pp. 211–225, 2001.
143. V. C. Gungor and G. P. Hancke, "Industrial wireless sensor networks: Challenges, design principles, and technical approaches," *IEEE Transactions on industrial electronics*, vol. 56, no. 10, pp. 4258–4265, 2009.
144. H. Yuan, X. Chen, J. Wang, J. Yuan, H. Yan, and W. Susilo, "Blockchain-based public auditing and secure deduplication with fair arbitration," *Information Sciences*, vol. 541, pp. 409–425, 2020.
145. R. U. Rasool, H. F. Ahmad, W. Rafique, A. Qayyum, and J. Qadir, "Security and privacy of internet of medical things: A contemporary review in the age of surveillance, botnets, and adversarial ml," *Journal of Network and Computer Applications*, vol. 201, p. 103332, 2022.
146. X. Li, H. Zhao, and W. Deng, "Bfod: Blockchain-based privacy protection and security sharing scheme of flight operation data," *IEEE Internet of Things Journal*, 2023.
147. A. Doricchi, C. M. Platnich, A. Gimpel, F. Horn, M. Earle, G. Lanzavecchia, A. L. Cortajarena, L. M. Liz-Marzán, N. Liu, R. Heckel, et al., "Emerging approaches to dna data storage: Challenges and prospects," *ACS nano*, vol. 16, no. 11, pp. 17552–17571, 2022.
148. C. Bhowmik, S. Bhowmik, A. Ray, and K. M. Pandey, "Optimal green energy planning for sustainable development: A review," *Renewable and Sustainable Energy Reviews*, vol. 71, pp. 796–813, 2017.
149. Y. Zhang, M. Qiu, C.-W. Tsai, M. M. Hassan, and A. Alamri, "Health-cps: Healthcare cyber-physical system assisted by cloud and big data," *IEEE Systems Journal*, vol. 11, no. 1, pp. 88–95, 2015.
150. M. Cheminod, L. Durante, and A. Valenzano, "Review of security issues in industrial networks," *IEEE transactions on industrial informatics*, vol. 9, no. 1, pp. 277–293, 2012.

Chapter 5
Data Integration and Fusion in Cyber Physical Systems

Abstract With the advent of data-driven cyber physical systems (DDCPS), there has been a fundamental shift in the way computers interact with the physical environment In this section, we begin by elucidating the concepts of data integration and fusion of DDCPS. Subsequently, we provide an overview of the various methods employed for data fusion, which can be categorized into three groups: stage-based, semantic-based, and artificial intelligence-based approaches. These techniques emphasize the fusion of knowledge rather than mere pattern mapping and data, distinguishing them from the conventional data integration methods in database research. Then, case studies of cloud-integrated cyber physical systems (CPS) for smart cities and big data fusion framework for smart healthcare are proposed. Finally, the existing challenges of CPS data integration and fusion are summarized, and future research directions are discussed.

This chapter provides an overview of the concepts, methods, case studies, and future directions of cyber physical systems (CPS) dataset integration and fusion, which enables readers to understand systematically. Understanding the characteristics of CPS data and the concepts of data integration and fusion is a prerequisite for achieving CPS data fusion. The summarized different types of fusion methods help readers to comprehensively learn about data fusion knowledge and provide a foundation for practical applications. Finally, the related challenges and future work provide readers with directions for research.

5.1 Data Integration and Fusion

The widespread adoption of CPS has resulted in the generation of vast amounts of heterogeneous data from various sources, including user feedback, sensor observation, and measuring devices. CPS data are critical for ensuring the quality of services in a pervasive environment such as smart homes and societies. To ensure the CPS applications, mining and using massive data are important [1, 2].

F. Li et al., *Data-Driven Cyber Physical Systems*,
https://doi.org/10.1007/978-981-96-8709-1_5

Effective data integration and fusion are essential components of CPS, as they enable the integration and analysis of diverse data sets to derive meaningful insights and improve service delivery [3]. Traditional data fusion techniques are typically designed to process data from the physical space, which is structured data such as sensor data [4, 5]. However, in CPS, data are collected from different spaces, namely the physical space and cyberspace. These spaces have different distributions, densities, scales, representations, and modalities, making it challenging to integrate and analyze data from all three spaces [6]. In particular, the data in CPS are complex and diverse, requiring special consideration in terms of representation, organization, processing, indexing, retrieval, and dissemination.

5.1.1 CPS Data Characteristics

CPS data include sensor data, control data, communication data, and system logs. Sensor data provide real-time information on the physical information on the physical environment and processes. The physical sensors embedded in CPS collect data on various parameters such as temperature, pressure, humidity, and motion. This data are then transmitted to the control systems for processing and analysis. The control data of CPS provide information on the state of the system, actions taken by the control systems, and feedback received from the physical sensors. Data are generated by the control systems that manage the physical processes in CPS and play a crucial role in ensuring the smooth operation of the system. Communication data provide information on the communication networks that connects the various components of the system. This data include information on the network topology, the traffic flow, and the communication protocols used. System logs provide a detailed record of the events and activities that occur within the system. These logs are generated automatically by CPS and include information on system errors, warnings, and performance metrics. In conclusion, the data generated by CPS have characteristics as follows:

Inter-space sensing. Inter-space sensing, also referred to as augmented sensing or collaborative sensing, plays a vital role in data acquisition [7]. CPS rely on three distinct sensing resources, each possessing unique perception capabilities. The coordination of these diverse perceptions is necessary to achieve interspace sense. In terms of signal and system dynamics, interspace sensing can be divided into two components: input and output [8]. The input involves multiple sensing sources, while the output corresponds to the associated data. This approach enables CPS to collect and analyze data from multiple sources, resulting in more accurate and comprehensive insights [9].

Heterogeneity. Heterogeneity is a main characteristic of data, resulting from differences in expression (i.e., modalities or views) and source (i.e. cyberspace, and physical space). Within the context of modern data fusion, heterogeneity manifests in the diversity of order and dimensionality, also known as expression differences. For instance, some data may exist as higher-order tensors while others are represented as matrices. Factors like the construction of data space, implementation of data management systems, technology, and other artificial elements lead to heterogeneity [10]. By employing data fusion techniques, CPS integrate diverse data from both physical space and cyberspace to improve data quality, resulting in enhanced decision-making capabilities [8].

In order to process the collected CPS data with different characteristics, understanding the corresponding definitions of data fusion, data integration, and the framework of CPS is an important step.

5.1.2 CPS Data Integration

CPS are intelligent platforms that utilize embedded technologies and physical systems to integrate both physical and computational elements. These systems leverage software and multi-agent systems to achieve specific tasks [11]. The current architectures of CPS comprise communications, interconnections, sensors, actuators, and controllers [12]. In the future, CPS will encompass millions of interconnected smart devices, including actuators and sensors, connected to thousands of central processing units.

As complex systems, CPS are manufactured by different companies, resulting in variations in data models and interfaces. The heterogeneity among building blocks poses a crucial challenge that needs to be addressed for successful implementation in industries [13]. With the involvement of external suppliers delivering many subsystems, organizations face the necessity for both horizontal integration within their own structures and vertical integration between sub-system suppliers and suppliers of complete systems. The integration of computation and physical processes is a fundamental aspect [14]. CPS are commonly integrated into complex systems to enhance their capabilities. Each system adheres to a specific data model derived from the nature of the corresponding physical processes. CPS systems provide data through interfaces to other CPS or external systems and, conversely, consume data from surrounding systems to enhance control over physical processes. Moreover, CPS utilize shared data storage, which can be either locally distributed or stored in the cloud. This necessitates a multidisciplinary development process, where designers need to consider not only the individual physical and computational components, but also their interactions across different disciplines. Such interactions often give rise to disruptions and compatibility issues during the design phase [15, 16].

Low-level integration. Low-level integration means the interconnections among components of CPS. CPS are composed of two main components: a physical part and a cyber part, as presented in Fig. 5.1. The physical part involves the tangible processes and objects that enable process control. The cyber part can be further divided into two layers for better understanding. The first layer, known as the platform layer,

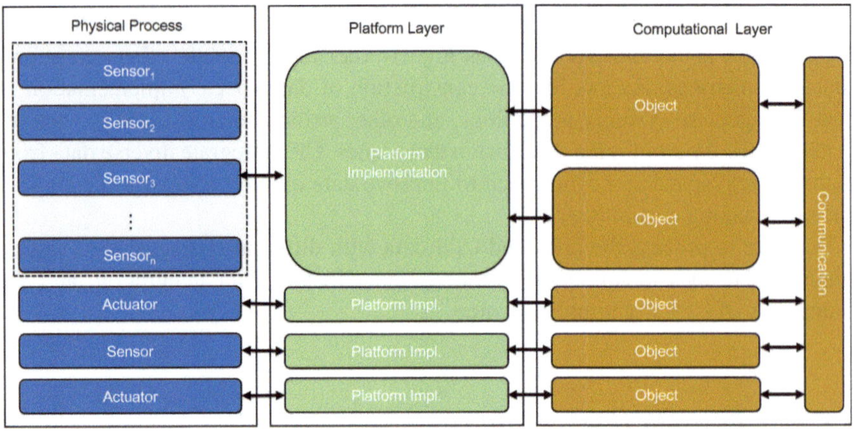

Fig. 5.1 Low-level CPS integration schema. The schema consists of a physical part and a network part that combines a platform layer and a computing layer

integrates various devices from different manufacturers. The second layer, known as the computational layer, is responsible for the computational process that controls the physical process based on pre-defined logic. The described process involves initially creating a model of the physical process using a physical layer abstraction. Then, a corresponding control system, represented by a computational (software) layer abstraction, is developed and implemented. Lastly, the control system is deployed on the computation platform, which is modeled using a platform layer abstraction.

High-level integration. High-level integration refers to the connection and integration of multiple CPS to create the Internet of Things (IoT). The integration process involves two key aspects: unifying CPS interfaces and unifying corresponding data models, as shown in Fig. 5.2. CPS integration can be approached from two dimensions. The initial step involves platform unification, addressing the disparate

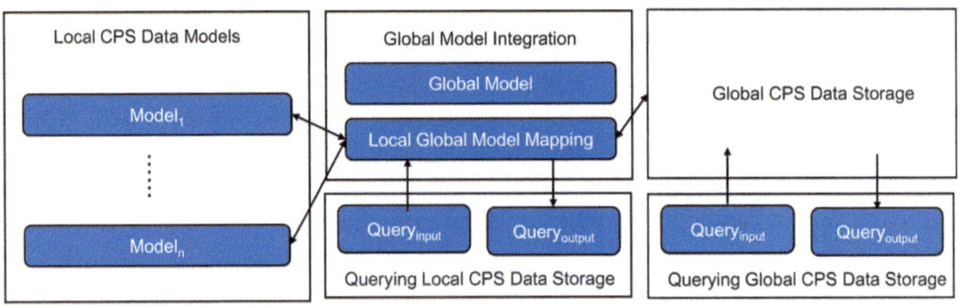

Fig. 5.2 High-level CPS integration schema. The schema consists of the unification of CPS interfaces as well as the unification of corresponding data models

interfaces offered by different manufacturers. The subsequent step is model integration, encompassing the identification of relevant concepts, their relationships, and the contextual framework.

5.1.3 CPS Data Fusion

Data fusion refers to a field of interdisciplinary research where the integration and correlation of data from various sources, such as information fusion, sensor fusion, and data integration. There are multiple definitions and explanations proposed by researchers from different fields or perspectives. In joint directors of laboratories workshop, data fusion or sensor fusion was defined as a multi-level procedure that involves combining data and information from single or multiple sources to achieve a distinguished position and timely assessments of situations, threats, and their significance [17]. In Ref. [18], data fusion was defined as the collaborative utilization of information derived from multiple sensors to assist in performing a specific function. In addition, data fusion algorithms involve combining data from multiple sources to enhance the performance of mobile robots [19]. Hall and Llinasd defined data fusion as the process of integrating data from multiple sensors and associated information from databases to achieve improved accuracy and more precise inferences compared to using a single sensor alone [20]. Furthermore, researchers conducted surveys on definitions of data fusion [21]. Another perspective on data fusion is the fusion of multiple datasets, which is referred to as modern data fusion.

Here, we define CPS data fusion as the integration of data from the physical and cyberspace of CPS spaces, allowing for the interaction and mutual information of corresponding data sets to provide services for the CPS space. In essence, CPS data fusion can be likened to how humans or animals gather information through multiple senses and comprehend the objective world through comprehensive analysis performed by the brain. Figure 5.3 illustrates the definition of CPS data fusion.

5.1.4 Data Integration and Fusion Framework

Data integration and fusion framework primarily pertains to the structure or architecture of data integration and fusion. CPS frameworks succeed embedded systems by incorporating computing and communication technology, as well as control methodology. Specifically, CPS frameworks are typically focused on sensing and controlling physical phenomena through networks of interconnected devices to achieve specific goals. Researchers have put forward various information fusion frameworks over time, as shown in Fig. 5.4. In 1990, Waltz and Llinas introduced the differentiation between high-level fusion (HLIF) and low-level fusion (LLIF). LLIF processes focus on target classification, identification, and tracking using numerical data, whereas HLIF processes handle abstract symbolic information related to situation, impact, and

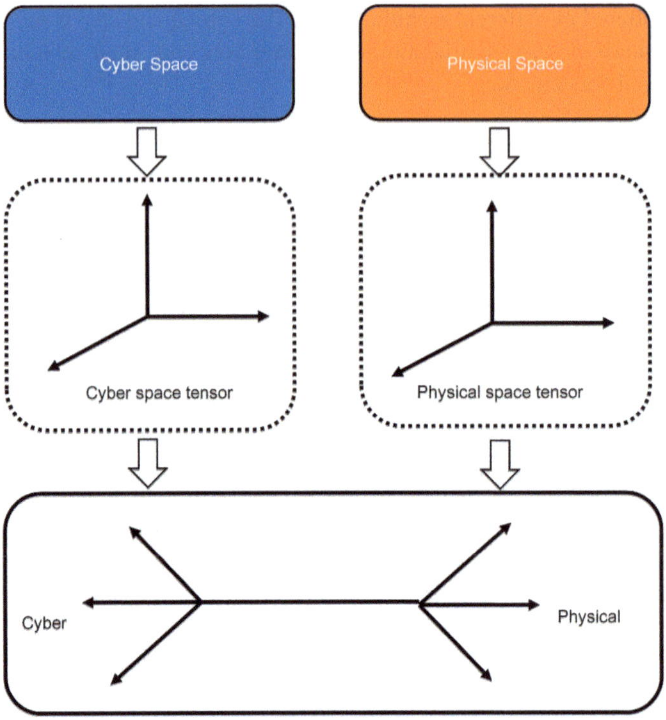

Fig. 5.3 The definition of CPS data fusion. CPS data fusion refers to the process of integrating data from both the cyber space and physical space. This integration allows the corresponding data sets from these two realms to interact and mutually inform each other, ultimately providing valuable information for the CPS space

fusion process refinement [22]. Subsequently, the Data Fusion Information Group (DFIG) model, an update to the Joint Director Laboratories (JDL) model, was introduced in a meeting held in 2004, in accordance with [23]. The information fusion community has embraced the Data Fusion Information Group (DFIG) model, which replaces the Joint Director Laboratories (JDL) model, as outlined in [24]. In 2006, Lambert et al. introduced the state transition data fusion (STDF) model, which provides a comprehensive framework for handling uncertainty [25]. Blasch et al. proposed the cognitive observe-orient-decide-act (C-OODA) model within the context of the DFIG information fusion model, offering a method for user and team analysis [3]. Fantini et al. suggested that CPS could benefit from employee participation and advocated for a scenario-based approach and multidimensional analysis framework [26]. Han et al. highlighted the role of user context in data fusion and integrated target models and problem frameworks [27].

Among them, the JDL framework and its variations have gained more popularity. Originally developed for military applications, the JDL framework comprises four levels: (a) target assessment, (b) situation assessment, (c) threat assessment, and (d) process optimization, as shown in Fig. 5.5.

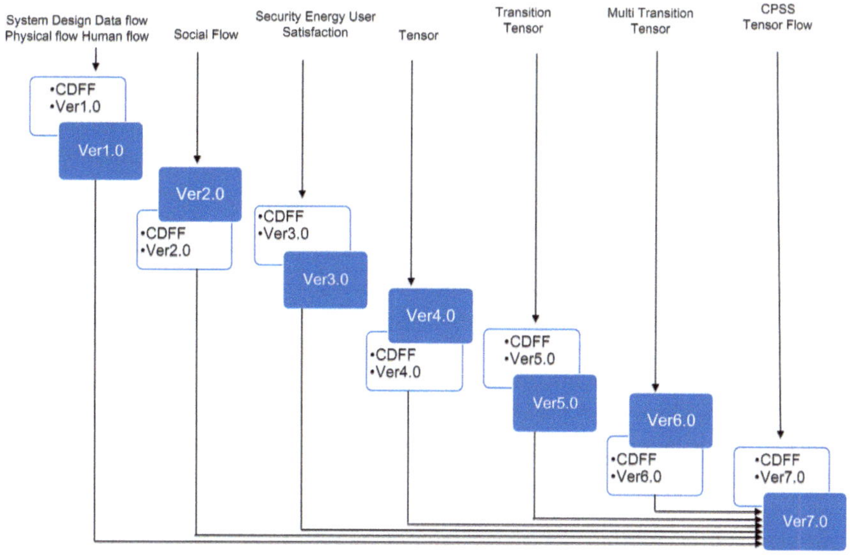

Fig. 5.4 The data integration and fusion framework timeline. The timeline of the data integration and fusion framework can be divided into three stages and seven versions. The initial stage includes Ver1.0, Ver2.0, and Ver3.0. The evolutionary stage includes Ver4.0, Ver5.0, and Ver6.0. The future stage is represented by Ver7.0

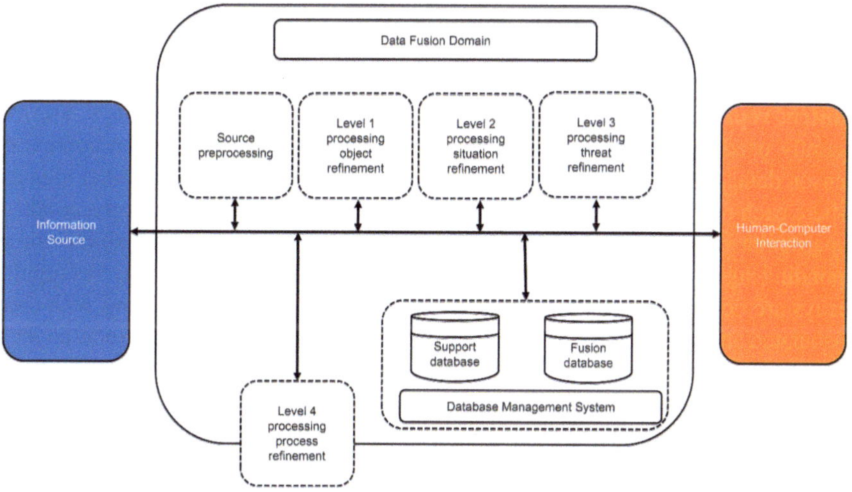

Fig. 5.5 The architecture of the JDL fusion framework. The JDL framework was proposed for military applications and consisted of four levels: target assessment, situation assessment, threat assessment, and process optimization

5.1.5 Data Representation

Data representation plays a fundamental role in the fusion and integration of various types of data in CPS, enabling effective data mining and machine learning. In recent years, there has been a proliferation of research on data representation, with numerous approaches being proposed.

Graph representation. Graph representation was introduced by [28], and has become a popular tool for analyzing and visualizing data. Graph representation is a technique used to represent data that can be easily expressed in matrix form [29]. A graph comprises nodes, some of which are linked by edges. An adjacency matrix is a type of matrix in which rows and columns correspond to the nodes of a graph, and the matrix elements represent the number of edges between any two nodes. By using matrices to represent graphs, researchers can easily apply mathematical and computational techniques to study the structure and properties of these networks.

Ontology representation. Ontology refers to a set of concepts that possess four essential characteristics: sharing, reusability, hierarchy, and reasoning [30]. These attributes allow ontology to represent objects as semantic conceptual models. The construction of an ontology involves two main steps: (1) mapping the real-world entities, attributes, and processes into a set of concepts; and (2) extracting the relationships between these concepts. Ontology provides a structured and organized way of representing knowledge, which can be used for a variety of purposes, such as information retrieval, data integration, and knowledge management [31].

Matrix representation. Matrices, which are also called two-way arrays, are frequently used to capture variations in time and space, especially in the context of sensor data. A matrix provides a natural representation for multichannel signals, where each row represents a specific sensor or channel, each column corresponds to a particular measurement time, and the values within the matrix indicate the corresponding signal values. In the field of data mining and machine learning, rectangular arrays are commonly employed to depict samples or observations along with their associated attributes or properties. In this representation, each row represents an individual sample or observation, while each column corresponds to a set of attributes or properties associated with that particular sample. The use of matrices in data analysis and machine learning allows for efficient computation and manipulation of large datasets. It enables the application of various mathematical and statistical techniques for data analysis and modeling.

Tensor representation. A tensor is a mathematical construct that expands the concept of a vector to encompass multiple dimensions [32]. Tensors are essentially arrays with multiple dimensions, where each element of the tensor is identified by multiple indices, each corresponding to a specific mode or order. Tensors are highly adaptable and versatile mathematical tools that can be used to represent a wide range of complex, multidimensional data with diverse attributes [33].

5.2 Techniques for Fusing Data from Multiple Sources

The data fusion in CPS poses significant challenges due to its inherent complexity, which is closely tied to multidisciplinary technologies. The section aims to introduce prominent data fusion methods utilized in CPS, with the goal of achieving system objectives or enhancing situation awareness. These fusion techniques enable collaboration between computers, smart devices, and individuals, ensuring that the system operates as a collective effort rather than solely relying on human users [34]. Furthermore, data fusion is an essential prerequisite for extracting valuable insights from the voluminous and noisy CPS big data, which is characterized by high dimensionality and semantic diversity.

Due to the interdisciplinary nature of data fusion, it is challenging to establish a strict and definitive classification. Various classification criteria exist, including the relationships among different data sources [35], the types and properties of input/output data [36], the levels of abstraction [37], the levels of data fusion [38], and the architectural type. A classification framework consists of three major categories: stage-based approaches, feature-level-based approaches, and semantic meaning-based approaches was proposed [39]. The categories serve as a guideline for organizing and understanding different data methods. In Ref. [40], data fusion methodologies were classified into cyber-physical, cyber-social, and physical-social system groups. They introduced their model in CPS, called the tensor-based model by extending the Markov chain (MC) model to the transition matrix of MC. According to the existing fusion techniques, we divide the data integration and fusion of CPS into stage-based, semantic meaning-based, and artificial intelligence (AI)-based methods, as shown in Fig. 5.6.

5.2.1 Stage-Based Data Fusion Methods

Stage-based data fusion entails partitioning a problem into distinct stages that address specific issues by utilizing relevant data. The results obtained from each stage are then aggregated to solve the overarching problem. Stage-based methods enable loosely linked data sets, without enforcing strict requirements for the consistency of their modalities. A map segmentation technique was employed to divide the city into regions based on major roads [41]. GPS trajectories of taxicabs were then superimposed onto these regions, generating a region graph. Each node in the map partition denoted a particular region, while the edges represented the combination of taxi commutes between two regions. Through an analysis of this region graph, the researchers

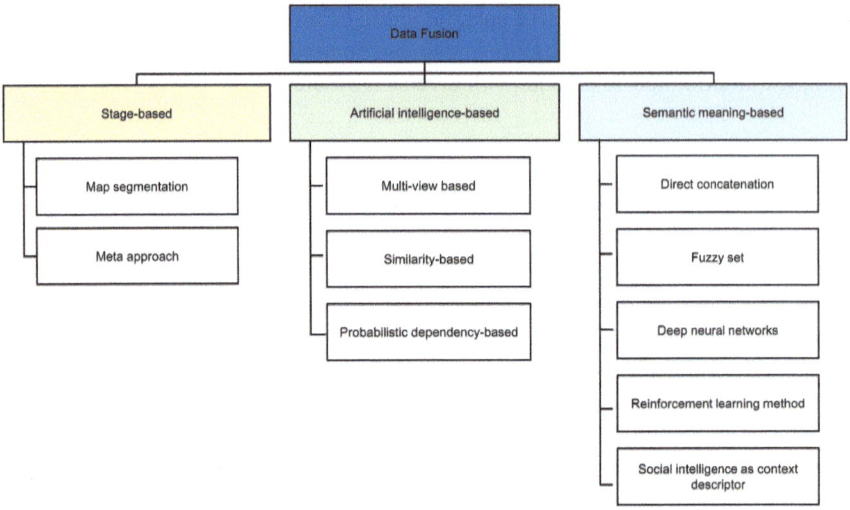

Fig. 5.6 The taxonomy of data fusion techniques. Different data fusion algorithms can be roughly categorized into three classes, namely, stage-based, artificial intelligence-based, and semantic meaning-based based on the model used

could identify problematic road networks, detect and diagnose traffic anomalies, and recognize urban functional regions. In Ref. [42], the objective was to identify similar users through friend recommendations. The researchers initially modeled a user's GPS trajectories using a semantic location history. They then estimated the similarity between two users based on their non-overlapping geographic spaces. For example, if a specific stay point was surrounded by five restaurants, one shopping mall, and one gas station, the feature vector would capture this contextual information. In other words, the measure of similarity between visited places can be determined by the disparity among their respective feature vectors. By analyzing GPS trajectories of vehicles and road network data, traffic anomalies were detected [43]. An anomaly was identified as a subset of the road network where drivers' routing behaviors significantly deviated from their normal patterns. However, there are various considerations to take into account when employing this approach. Firstly, dividing the problem into stages runs the risk of potential disconnection between different stages. Secondly, it is essential to determine the appropriate level of granularity for each stage. Lastly, optimizing the combination of solutions across various stages is necessary. Different data fusion methods can yield different effects on the combined outcomes. Stage-based data fusion techniques can be employed as a meta-approach along with other data fusion methods. For example, road network data and taxi trajectories can be used to construct a region graph, which can then be fused with information about points of interest and knowledge of the region graph [44].

5.2.2 Semantic Meaning-Based Data Fusion

Semantic meaning-based methods consider the semantics and relationships between features across different datasets [45]. Semantic reasoning methods (SR) are commonly used in representing CPS, as they are effective in detecting and coordinating services or agents [46, 47]. SR refers to the process of using knowledge representations and reasoning techniques to draw logical conclusions from semantic data, which involves analyzing the meaning and relationships between concepts and entities in order to make inferences and deductions. In [48], semantic web-based methods were utilized to convert different types of data into interconnected triples in the resource description framework format and allowed for the mapping of these data sources and enabling them to be queried using SPARQL. The data types included proprietary relational datasets, environment monitoring data streams, and participatory sensing data. By utilizing semantic web technologies, the wi-city plus system is able to integrate and analyze diverse data sources in a more efficient and effective manner. In Ref. [49], semantic rule-based reasoning was utilized in a smart airport setting. Specifically, device and environment capabilities were modeled as social objects and were combined with user preferences to form a smart services framework within a CPS context (called dynamic social structure of things). By leveraging semantic technologies and rule-based reasoning, it was possible to create a flexible and adaptive system that could dynamically adjust its services based on user preferences and environmental context. The method allowed for more personalized and efficient service delivery in the airport setting. SR is also utilized to describe system interfaces and generate applications [50]. The W3C semantic sensor network ontology offers similar capabilities, providing a consistent semantic model of sensors that can be used by backend systems regardless of the underlying technology. Compared to other metadata models like XML, semantic models offer more structure, flexibility, and extensibility. Thus, the semantic meaning-based data fusion methods can be divided into four groups.

5.2.2.1 Multi-view-based Data Fusion

The multi-view-based data fusion method explores an object by considering multiple perspectives or views. When studying an object, this method is divided into several techniques: subspace learning method, multi-kernel learning method, and cross-space data fusion through correlation.

Subspace learning. The subspace learning method focuses on extracting relevant features from different views of the object. It aims to identify subspaces within each view that capture the most informative aspects of the object, as shown in Fig. 5.7. After obtaining the subspace through the multi-view-based data fusion method, further tasks such as classification and clustering can be carried out. The subspace,

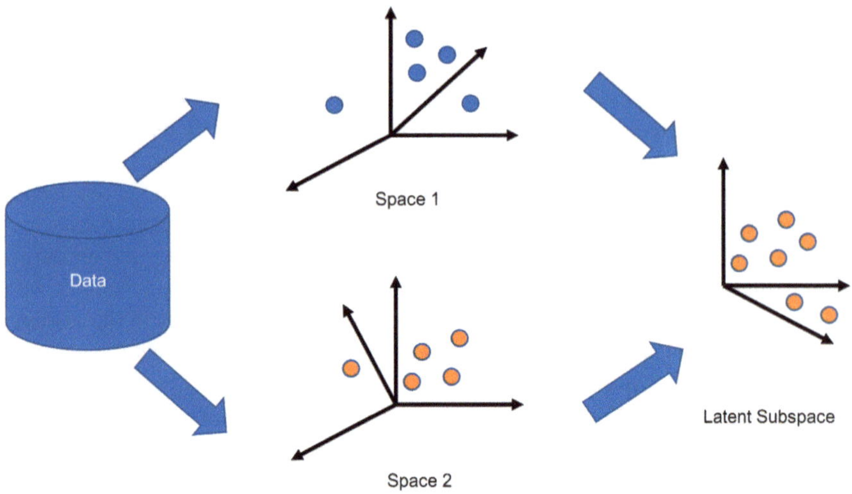

Fig. 5.7 Concept of subspace learning, which generates a potential subspace from different views of data

which is constructed based on the input views, often has a lower dimensionality compared to the original data. This characteristic provides some relief to the "curse of dimensionality problem".

Subspace learning-based approaches have gained popularity in recent years due to their ability to handle multi-view data. By leveraging the shared latent subspace, subspace learning-based approaches effectively capture the underlying structure of the data and extract meaningful features. The Gaussian process as a means of constructing a latent variable model was proposed, which has gained popularity due to its ability to capture complex relationships between variables and model uncertainty [51]. Based on the Gaussian process, obtaining a low-dimensional representation of high-dimensional data has been possible that preserves the underlying structure. A statistical framework for learning a predictive subspace was proposed [52]. The framework was achieved by using a generic multi-view latent space Markov network, which was capable of modeling complex dependencies between views to extract meaningful features from multi-view data and perform subsequent tasks such as classification and clustering. The Gaussian process and statistical framework have been applied in a wide range of fields, including computer vision, natural language processing, and bioinformatics. However, there are still several challenges that need to be addressed, such as the selection of appropriate regularization parameters and the determination of the optimal subspace dimensionality. Despite their advantages, subspace learning-based approaches also face certain challenges, such as the determination of the optimal subspace dimensionality and the selection of appropriate regularization parameters. Addressing these challenges is an active area of research in the field of subspace learning.

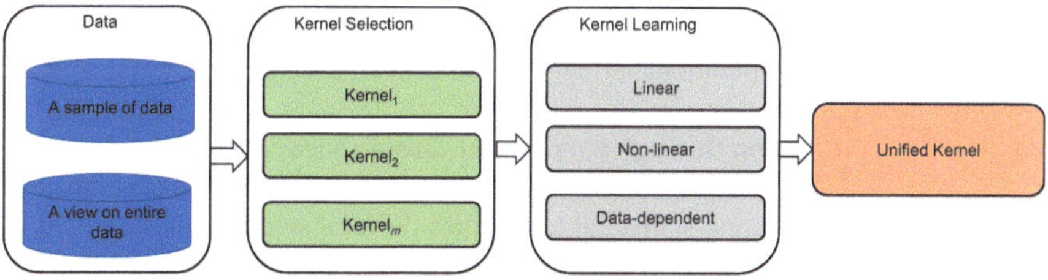

Fig. 5.8 Procedure of multi-kernel learning by using different kernel in different machine learning methods to learn

Multi-kernel learning. Multi-kernel learning (MKL) leverages multiple kernels within various machine learning techniques to enhance learning capabilities. In MKL, a kernel represents a hypothesis or concept about the data, which can encompass similarity measures, classifiers, or regressors. By incorporating multiple kernels, MKL aims to improve the overall accuracy and performance of the learning process. The procedure of multi-kernel learning involves several steps, as shown in Fig. 5.8.

MKL algorithms are commonly divided into two main groups: one-step methods and two-step methods [53]. The first group, one-step methods, aims to determine the parameters of the combination function and base learners in a single pass. The one-step methods can be accomplished through two different approaches: sequential and simultaneous. In the sequential approach, the combination function parameters are determined first, followed by the training of a kernel-based learner using the combined kernel. This allows for a step-by-step process where each parameter is updated sequentially. On the other hand, the simultaneous approach involves learning both sets of parameters at the same time, considering them as an integrated entity. The second group of MKL algorithms consists of two-step methods that adopt an iterative approach. These algorithms update the combination function parameters and the base learner parameters in separate steps. During each iteration, the parameters of the combination function are updated while keeping the base learner parameters fixed, which means that the combination function is optimized independently at each step. After that, the parameters of the base learners are updated while keeping the combination function parameters fixed, which enables the base learners to be fine-tuned in each iteration. These two steps are repeated until the algorithm converges, ensuring the optimal parameters for both the combination function and the base learners are obtained.

Both one-step and two-step methods have their advantages and disadvantages. One-step methods are computationally efficient and require less memory, making them suitable for large-scale problems. However, one-step methods may not always converge to the optimal solution. In contrast, two-step methods are more computationally expensive and require more memory, but they are more likely to converge to the optimal solution [54]. The choice of training methodology depends on the specific problem at hand and the available computational resources. Researchers continue to explore new MKL algorithms and training methodologies that strike a balance between computational efficiency and convergence to the optimal solution.

Cross-space data fusion through correlation. Cross-space data fusion refers to the process of integrating and analyzing numerical data streams obtained from both physical and cyber spaces using statistical techniques. Several studies have explored the correlation between different data sources to extract meaningful insights. Komninos et al. conducted a study in Paras, Greece, where they analyzed Foursquare check-in data along with diurnal pollutant levels and traffic volume [55]. By calculating the correlation and p-value between the diurnal cumulative averages of pollutant levels and Foursquare check-ins, they discovered a close alignment between the check-in patterns and pollution levels. Another study focused on correlating traffic behavior with temperature in the city of Santander [56]. The study discovered a detailed correlation in the evolution of both traffic flow and temperature, highlighting the interdependencies between these two factors. Addressing the challenge of inconsistent semantics derived from cross-space, multi-modal sensory data (such as audio clips, images, videos, and texts), a proposed DDCPS approach involved aggregating heterogeneous and complementary data from various sources to provide a more comprehensive characterization of the sensing target [57]. DDCPS method effectively managed the complexity arising from inconsistent semantics in cross-space data fusion. These above studies demonstrate the value of cross-space data fusion in uncovering correlations and extracting insights from diverse data sources. By leveraging statistical techniques and exploring the relationships between different data streams, it becomes possible to gain a deeper understanding of complex phenomena in both physical and cyber spaces.

5.2.2.2 Similarity-Based Data Fusion

The similarity-based data fusion method commonly involves measuring the correlation or similarity between multiple sources of data. The similarity-based method aims to quantify the degree of similarity and construct a similarity matrix for the data sources.

Corrective matrix factorization Corrective matrix factorization (CMF) is a data fusion technique that is based on matrix decomposition. It is designed to handle multiple data sets from different sources collectively. The fused matrices generated using CMF have one or more common modes, which means that they share certain dimensions or features. The main objective of CMF is to jointly factorize multiple relation matrices from diverse sources, allowing for the discovery of hidden or latent associations. By transferring information through a common model, CMF facilitates the integration of different matrices and reveals the underlying relationships between them through a common model. An illustrative example of two matrices coupling in a common mode is shown in Fig. 5.9. The CMF fusion method is widely used in various applications due to its simplicity and practicality. For example, Klami

Fig. 5.9 An illustration example of two matrices couple **X** and **Y** in a common mode, where $\mathbf{X} \in \mathbb{R}^{I \times J}$ and $\mathbf{Y} \in \mathbb{R}^{I \times K}$

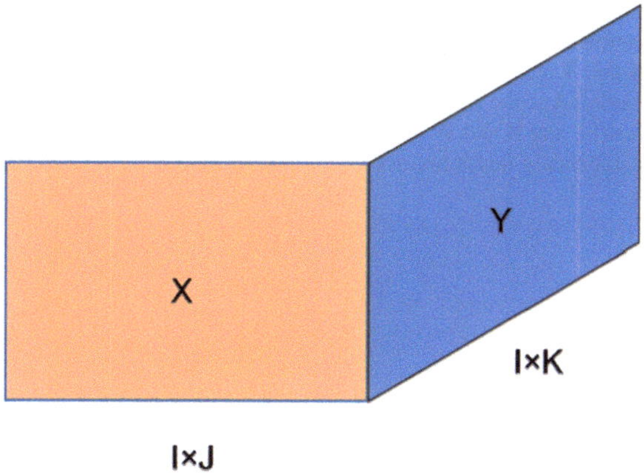

et al. proposed a novel CMF solution that allows each matrix to have a separate low-rank structure that is independent of other matrices, as well as structures that are shared only by a subset of them [58]. The approach enables the identification of both global and local patterns across multiple data sources. A unified view of matrix factorization was proposed to frame the differences among popular methods in terms of some modeling choices [59]. By framing these differences in a unified manner, they provide a clearer understanding of how various methods relate to each other and identify common principles underlying these methods. Furthermore, fusing multiple data sources not only increases accuracy but also provides a more comprehensive understanding of the target being studied [60]. The main reason is that different data sources may capture unique aspects of the target, and combining them can result in a more complete picture.

Coupled matrix and tensor factorization. The CMF method is limited in its ability to address the fusion problem between matrices and tensors. In practical applications, data often come in highly heterogeneous forms, with both matrices and tensors being present in the datasets. To address this issue, the coupled matrix and tensor factorization (CMTF) approach has been proposed as a natural extension of the CMF method. CMTF can jointly analyze one tensor with multiple matrices simultaneously, by identifying common modes among them, as shown in Fig. 5.10. CMTF allows for the integration of different sources of data, enabling the discovery of hidden patterns and relationships. Tensor-based methods have been employed in data fusion techniques to uncover concealed information, often involving the analysis of user behavior. Group-centric data fusion has been executed using the approximate tensor, where each element in the approximate tensor indicates the prevalence of the corresponding behavior in the group [61]. The CMTF method has made significant improvements over the CMF method and is widely used in big data analysis fields. It allows for the simultaneous analysis of multiple heterogeneous data sources and has been successfully applied in various domains, such as social network analysis, recommender systems, and biomedical data analysis [62, 63].

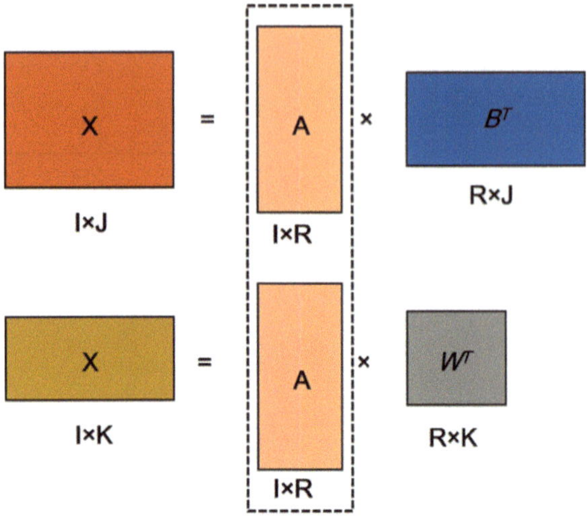

Fig. 5.10 Visualization of coupled matrix factorization, **X** and **Y** share the common model, **A** is the shared factor matrix, and **B** and **W** are the independent factor matrices

5.2.2.3 Probabilistic Dependency-Based

Probabilistic dependency-based data fusion method is a graph-based approach to information fusion. This method involves representing different data sources as nodes in a graph and measuring the relationships between them using edges. The edges can be either directed or undirected, depending on the structure of the graph being used. The key idea behind this method is to model the dependencies between different data sources and use this information to make more accurate predictions or decisions. The probabilistic dependency-based data fusion method includes the weighted average, the least squares, the maximum likelihood, and so on.

Weight average. Weighted average method is a simple and widely used approach to information fusion. The method involves directly combining the signals from different sensors in the measuring environment by assigning weights to each signal based on its reliability [64]. For example, if one sensor has a higher resolution or is less prone to noise, its signal may be given a higher weight in the fused signal. The weighted average method has several advantages, including its simplicity and ease of implementation, and has been used in various fields, including sensor networks, image processing, and speech recognition. However, the weighted average method also has some limitations. (1) The method assumes that the signals from different sensors are independent and identically distributed, which may not always be the case in practice. (2) The method does not take into account the correlations between different signals or the uncertainties associated with each sensor. To address these limitations, researchers have developed various extensions and modifications to the weighted average method, such as the Kalman filter and the particle filter. These methods can handle more complex dependencies between signals and can incorporate uncertainty and noise more effectively.

Least squares. Least squares method involves finding the best estimate match by minimizing the sum of squares of the error between the obtained data and the actual data. The key idea behind the least squares method is to obtain the unknown data by minimizing the error sum of squares, which can improve the reliability of information system data [65]. There are several variants of the least squares method including partial least squares fusion, weighted least squares fusion, and least squares combined with a support vector machine are commonly used for information fusion [66, 67]. Partial least squares fusion is a variant of the least squares method that is used when the number of variables is larger than the number of samples [68]. Weighted least squares fusion is a variant of the least squares method that assigns weights to each sensor based on its reliability or accuracy. Least squares combined with support vector machine is a variant of the least squares method that uses support vector machine to improve the classification accuracy of the fused data.

Maximum likelihood. Maximum likelihood estimation (MLE) is a statistical method that uses probability theory to estimate model parameters based on observed data. The main objective of MLE is to find the parameter values that maximize the likelihood of the observed data. This approach has been widely used in various applications, including information systems and data fusion. One significant application of MLE is in information systems, where it can accurately estimate and correct errors within the system. By maximizing the likelihood of the observed data, MLE enables the identification of the most probable parameter values, allowing for effective error detection and correction. In the context of data fusion, MLE is utilized to solve the registration problem. The registration problem involves aligning data from different sources to create a unified representation. MLE helps identifying the parameter values that maximize the probability of achieving a consistent and accurate registration, improving the quality of the fused data. MLE offers several advantages in information fusion tasks. It can handle complex dependencies between data sources, capturing the relationships and interdependencies effectively. Additionally, MLE can incorporate prior knowledge about the system, providing a flexible framework for integrating existing knowledge into the estimation process. There are various variants of the MLE method employed in information fusion, and they have been extensively studied and utilized. For instance, researchers like Myung et al. [69] and Barlow et al. [70] have proposed and explored different variants of MLE for information fusion. An example of MLE in action is the work of Chen et al. [71], who employed maximum likelihood estimation for joint image registration and fusion. By maximizing the likelihood of the observed data, they successfully aligned and fused images from multiple sources, improving the overall image quality.

5.2.3 *Artificial Intelligence-Based Data Fusion*

Artificial intelligence-based data fusion methods employ AI techniques to extract features of the same dimension from diverse datasets, and the features are directly concatenated or employed using deep learning approaches to perform classification or prediction tasks.

5.2.3.1 Direct Concatenation

Direct concatenation treats features extracted from different data sets equally and concatenates features sequentially into a feature vector, which is used in clustering and classification tasks [72]. However, direct concatenation has been criticized by several studies [73, 74] due to limitations that arise from differences in the representation, distribution, and scale of different datasets. (1) Concatenation can lead to over-fitting when the training sample size is small, and it ignores the specific statistical properties of each view. (2) Uncovering highly non-linear connections between low-level characteristics across diverse modalities poses a challenge. (3) There could be redundancies and interdependencies among attributes derived from distinct datasets that display correlation. Thus, when using the direct feature combination method, there are several issues that need to be considered. (1) It is necessary to remove any duplicate features when merging the data. (2) Some features with different dimensions that contribute to the model's performance may be lost during the direct merging process. (3) Direct feature merging may lead to overfitting.

5.2.3.2 Fuzzy Set

Fuzzy set theory, initially proposed in the 1960s, is a classification-based approach that allows for more flexible modeling of data fusion compared to probability theory. It provides a means to address conflicts between information and decision-making [75]. Fuzzy theory is often combined with other methods to enhance the accuracy of information fusion [76]. The concept of a membership function is utilized in fuzzy theory to handle fuzziness. However, determining the degree of membership is typically subjective and relies on experience or expertise from domain specialists, as shown in Fig. 5.11. Due to this inherent subjectivity, fuzzy set theory is not widely used in information fusion. Nevertheless, it still finds application in certain IoT scenarios where obtaining the degree of membership is relatively straightforward.

Despite its limitations, fuzzy set theory has been applied in various fields, including control systems, decision-making, and pattern recognition [77, 78]. Fuzzy logic has been used to model complex systems and make decisions based on imprecise or uncertain information. Moreover, fuzzy logic has been used in combination with other methods, such as neural networks and genetic algorithms, to improve the accuracy of predictions and decision-making. However, the subjective nature of fuzzy

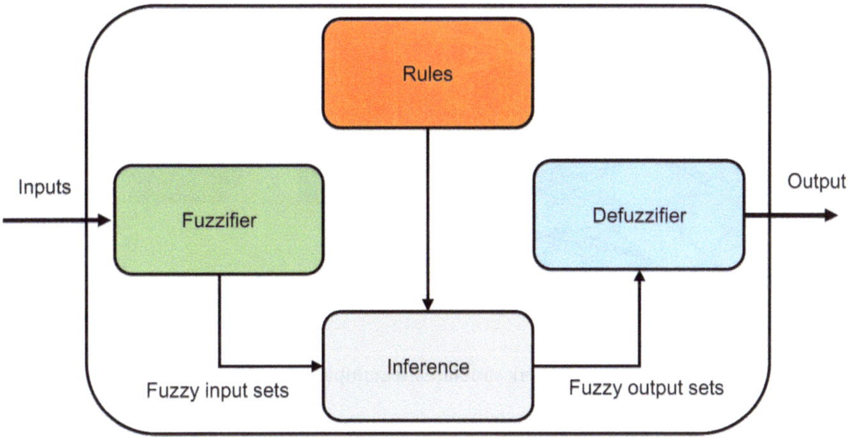

Fig. 5.11 Fuzzy logic system. The fuzzy logic system infers fuzzy conclusions by applying fuzzy logic operations and inference methods based on fuzzy inference rules and fuzzy sets generated by fuzzy generators

set theory remains a challenge, and researchers continue to explore new methods for information fusion that can handle uncertainty and imprecision in a more objective manner.

5.2.3.3 Deep Neural Networks

Recently, advanced techniques have been developed to leverage deep neural networks (DNN) to learn a unified feature representation from diverse datasets [79, 80]. While DNN is not a new concept in the field of AI, it refers to a multi-layer neural network with a large number of parameters, as shown in Fig. 5.12a. Neural networks were trained using a back-propagation algorithm, which exhibited limitations when dealing with networks comprising numerous hidden layers. However, novel learning algorithms have emerged to address the challenges of training DNNs with many hidden layers. Some of these algorithms include the autoencoder as shown in Fig. 5.12b, restricted Boltzmann machines (RBM), and convolutional neural networks (CNN) shown in Fig. 5.12c. The above networks enable the learning of DNN layer by layer, using supervised, unsupervised, and semi-supervised approaches. Deep learning can acquire multiple levels of representation and abstraction, enabling the interpretation of various data types such as images, sound, and text. In addition to its predictive capabilities, DNN can also learn new feature representations that can be subsequently used in other classification or prediction models. Such learned feature representations have proven to be more effective than manually crafted features in domains like image recognition and speech translation [81, 82].

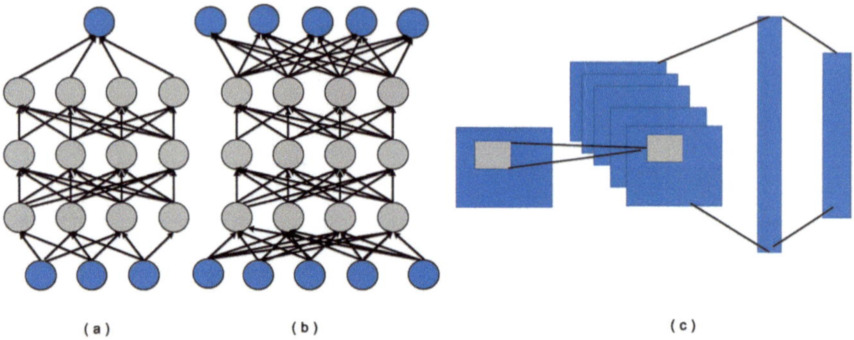

Fig. 5.12 Three generic neural network structures, **a** multiple-layer neural network, **b** autoencoder, **c** convolutional neural network

5.2.3.4 Reinforcement Learning

Reinforcement learning (RL), a machine learning approach and methodology, addresses the challenge of enabling agents to learn strategies for achieving specific objectives or maximizing returns through interactions with their environment. RL utilizes learning algorithms to optimize the decision-making process of these agents, allowing them to adapt and improve their performance over time. Recently, RL has been utilized in various applications. For instance, RL was combined with semantic fusion to provide suggestions for decision-making [83] and a framework based on reinforcement learning and human-in-the-loop was presented to optimize decision-making [84]. In Ref. [85], ontology and reinforcement learning were combined for zero-shot classification. Similarly, an ontology-based multi-agent reinforcement learning method was designed to optimize the scheduling of a manufacturing system [86]. Blockchain technology and deep reinforcement learning were combined to create a secure and reliable environment [87]. Multi-granularity reinforcement learning models were proposed to accelerate the learning process and adapt to dynamic environments [88]. These studies indicate the growing utilization of reinforcement learning in ontology-related computations, although more in-depth research is needed. Based on the reinforcement learning algorithm, a ternary data fusion model was proposed in Ref. [8]. This model effectively merges internet data, IoT data, and other data sources to derive better and faster decisions. The data fusion approach integrates knowledge and expertise from experts and managers, traditional information systems, and sensor data, enabling more comprehensive and efficient decision-making processes.

5.3 CPS Data Integration and Fusion Case Studies

The definitions of data integration and fusion have been introduced in Sect. 5.1 and the relevant data fusion methods have been summarized in Sect. 5.2. To explain data integration and fusion of CPS data comprehensively and systematically, two case studies about cloud-integrated CPS and CPS big data fusion are introduced.

5.3.1 Cloud-Integrated CPS for Smart Cities Case Study

Cloud-integrated CPS enables the creation, deployment, management, and efficient control of application scenarios. A complex industrial application (CIA) CPS consists of three levels: perception, decision-making, and execution, as shown in Fig. 5.13.

Physical part. The physical part includes hardware and its integration, as shown in Fig. 5.13. The cloud-integrated CPS is compliant with the ISA' 95 automation pyramid view but complements it with a flat information-driven modern system and enhances the integrality via modern software engineering practices [89]. By expressing functionalities as services, CPS take advantage of the benefits of service-oriented architecture (SOA), which enables CPS to be more modular, flexible, and scalable, as services can be easily added or removed as needed. Additionally, services can be

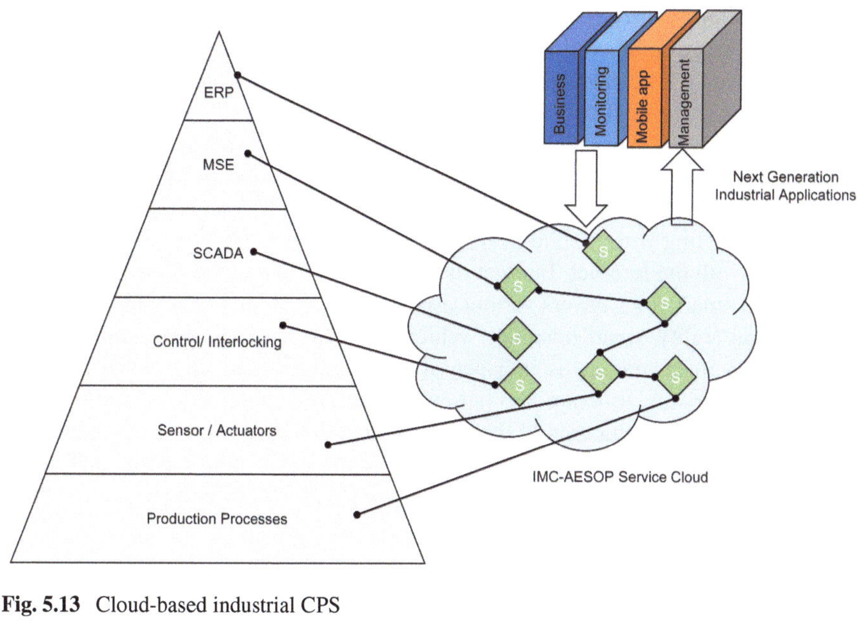

Fig. 5.13 Cloud-based industrial CPS

realized on-device or in-network, providing greater flexibility in terms of where and how they are executed. Furthermore, by using services, CPS are easily integrated with other systems and technologies. This approach not only simplifies the integration process but also enables CPS to take advantage of cloud computing, providing greater scalability and cost-effectiveness.

Cyber part. The cyber part includes software and more importantly capability exposure as services, which splits into services on-device and in networks running outside CPS. By implementing certification, CPS take advantage of modern software engineering practices, such as service-driven integration to offer several benefits for the design, development, integration, operation, and management of complex and large-scale systems.

Service-driven integration enables CPS to be more agile and responsive to changing requirements, as new services can be easily added or removed as needed. Furthermore, the move towards a modular and service-based infrastructure enables easier evolution over time, allowing CPS to be adapted to future needs and facilitating the development of new functionalities and capabilities. In addition to these benefits, service-driven integration also simplifies the integration process and enables better management of complex and large-scale systems. By breaking down functionalities into services, CPS are more easily managed and maintained, as services can be individually monitored and updated as needed.

By integrating CPS, third-party services, and APIs, CPS can modularize its functionalities and provide services to external parties, while also benefiting from independent services regardless of their physical location. This approach enhances the flexibility of CPS, enabling advanced and adaptable CPS systems in various industrial applications. Through modularization, CPS can offer services to third-party entities, fostering collaboration and integration with diverse systems and technologies. Moreover, CPS can leverage independent services, regardless of their physical location, thus ensuring flexibility in the execution of services.

For example, to ensure the well-being and economic prosperity of their citizens, decision-makers and city administrations actively explore solutions for smart cities. These cities rely on an infrastructure that utilizes sensor and actuator network technology, connecting with mobile devices like smartphones, tablets, and laptops, and integrating with the Internet, Internet of Things (IoT), and cloud services. The data collected in smart cities covers various aspects such as air and water quality, traffic, parking availability, smart hospitals, vehicular networks, and other essential facilities. A CPS infrastructure is indispensable to enable smart city technology and services, comprising software and hardware platforms equipped to handle mobility and process large amounts of data. In essence, the infrastructure of smart cities encompasses the abilities of sensing, communication, decision-making, and actuation. Cloud computing and IoT technologies play a pivotal role in facilitating these capabilities.

5.3.2 Data Fusion Framework for Smart Healthcare Case Study

Data fusion framework for smart factory based on tensor decomposition is shown in Fig. 5.14. The framework is composed of three layers: data perception layer, cloud data processing, and application layer [90].

Data perception layer. The data perception layer is a critical component of large CPS that consists of numerous local CPS. These local CPS generate vast amounts of data on a daily basis, which need to be collected and organized efficiently. To represent the high-order characteristics of CPS data, tensors are utilized. Depending on the specific circumstances, various tensor decomposition methods have been employed to preprocess the collected data in different local CPS. By leveraging the preprocessing, noisy and redundant data from raw CPS data are eliminated and higher-quality data are extracted. Then, the processed data are uploaded to the cloud for integration and fusion processing [91].

In previous research, we have discussed the representation, preprocessing, and computing of multi-sources and heterogeneous big data using tensor and tensor decomposition. We have also described the tensor representation of various heterogeneous data. By utilizing tensor decomposition methods, we can effectively clean and reduce invalid CPS data. Distributed and incremental tensor decomposition approaches can also be utilized to improve the efficiency of data processing in the data perception layer. Distributed tensor decomposition involves breaking down a large tensor into smaller sub-tensors that can be processed simultaneously on multiple computing nodes. This can help reducing the computational load and improving efficiency. Incremental tensor decomposition involves processing data in small batches, which can help reducing memory requirements and improving scalability. Furthermore, the data perception layer can also leverage various machine learning algorithms to preprocess and analyze CPS data. For example, deep learning algorithms such as CNN and recurrent neural networks (RNN) can be used to extract features from raw data and improve the accuracy of data analysis.

Therefore, the data perception layer plays a critical role in the efficient processing of CPS data. By utilizing tensor decomposition methods, distributed and incremental computing approaches, and machine learning algorithms, we can effectively preprocess and analyze CPS data, extract high-quality information, and improve the accuracy and efficiency of data fusion in the cloud data processing layer.

Cloud data processing layer. In the cloud data processing layer, the data acquired from the data perception layer undergoes a refinement process. The refined data, represented as local tensor data, is then transformed into tensor train (TT) format tensor networks using fundamental tensor operations. This conversion process comprises four primary operations: tensor contractions, tensor reshaping, tensor singular value decomposition (SVD), and tensor rounding. Tensor contractions involve the multiplication of two or more tensors to produce a new tensor. Tensor reshaping involves

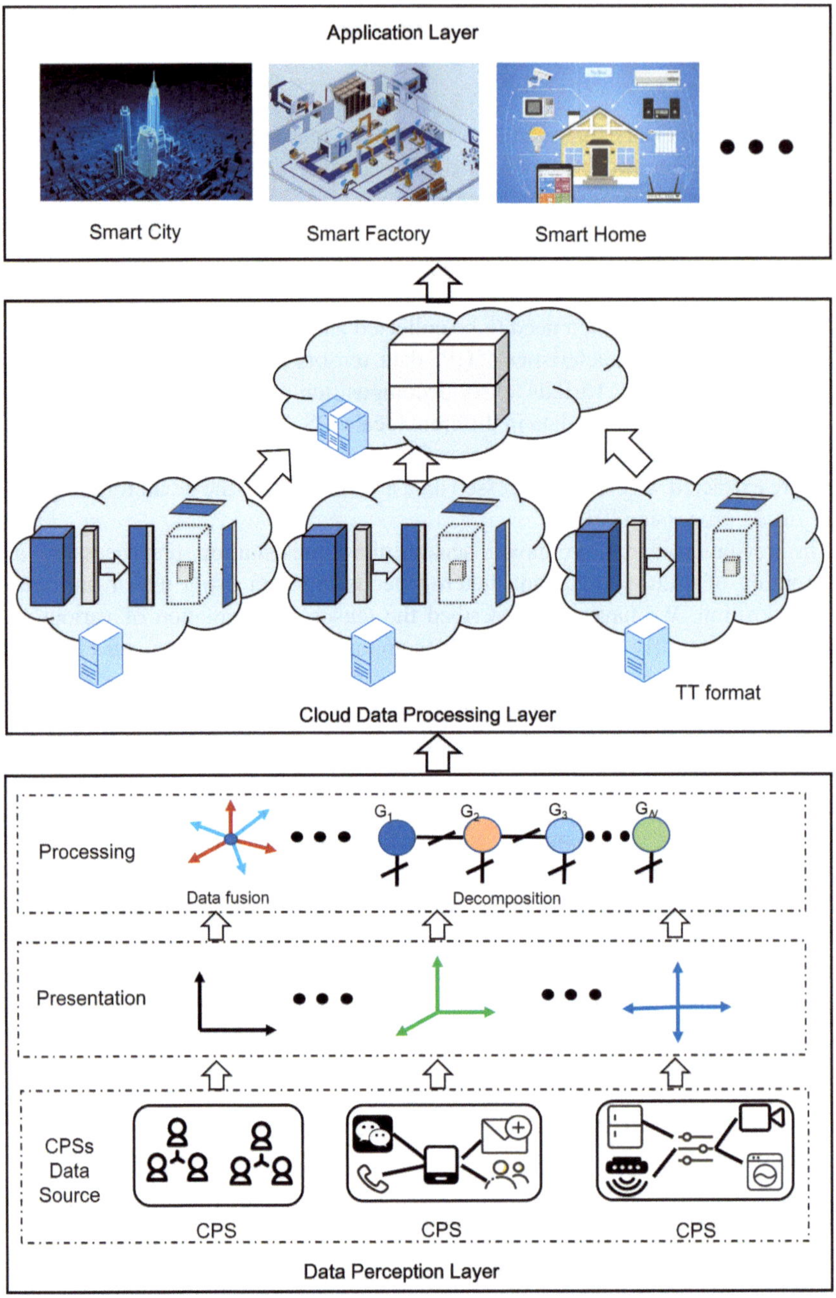

Fig. 5.14 Data fusion framework for the smart factory. The framework consists of three layers: the data perception layer, the cloud data processing layer, and the application layer

changing the shape of a tensor while maintaining the same number of elements. Tensor SVD is a decomposition technique that breaks down a tensor into a set of lower-dimensional tensors, which can be used to reduce the size of the tensor and extract important features. Tensor rounding involves rounding the values of a tensor to a certain number of decimal places, which can help reducing the size of the tensor and improving computational efficiency.

Through the utilization of tensor networks and associated tensor operations, it is possible to convert tensor data in various formats into TT format tensors. This enables the fusion of tensor data conversions by utilizing tensor join or union operations. Nonetheless, when dealing with global TT format tensor networks, the size of the tensors can become considerably large, leading to increased computational overheads during the conversion process. To tackle this problem, several techniques can be employed, including parallel distributed computing, incremental computing, collaboration computing between edge cloud and fog, and more [92]. Parallel distributed computing involves breaking down a large computation task into smaller sub-tasks that can be processed simultaneously on multiple computing nodes. Incremental computing involves processing data in small batches, which can help reduce the computational load and improve efficiency. Edge cloud and fog collaboration computing involve distributing computation tasks between edge devices and cloud/fog servers, which can help reduce latency and improve response time. By employing these techniques, we can significantly reduce the time complexity of the conversion process and accelerate the TT decomposition process.

For the representation, computation, and storage of higher-order CPS data, the TT format tensor networks offer exceptional effectiveness and efficiency. To enhance the processing of large-scale CPS data using these networks, we can leverage a range of techniques including incremental computing, parallel distributed computing, and collaborative computing between edge cloud and fog. With the application of these methods, the efficiency and effectiveness of the TT format tensor networks can be further optimized.

Application layer. In order to cater to specific application scenarios, the application layer within the CPS data fusion framework is specifically designed to offer personalized and proactive services. This is accomplished by employing mapped tensor network models derived from preceding layers within the framework. Given that the CPS data fusion framework is grounded in tensor decomposition and transformation, it becomes imperative to reconfigure data analysis and processing methodologies for tasks like prediction and recommendation, effectively harnessing the framework's inherent capabilities. This redesign effort enables robust support for diverse implementations within the application layer [93].

For example, In the realm of intelligent healthcare, the utilization of the data fusion framework can be observed in the provision of personalized and proactive health monitoring services for patients. A fine instance exists within the data perception layer, where the CPS data fusion framework assimilates insights from diverse sensors and devices, ultimately generating a mapped tensor network model representing the patient's state of health. This layer encompasses an array of sensors interconnected via wires, facilitating seamless connectivity with a wearable wireless data transmitter. These sensors encompass both wearable and implantable devices, substantially enhancing continuous patient monitoring by providing comprehensive contextual information about abnormal conditions. Furthermore, implantable sensors, forming an integral part of body sensor networks, enable metabolite level measurements independent of the patient's physiological state (such as rest or sleep). A handheld device, akin to an implantable sensor, further enhances data collection by capturing biometric markers like handwriting, signatures, and fingerprints, as well as leveraging cameras for video footage, facial recognition, and microphones for sound capture. To handle multi-sources and diverse data, tensor decomposition techniques are employed. In addition, incremental TT decomposition methods can effectively cleanse and minimize erroneous data in CPS. Within the cloud data processing layer, the different tensor decomposition methods are converted into the TT format. This enables integration and transformation of the data. By leveraging tensor networks and their associated tensor operations, it is possible to convert tensor data from various formats into TT (Tensor Train) format tensors. The TT format is a compressed representation of high-dimensional tensors that allows for efficient storage and manipulation. With the use of tensor join or union operations, tensor data conversion fusion can be achieved, enabling seamless integration and transformation of data across different tensor formats. This approach not only facilitates efficient data processing but also enables the exploration of complex relationships and patterns within the tensor data, leading to enhanced insights and decision-making capabilities. In the application layer, this model is used to provide personalized recommendations for diet and exercise, as well as to predict potential health issues before they occur.

In conclusion, the application layer of CPS data fusion framework provides a powerful platform for the development of personalized and proactive services in a variety of application scenarios. By leveraging the capabilities of tensor decomposition and transformation, the framework can provide highly accurate and efficient data analysis and processing, enabling the development of innovative and effective solutions for a wide range of industries.

5.4 Challenges and Future Work Opportunities

In the section, we summarize the challenges of DDCPS data integration and fusion and analyze some future work opportunities.

5.4.1 Integrated Models Challenges

The design process for integrated models of CPS requires a multidisciplinary approach, where designers must focus on integration and interaction rather than just computational and physical components. As a result, the challenges associated with integrated systems are related to physical processes, computation, and integration. Compared to other complex systems, environmental influences play a significant role in the design process of integrated CPS. This is because integrated CPS must consider the impact of environmental factors on the system's performance and reliability [94].

From a computational perspective, integrated CPS are developed based on embedded systems. This means that the system's computing capabilities are integrated into the physical components, allowing for real-time processing and control. This approach enables integrated CPS to respond quickly to changing conditions and make decisions based on real-time data.

The design of integrated CPS requires the integration of physical and computational processes within the collaborative design process of CPS. However, the interaction between computational and physical components across different disciplines can lead to interruptions and incompatibilities during the design process.

One of the major challenges in designing integrated CPS is the lack of documented specifications, interactions, and interfaces between the various disciplines and components involved. This can lead to misunderstandings and communication issues, which are critical issues that must be addressed, as shown in Fig. 5.15.

Network communications. Integrated CPS systems often involve multiple heterogeneous network technologies. These systems typically include applications that require autonomous transmission of collected data, which can lead to a significant increase in bandwidth consumption. As a result, ensuring efficient management to enable continuous transmission and optimize bandwidth usage remains a critical challenge. Most integrated CPS require reliable and continuous data transmission, as well as fault tolerance from the edge devices to the cloud. This is essential to ensure that data are transmitted securely and without interruption, even in the event of network failures or other disruptions. Researchers have identified this as a critical challenge in the design of integrated CPS, and have proposed various solutions to address it, including the use of advanced networking technologies and fault-tolerant communication protocols [95, 96].

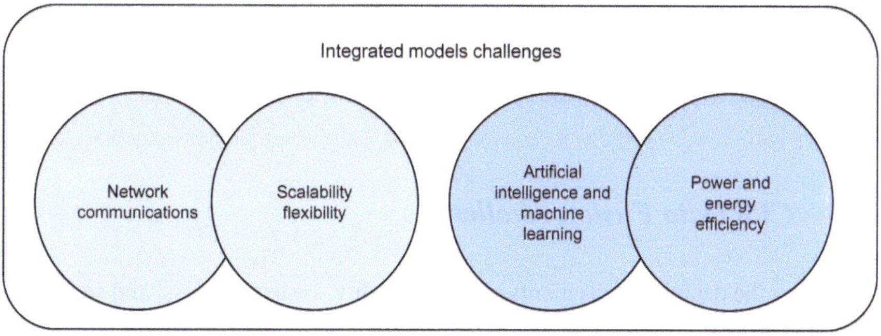

Fig. 5.15 The challenges of integrated models

Scalability and flexibility. To effectively integrate data and events in complex CPS, efficient mechanisms for managing applications and services are required. In order to ensure scalability with respect to the number of connected devices and users, flexible event management and subscription schemes are necessary [97]. However, developing such mechanisms remains a significant challenge. The challenge lies in designing systems that can handle the large volume of data generated by CPS, while also providing flexible and scalable event management capabilities. This requires the development of advanced algorithms and protocols that can efficiently manage events and subscriptions in real-time. By addressing this challenge, we can enable the development of more sophisticated CPS applications and services that can improve our lives in many ways, from smart homes to intelligent transportation systems.

Artificial intelligence and machine learning. AI and data mining techniques are essential for processing big data because intelligent algorithms, systems, and services are necessary to manage the vast amounts of data involved. These technologies enable the understanding of complex relationships between different variables, the processing of diverse requirements, and the presentation of outputs in a meaningful way. The next challenge is to develop predictive systems that can detect diseases at an early stage, particularly in healthcare applications. This will require the integration of advanced AI and data mining techniques with medical knowledge and expertise to create accurate and reliable diagnostic tools. Such systems have the potential to revolutionize healthcare by enabling early detection and treatment of diseases, leading to better health outcomes for patients [98].

Power and energy efficiency. Integrated CPS involve the transmission of data from various devices, such as IoT and cloud applications, to the cloud. However, this process can consume significant amounts of battery capacity on both the devices and the gateway, which can limit their continuous operation throughout the day. As a result, improving energy efficiency in both data processing and transmission is a critical challenge that needs to be addressed [99].

In addition to energy efficiency, there are many other critical research issues that need to be addressed in order to build resilient CPS. However, these issues have not received enough attention in academia. It is urgent to address these issues, as they are essential for the development of robust and reliable CPS. By addressing these challenges, we can create more efficient and effective CPS that can improve our lives in many ways, from smart homes to intelligent transportation systems.

5.4.2 CPS Data Fusion Challenges

To achieve the design requirements such as stability, systematicness, and security in CPS, we summarize some challenges of CPS data fusion, as shown in Fig. 5.16.

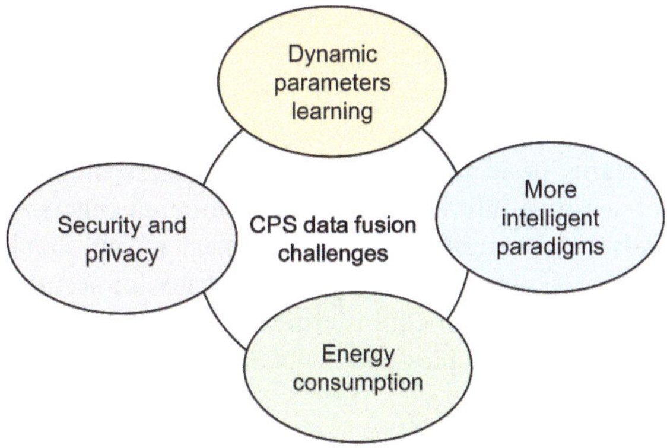

Fig. 5.16 The challenges of CPS data fusion

5.4.2.1 Security and Privacy

Data security. The collection of CPS data from a diverse range of mobile phones, smart terminals, and wearable devices has introduced new threats, such as Trojan viruses, to computing systems. As a result, ensuring the security of computing systems has become a new challenge that needs to be addressed. It is imperative to find ways to prevent computing systems from being vulnerable to these threats and to safeguard the integrity and confidentiality of CPS data. This requires a comprehensive approach that incorporates advanced security measures and protocols to mitigate the risks associated with the collection and analysis of CPS data.

Privacy. The issue of privacy is a significant concern in the context of CPS data fusion. Interspace data are often beyond the control of data owners, making access control a challenging task. This is further complicated by the fact that different spaces are often untrusted, which can lead to potential breaches of user privacy. When data are outsourced to another space, securely accessing and computing user data becomes a significant challenge. Additionally, designing a privacy-aware architecture that can effectively preserve the privacy of a large number of CPS participants is also a major challenge [100].

To address these issues, a privacy-aware fusion system is required. One such effort is the AnonySense architecture proposed in Ref. [101], which focuses on privacy-aware applications in crowd sensing. This architecture aims to ensure that user privacy is protected while still allowing for effective data access and computation. Advanced security measures and protocols that enable secure and effective data querying and searching are incorporated. Overall, developing a privacy-aware architecture is crucial for safeguarding user privacy in the context of CPS data fusion.

Transmission security. The transmission of sensitive information in CPS is becoming increasingly common, and this has led to a growing concern about transmission security. With the use of wireless sensor networks (WSN) systems, the transmission of data has become more vulnerable to cyber-attacks. While some research has been conducted on this issue, such as the works studied in Ref. [92], these studies mainly focus on traditional cyberattacks.

To ensure the security of transmission, it is necessary to consider a range of security measures, including authentication, key agreement, data confidentiality, and data integrity protection. These measures are essential for safeguarding the transmission of sensitive information and preventing unauthorized access or tampering of data. Advanced security protocols and mechanisms should be developed to address the unique challenges posed by CPS transmission security. Overall, ensuring the security of transmission in CPS is critical for protecting the integrity and confidentiality of sensitive information and maintaining the trust of CPS participants [102].

5.4.2.2 Dynamic Parameters Learning

In large-scale application scenarios, the data flows are constantly evolving and require on-the-fly re-estimation. However, the challenge lies in the fact that CPS data flows are often captured by fixed specifications, making it difficult to adapt the system to changing conditions. This is further complicated by the lack of knowledge about environmental parameters in advance. Therefore, future work should focus on learning and understanding the dynamic parameters [103].

To address this challenge, it is crucial to develop methodologies that can adaptively learn and update the dynamic parameters. This can be achieved through data-driven approaches that leverage real-time data collected from the system. By analyzing this data using machine learning algorithms, such as neural networks or reinforcement learning, it becomes possible to extract valuable insights and predict the changing parameters [104]. Moreover, researchers develop systems that are capable of learning from data, identifying patterns and trends, and making decisions based on this information. This will enable CPS to be more flexible and adaptable, providing significant benefits for a wide range of applications

Additionally, online estimation techniques can be employed to continuously update and re-estimate the parameters based on the incoming data. These techniques involve using adaptive algorithms and statistical methods to dynamically adjust the system's behavior and adapt to the changing environment parameters

5.4.2.3 More Intelligent Paradigms

CPS data spaces have undergone a significant transformation from a single space to a multi-space, which has resulted in the emergence of new security concerns. Data heterogeneity is a significant challenge, as it can lead to inconsistencies and inaccuracies in the data, making it difficult to integrate and analyze. Data from different sensors and devices may have different formats, precision, sampling rates, and timestamps, where data are generated from a wide range of sources, including sensors, actuators, and other devices [105]. To ensure that the data are consistent and accurate, data format conversion and calibration are necessary. This is particularly challenging in

CPS. To address this challenge, researchers are developing advanced multi-source data fusion algorithms that can handle data heterogeneity and inconsistencies [104]. These algorithms use techniques such as machine learning, data mining, and statistical analysis to identify patterns and relationships in the data and fuse the data from multiple sources. Machine learning techniques can be used to learn the relationships between the data from different sources and predict missing values, while data mining techniques can be used to identify patterns and anomalies in the data. Statistical analysis techniques can be used to model the data and estimate the uncertainty in the results.

5.4.2.4 Energy Consumption

Energy consumption plays an important role in CPS design, especially in the context of next-generation smart energy systems. To achieve sustainable development, it is essential to incorporate energy-aware and data-driven methodologies and models into the design of CPS. By incorporating energy-aware methodologies, designers can optimize the energy consumption of CPS components, such as sensors, actuators, and communication modules, which can be achieved through techniques like power management, energy harvesting, and energy-efficient algorithms. Furthermore, designers can gain valuable insights into energy usage patterns and identify areas for improvement based on large volumes of data generated by CPS components. The development of predictive models can optimize energy consumption based on real-time data, environmental conditions, and user behavior. Incorporating energy-aware and data-driven methodologies into CPS design not only reduces energy consumption but also enhances the overall performance and reliability of the system.

One approach to designing energy-aware CPS is to leverage computational methodologies that optimize energy consumption. This can involve techniques such as energy-efficient scheduling, resource allocation, and power management strategies. By considering energy consumption as a key design criterion, CPS can be developed to minimize energy usage while still meeting performance requirements.

5.4.3 Future Work Opportunities

Despite the feasibility of the automatic design of CPS enabled by the latest technologies, further research efforts are necessary. A system-level design methodology may offer a promising solution for enhancing the design of CPS. Nevertheless, several challenges need to be tackled. Some aspects of issues have a significant impact on future design.

Blockchain technology. Blockchain technology has emerged as an innovative and decentralized solution for secure transactions and data storage, eliminating the need for a centralized authority [106]. Initially associated with Bitcoin, blockchain has now gained widespread acceptance in various industries, including finance, healthcare, and logistics. Its public accessibility and absence of centralized control have made it popular among users. However, security and privacy concerns have been raised.

Existing blockchain protocols [100] have shown limitations in addressing these concerns. To tackle these issues, various privacy strategies have been proposed, focusing on anonymity, encryption, and perturbation. Machine learning techniques have been employed to enhance perturbation-based strategies, making them more intelligent. We believe that combining algorithms based on differential privacy with blockchain can effectively address privacy-related challenges. Differential privacy, known for its lightweight nature and mathematical foundation, enables the preservation of privacy in data storage and transactions within blockchain technology. Therefore, researchers should focus on the integration of these two techniques to achieve effective results [107].

Game theory. Game theory is a branch of mathematics that deals with situations where multiple participants with different goals compete against each other [106]. It provides a framework for analyzing and understanding strategic interactions and decision-making processes. By considering a wide range of possible scenarios, game theory can assist administrators in making informed decisions.

When applying game theory-based algorithms, it is crucial to take into account important security and privacy considerations. In the context of differential privacy, one critical issue to address is the trade-off between privacy and utility. This trade-off has become increasingly complex and can be seen as a game problem in itself. Researchers are now focusing on developing contemporary differential privacy methods that effectively manage the trade-off between utility and privacy through algorithms based on game theory [108].

We believe that combining game theory with differential privacy can provide a powerful approach to address privacy concerns in applications of CPS. By leveraging the principles of game theory, administrators can make strategic decisions that balance privacy and utility effectively. This integration of game theory and differential privacy has the potential to enhance the privacy protection of CPS applications [109].

Intelligent archetypes. With the progress of AI, CPS have been able to advance into more advanced and intelligent stages. To maximize the potential of CPS, researchers should concentrate on equipping these systems with the intelligence they require. The emergence of the cyber-physical thinking (CPST) paradigm holds great significance for the future of CPS [107]. However, this paradigm presents new features that require the development of new data fusion and learning methods. Intelligent archetypes are essential for the development of CPS that can adapt to the dynamic and complex nature of modern applications. These archetypes must be capable of learning from data, identifying patterns and trends, and making decisions based on this information. To achieve this, researchers must develop new data fusion and learning methods that can handle the diverse and heterogeneous data generated by CPST.

The CPST paradigm requires a holistic approach to data integration and analysis, which can be achieved through the development of intelligent archetypes. These archetypes must be able to integrate data from multiple sources, including physical sensors, and other digital sources. Additionally, they must be capable of learning from this data to identify patterns and trends that can inform decision-making in real-time scene [110].

Real-time computing and ubiquitous sensing. Fog/edge computing plays a pivotal role in CPS, enabling rapid real-time computing and extensive sensing capabilities. Nevertheless, the privacy and reliability challenges associated with fog/edge-based applications need to be tackled. To effectively address these concerns, the adoption of differentially private deep tensor computation becomes essential. By incorporating this advanced technique, sensitive data can be protected while preserving the integrity and accuracy of computations for enhanced privacy and trustworthy fog/edge computing. This approach not only avoids privacy risks but also ensures the reliable and secure operation of applications in diverse CPS scenarios.

To ensure privacy in fog computing, tensor computations are employed. This involves utilizing tensor computations for differentially private fog computing, cloud-fog schemes, cloud-edge schemes, and cloud-fog-edge schemes. Differential privacy is crucial for safeguarding sensitive data in fog/edge computing [111]. Tensor computations provide a powerful tool for achieving differential privacy by allowing for the efficient processing of large amounts of data while preserving privacy. By using differentially private tensor computations, fog/edge computing applications can ensure that sensitive data are protected while still providing accurate and reliable results.

Embedding techniques. There is a requirement for innovative approaches to embedding new hardware and software components into tangible objects. These cyber components should have the ability to detect and monitor the complete status of the physical objects [109]. One of the primary challenges is the physical integration of the new components into the existing objects. This involves designing and implementing techniques that ensure a seamless and efficient incorporation of the cyber components without compromising the functionality or appearance of the objects. It requires careful consideration of factors such as size, shape, and compatibility with the physical structure [112]. Another challenge lies in enabling the cyber components to accurately perceive and understand the entire status of the physical objects. This necessitates the development of advanced sensing technologies and algorithms that can capture and interpret various parameters, such as position, orientation, temperature, and other relevant data. The integration of these sensing capabilities into physical objects can be complex and requires careful calibration and synchronization. Furthermore, ensuring the interoperability and compatibility of the embedded cyber components with existing systems and networks is a significant challenge. This involves developing standardized protocols and interfaces that facilitate seamless communication and data exchange between the physical objects and the cyber components. It requires addressing issues related to data formats, communication protocols, and system integration.

5.5 Summary

Data integration and data fusion are two critical processes in CPS that enable the collection and processing of data from multiple sources to support intelligent decision-making and control. Data integration involves combining data from different sources, such as sensors, databases, and other systems, to create a unified view of the data. Data fusion involves combining data from multiple sources to create a more complete and accurate picture of the system being monitored. However, these processes present several challenges, including data heterogeneity and data quality. To address these challenges, researchers are developing advanced techniques such as stage-based, semantic meaning-based, and AI-based to ensure the accuracy and reliability of the results. These techniques can help identifying patterns and relationships in the data and improving the quality of the data by filling in missing values or correcting errors. To better illustrate the integration and fusion of CPS data, two case studies are provided cloud-integrated CPS for complex industrial applications and big data fusion framework for the integration of things. Finally, the challenges and future works opportunities are summarized. By addressing these challenges and advancing research in these areas, CPS can unlock its full potential and enable new applications and services that can transform the world around us.

Competing Interests

1. The research is supported by the National Key Research and Development Project under Grant 2023YFB3307300, the National Natural Science Foundation of China under Grants 62373014, 62422301, 62125301, and 92467205.
2. The authors have no conflicts of interest to declare that are relevant to the content of this chapter.

References

1. P. Jiang, K. Ding, and J. Leng, "Towards a cyber-physical-social-connected and service-oriented manufacturing paradigm: Social manufacturing," *Manufacturing Letters*, vol. 7, pp. 15–21, 2016.
2. X. Wang, L. T. Yang, X. Xie, J. Jin, and M. J. Deen, "A cloud-edge computing framework for cyber-physical-social services," *IEEE Communications Magazine*, vol. 55, no. 11, pp. 80–85, 2017.
3. E. P. Blasch, R. Breton, P. Valin, and E. Bosse, "User information fusion decision making analysis with the c-ooda model," in *14th International Conference on Information Fusion*, pp. 1–8, IEEE, 2011.
4. D. Lahat, T. Adali, and C. Jutten, "Multimodal data fusion: an overview of methods, challenges, and prospects," *Proceedings of the IEEE*, vol. 103, no. 9, pp. 1449–1477, 2015.
5. L. Kuang, L. T. Yang, and Y. Liao, "An integration framework on cloud for cyber-physical-social systems big data," *IEEE Transactions on Cloud Computing*, vol. 8, no. 2, pp. 363–374, 2015.

6. J. Ma, L. T. Yang, B. O. Apduhan, R. Huang, L. Barolli, and M. Takizawa, "Towards a smart world and ubiquitous intelligence: a walkthrough from smart things to smart hyperspaces and ubickids," *International Journal of Pervasive Computing and Communications*, vol. 1, no. 1, pp. 53–68, 2005.
7. K. L. Huang, S. S. Kanhere, and W. Hu, "Are you contributing trustworthy data? the case for a reputation system in participatory sensing," in *Proceedings of the 13th ACM international conference on Modeling, analysis, and simulation of wireless and mobile systems*, pp. 14–22, 2010.
8. S. Chen, J. Wang, H. Li, Z. Wang, F. Liu, and S. Li, "Top-down human-cyber-physical data fusion based on reinforcement learning," *IEEE Access*, vol. 8, pp. 134233–134245, 2020.
9. A. Jones, E. Subrahmanian, A. Hamins, and C. Grant, "Humans' critical role in smart systems: A smart firefighting example," *IEEE Internet Computing*, vol. 19, no. 3, pp. 28–31, 2015.
10. J. Zeng, L. T. Yang, and J. Ma, "A system-level modeling and design for cyber-physical-social systems," *ACM Transactions on Embedded Computing Systems (TECS)*, vol. 15, no. 2, pp. 1–26, 2016.
11. A. Darwish and A. E. Hassanien, "Cyber physical systems design, methodology, and integration: the current status and future outlook," *Journal of Ambient Intelligence and Humanized Computing*, vol. 9, pp. 1541–1556, 2018.
12. P. Hehenberger, B. Vogel-Heuser, D. Bradley, B. Eynard, T. Tomiyama, and S. Achiche, "Design, modelling, simulation and integration of cyber physical systems: Methods and applications," *Computers in Industry*, vol. 82, pp. 273–289, 2016.
13. J. Gama, I. Žliobaitė, A. Bifet, M. Pechenizkiy, and A. Bouchachia, "A survey on concept drift adaptation," *ACM computing surveys (CSUR)*, vol. 46, no. 4, pp. 1–37, 2014.
14. E. A. Lee, "Cyber physical systems: Design challenges," in *2008 11th IEEE international symposium on object and component-oriented real-time distributed computing (ISORC)*, pp. 363–369, IEEE, 2008.
15. J. Lefèvre, S. Charles, M. Bosch-Mauchand, B. Eynard, and R. Padiolleau, "Multidisciplinary modelling and simulation for mechatronic design," *J of Design Research*, vol. 12, no. 1-2, pp. 127–144, 2014.
16. M. Grimm, R. Anderl, and Y. Wang, "Conceptual approach for multi-disciplinary cyber physical systems design and engineering," 2014.
17. F. E. White, "Data fusion lexicon," *data fusion lexicon*, 1991.
18. C. Federico, "A review of data fusion techniques," *Scientific world journal*, vol. 2013, p. 704504, 2013.
19. H. F. Durrant-Whyte, "Sensor models and multisensor integration," *The International Journal of Robotics Research*, 1988.
20. J. Llinas and D. L. Hall, "Introduction to multi-sensor data fusion," *Proceedings of the IEEE*, 1998.
21. H. Boström, S. F. Andler, M. Brohede, R. Johansson, A. Karlsson, J. Van Laere, L. Niklasson, M. Nilsson, A. Persson, and T. Ziemke, "On the definition of information fusion as a field of research," 2007.
22. E. Waltz, J. Llinas, et al., *Multisensor data fusion*, vol. 685. Artech house Boston, 1990.
23. E. P. Blasch and S. Plano, "Jdl level 5 fusion model: user refinement issues and applications in group tracking," in *Signal processing, sensor fusion, and target recognition XI*, vol. 4729, pp. 270–279, SPIE, 2002.
24. E. Blasch, I. Kadar, J. Salerno, M. M. Kokar, S. Das, G. M. Powell, D. D. Corkill, and E. H. Ruspini, "Issues and challenges of knowledge representation and reasoning methods in situation assessment (level 2 fusion)," *Signal Processing, Sensor Fusion, and Target Recognition XV*, vol. 6235, pp. 355–368, 2006.
25. D. A. Lambert, "Stdf model based maritime situation assessments," in *2007 10th International Conference on Information Fusion*, pp. 1–8, IEEE, 2007.
26. P. Fantini, G. Tavola, M. Taisch, J. Barbosa, P. Leitão, Y. Liu, M. S. Sayed, and N. Lohse, "Exploring the integration of the human as a flexibility factor in cps enabled manufacturing environments: Methodology and results," in *IECON 2016-42nd Annual Conference of the IEEE Industrial Electronics Society*, pp. 5711–5716, IEEE, 2016.

27. D. Han, J. Xing, Q. Yang, J. Li, X. Zhang, and Y. Chen, "Integrating goal models and problem frames for requirements analysis of self-adaptive cps," in *2017 IEEE 41st Annual Computer Software and Applications Conference (COMPSAC)*, vol. 2, pp. 529–535, IEEE, 2017.
28. L. Sorber, "Data fusion: Tensor factorizations by complex optimization (data fusie: tensor factorisaties dmv complexe optimalisatie)," 2014.
29. S. T. Roweis and L. K. Saul, "Nonlinear dimensionality reduction by locally linear embedding," *science*, vol. 290, no. 5500, pp. 2323–2326, 2000.
30. S. Vigneshwari and M. Aramudhan, "Social information retrieval based on semantic annotation and hashing upon the multiple ontologies," *Indian Journal of Science and Technology*, vol. 8, no. 2, pp. 103–107, 2015.
31. T.-D. Cao, T.-H. Phan, and A.-D. Nguyen, "An ontology based approach to data representation and information search in smart tourist guide system," in *2011 Third International Conference on Knowledge and Systems Engineering*, pp. 171–175, IEEE, 2011.
32. M. Signoretto, L. De Lathauwer, and J. A. Suykens, "A kernel-based framework to tensorial data analysis," *Neural networks*, vol. 24, no. 8, pp. 861–874, 2011.
33. I. Kotsia and I. Patras, "Support tucker machines," in *CVPR 2011*, pp. 633–640, IEEE, 2011.
34. S. K. Sowe, E. Simmon, K. Zettsu, F. De Vaulx, and I. Bojanova, "Cyber-physical-human systems: Putting people in the loop," *IT professional*, vol. 18, no. 1, pp. 10–13, 2016.
35. H. F. Durrant-Whyte, "Sensor models and multisensor integration," *The international journal of robotics research*, vol. 7, no. 6, pp. 97–113, 1988.
36. B. V. Dasarathy, "Sensor fusion potential exploitation-innovative architectures and illustrative applications," *Proceedings of the IEEE*, vol. 85, no. 1, pp. 24–38, 1997.
37. R. C. Luo, C.-C. Yih, and K. L. Su, "Multisensor fusion and integration: approaches, applications, and future research directions," *IEEE Sensors journal*, vol. 2, no. 2, pp. 107–119, 2002.
38. F. E. White, "Data fusion lexicon," tech. rep., Joint Directors of Labs Washington DC, 1991.
39. Y. Zheng, "Methodologies for cross-domain data fusion: An overview," *IEEE transactions on big data*, vol. 1, no. 1, pp. 16–34, 2015.
40. P. Wang, L. T. Yang, J. Li, J. Chen, and S. Hu, "Data fusion in cyber-physical-social systems: State-of-the-art and perspectives," *Information Fusion*, vol. 51, pp. 42–57, 2019.
41. Y. Zheng, Y. Liu, J. Yuan, and X. Xie, "Urban computing with taxicabs," in *Proceedings of the 13th international conference on Ubiquitous computing*, pp. 89–98, 2011.
42. X. Xiao, Y. Zheng, Q. Luo, and X. Xie, "Finding similar users using category-based location history," in *Proceedings of the 18th SIGSPATIAL international conference on advances in geographic information systems*, pp. 442–445, 2010.
43. B. Pan, Y. Zheng, D. Wilkie, and C. Shahabi, "Crowd sensing of traffic anomalies based on human mobility and social media," in *Proceedings of the 21st ACM SIGSPATIAL international conference on advances in geographic information systems*, pp. 344–353, 2013.
44. N. J. Yuan, Y. Zheng, X. Xie, Y. Wang, K. Zheng, and H. Xiong, "Discovering urban functional zones using latent activity trajectories," *IEEE Transactions on Knowledge and Data Engineering*, vol. 27, no. 3, pp. 712–725, 2014.
45. R. C. Schank, "Conceptual dependency: A theory of natural language understanding," *Cognitive psychology*, vol. 3, no. 4, pp. 552–631, 1972.
46. P. Derler, E. A. Lee, and A. S. Vincentelli, "Modeling cyber–physical systems," *Proceedings of the IEEE*, vol. 100, no. 1, pp. 13–28, 2011.
47. A. M. Collins and M. R. Quillian, "Retrieval time from semantic memory," *Journal of verbal learning and verbal behavior*, vol. 8, no. 2, pp. 240–247, 1969.
48. A. Costanzo, A. Faro, D. Giordano, and C. Spampinato, "An ontological ubiquitous city information platform provided with cyber-physical-social-systems," in *2016 13th IEEE Annual Consumer Communications & Networking Conference (CCNC)*, pp. 137–144, IEEE, 2016.
49. D. Hussein, S. Park, S. N. Han, and N. Crespi, "Dynamic social structure of things: A contextual approach in cpss," *IEEE Internet Computing*, vol. 19, no. 3, pp. 12–20, 2015.
50. J. Ploennigs, A. Ahmed, B. Hensel, P. Stack, and K. Menzel, "Virtual sensors for estimation of energy consumption and thermal comfort in buildings with underfloor heating," *Advanced Engineering Informatics*, vol. 25, no. 4, pp. 688–698, 2011.

51. N. Lawrence, "Gaussian process latent variable models for visualisation of high dimensional data," *Advances in neural information processing systems*, vol. 16, 2003.
52. N. Chen, J. Zhu, and E. Xing, "Predictive subspace learning for multi-view data: a large margin approach," *Advances in neural information processing systems*, vol. 23, 2010.
53. J. He and R. Lawrence, "A graph based framework for multi-task multi-view learning.," in *ICML*, pp. 25–32, 2011.
54. T. K. Ho, "Random decision forests," in *Proceedings of 3rd international conference on document analysis and recognition*, vol. 1, pp. 278–282, IEEE, 1995.
55. A. Komninos, V. Stefanis, A. Plessas, and J. Besharat, "Capturing urban dynamics with scarce check-in data," *IEEE Pervasive Computing*, vol. 12, no. 4, pp. 20–28, 2013.
56. A. J. Jara, D. Genoud, and Y. Bocchi, "Big data for smart cities with knime a real experience in the smartsantander testbed," *Software: Practice and Experience*, vol. 45, no. 8, pp. 1145–1160, 2015.
57. B. Guo, Z. Yu, and X. Zhou, "A data-centric framework for cyber-physical-social systems," *It Professional*, vol. 17, no. 6, pp. 4–7, 2015.
58. A. Klami, G. Bouchard, and A. Tripathi, "Group-sparse embeddings in collective matrix factorization," arXiv:1312.5921, 2013.
59. A. P. Singh and G. J. Gordon, "A unified view of matrix factorization models," in *Machine Learning and Knowledge Discovery in Databases: European Conference, ECML PKDD 2008, Antwerp, Belgium, September 15-19, 2008, Proceedings, Part II 19*, pp. 358–373, Springer Berlin Heidelberg, 2008.
60. M. Žitnik and B. Zupan, "Data fusion by matrix factorization," *IEEE transactions on pattern analysis and machine intelligence*, vol. 37, no. 1, pp. 41–53, 2014.
61. Y. Zhang, "Grorec: a group-centric intelligent recommender system integrating social, mobile and big data technologies," *IEEE Transactions on Services Computing*, vol. 9, no. 5, pp. 786–795, 2016.
62. V. Zheng, B. Cao, Y. Zheng, X. Xie, and Q. Yang, "Collaborative filtering meets mobile recommendation: A user-centered approach," in *Proceedings of the AAAI Conference on Artificial Intelligence*, vol. 24, pp. 236–241, 2010.
63. Y.-R. Lin, J. Sun, P. Castro, R. Konuru, H. Sundaram, and A. Kelliher, "Metafac: community discovery via relational hypergraph factorization," in *Proceedings of the 15th ACM SIGKDD international conference on Knowledge discovery and data mining*, pp. 527–536, 2009.
64. F. Garcia, B. Mirbach, B. Ottersten, F. Grandidier, and A. Cuesta, "Pixel weighted average strategy for depth sensor data fusion," in *2010 IEEE International Conference on Image Processing*, pp. 2805–2808, IEEE, 2010.
65. H. Su and G. Zheng, "A partial least squares regression-based fusion model for predicting the trend in drowsiness," *IEEE Transactions on Systems, Man, and Cybernetics-Part A: Systems and Humans*, vol. 38, no. 5, pp. 1085–1092, 2008.
66. J. Hair and A. Alamer, "Partial least squares structural equation modeling (pls-sem) in second language and education research: Guidelines using an applied example," *Research Methods in Applied Linguistics*, vol. 1, no. 3, p. 100027, 2022.
67. G. Menduni, A. Zifarelli, A. Sampaolo, P. Patimisco, M. Giglio, N. Amoroso, H. Wu, L. Dong, R. Bellotti, and V. Spagnolo, "High-concentration methane and ethane qepas detection employing partial least squares regression to filter out energy relaxation dependence on gas matrix composition," *Photoacoustics*, vol. 26, p. 100349, 2022.
68. X. He, T. Wang, W. Liu, and T. Luo, "Measurement data fusion based on optimized weighted least-squares algorithm for multi-target tracking," *IEEE Access*, vol. 7, pp. 13901–13916, 2019.
69. I. J. Myung, "Tutorial on maximum likelihood estimation," *Journal of mathematical Psychology*, vol. 47, no. 1, pp. 90–100, 2003.
70. R. Barlow, "Extended maximum likelihood," *Nuclear Instruments and Methods in Physics Research Section A: Accelerators, Spectrometers, Detectors and Associated Equipment*, vol. 297, no. 3, pp. 496–506, 1990.

71. S. Chen, Q. Guo, H. Leung, and E. Bosse, “A maximum likelihood approach to joint image registration and fusion,” *IEEE Transactions on Image Processing*, vol. 20, no. 5, pp. 1363–1372, 2010.
72. J. B. Tenenbaum, V. d. Silva, and J. C. Langford, “A global geometric framework for nonlinear dimensionality reduction,” *science*, vol. 290, no. 5500, pp. 2319–2323, 2000.
73. J. Ngiam, A. Khosla, M. Kim, J. Nam, H. Lee, and A. Y. Ng, “Multimodal deep learning,” in *Proceedings of the 28th international conference on machine learning (ICML-11)*, pp. 689–696, 2011.
74. N. Srivastava and R. R. Salakhutdinov, “Multimodal learning with deep boltzmann machines,” *Advances in neural information processing systems*, vol. 25, 2012.
75. P. Manjunatha, A. K. Verma, and A. Srividya, “Multi-sensor data fusion in cluster based wireless sensor networks using fuzzy logic method,” in *2008 IEEE region 10 and the third international conference on industrial and information systems*, pp. 1–6, IEEE, 2008.
76. W. Su and T. C. Bougiouklis, “Data fusion algorithms in cluster-based wireless sensor networks using fuzzy logic theory,” in *Proceedings of the 11th WSEAS international conference on communications*, 2007.
77. H.-J. Zimmermann, “Applications of fuzzy set theory to mathematical programming,” *Readings in Fuzzy Sets for Intelligent Systems*, pp. 795–809, 1993.
78. J. de Andrés-Sánchez, “A systematic review of the interactions of fuzzy set theory and option pricing,” *Expert Systems with Applications*, p. 119868, 2023.
79. J. Gawlikowski, C. R. N. Tassi, M. Ali, J. Lee, M. Humt, J. Feng, A. Kruspe, R. Triebel, P. Jung, R. Roscher, et al., “A survey of uncertainty in deep neural networks,” *Artificial Intelligence Review*, pp. 1–77, 2023.
80. R. Kaur, R. Kumar, and M. Gupta, “Deep neural network for food image classification and nutrient identification: A systematic review,” *Reviews in Endocrine and Metabolic Disorders*, pp. 1–21, 2023.
81. Y. LeCun and M. Ranzato, “Deep learning tutorial,” in *Tutorials in international conference on machine learning (ICML’13)*, pp. 1–29, Citeseer, 2013.
82. Y. Bengio, A. Courville, and P. Vincent, “Representation learning: A review and new perspectives,” *IEEE transactions on pattern analysis and machine intelligence*, vol. 35, no. 8, pp. 1798–1828, 2013.
83. F. S. Gohari and M. J. Tarokh, “New recommender framework: combining semantic similarity fusion and bicluster collaborative filtering,” *Computational Intelligence*, vol. 32, no. 4, pp. 561–586, 2016.
84. H. Liang, L. Yang, H. Cheng, W. Tu, and M. Xu, “Human-in-the-loop reinforcement learning,” in *2017 Chinese Automation Congress (CAC)*, pp. 4511–4518, IEEE, 2017.
85. B. Liu, L. Yao, Z. Ding, J. Xu, and J. Wu, “Combining ontology and reinforcement learning for zero-shot classification,” *Knowledge-Based Systems*, vol. 144, pp. 42–50, 2018.
86. S. Qu, J. Wang, S. Govil, and J. O. Leckie, “Optimized adaptive scheduling of a manufacturing process system with multi-skill workforce and multiple machine types: An ontology-based, multi-agent reinforcement learning approach,” *Procedia Cirp*, vol. 57, pp. 55–60, 2016.
87. C. H. Liu, Q. Lin, and S. Wen, “Blockchain-enabled data collection and sharing for industrial iot with deep reinforcement learning,” *IEEE Transactions on Industrial Informatics*, vol. 15, no. 6, pp. 3516–3526, 2018.
88. B. Xin, K. Tang, L. Wang, and C. Chen, “Knowledge transfer between multi-granularity models for reinforcement learning,” in *2018 IEEE International Conference on Systems, Man, and Cybernetics (SMC)*, pp. 2881–2886, IEEE, 2018.
89. S. Karnouskos, A. W. Colombo, T. Bangemann, K. Manninen, R. Camp, M. Tilly, P. Stluka, F. Jammes, J. Delsing, and J. Eliasson, “A soa-based architecture for empowering future collaborative cloud-based industrial automation,” in *IECON 2012-38th Annual Conference on IEEE Industrial Electronics Society*, pp. 5766–5772, IEEE, 2012.
90. S. Zhang, L. T. Yang, J. Feng, W. Wei, Z. Cui, X. Xie, and P. Yan, “A tensor-network-based big data fusion framework for cyber–physical–social systems (cpss),” *Information Fusion*, vol. 76, pp. 337–354, 2021.

91. H. Liu, L. T. Yang, J. Ding, Y. Guo, and S. S. Yau, "Tensor-train-based high-order dominant eigen decomposition for multimodal prediction services," *IEEE Transactions on Engineering Management*, vol. 68, no. 1, pp. 197–211, 2019.
92. J. Feng, L. T. Yang, G. Dai, W. Wang, and D. Zou, "A secure high-order lanczos-based orthogonal tensor svd for big data reduction in cloud environment," *IEEE Transactions on Big Data*, vol. 5, no. 3, pp. 355–367, 2018.
93. R. Yu, B. Iung, and H. Panetto, "A multi-agents based e-maintenance system with case-based reasoning decision support," *Engineering applications of artificial intelligence*, vol. 16, no. 4, pp. 321–333, 2003.
94. X. Li, C. Qiao, A. Wagh, R. Sudhaakar, S. Addepalli, C. Wu, and A. Sadek, "A holistic approach to service delivery in driver-in-the-loop vehicular cps," *IEEE Journal on Selected Areas in Communications*, vol. 31, no. 9, pp. 513–522, 2013.
95. M. Krotofil, J. Larsen, and D. Gollmann, "The process matters: Ensuring data veracity in cyber-physical systems," in *Proceedings of the 10th ACM Symposium on Information, Computer and Communications Security*, pp. 133–144, 2015.
96. J. Biswas, J. Maniyeri, K. Gopalakrishnan, L. Shue, J. E. Phua, H. N. Palit, Y. S. Foo, L. S. Lau, and X. Li, "Processing of wearable sensor data on the cloud-a step towards scaling of continuous monitoring of health and well-being," in *2010 annual international conference of the IEEE engineering in medicine and biology*, pp. 3860–3863, IEEE, 2010.
97. E. K. Jones, N. Denis, and D. Hunter, "Hypothesis management for information fusion," *IEEE aerospace and electronic systems magazine*, vol. 18, no. 6, pp. 3–8, 2003.
98. G. Hackmann, W. Guo, G. Yan, C. Lu, and S. Dyke, "Cyber-physical codesign of distributed structural health monitoring with wireless sensor networks," in *Proceedings of the 1st ACM/IEEE international conference on cyber-physical systems*, pp. 119–128, 2010.
99. H. B. McMahan, D. Ramage, K. Talwar, and L. Zhang, "Learning differentially private recurrent language models," arXiv:1710.06963, 2017.
100. R. Henry, A. Herzberg, and A. Kate, "Blockchain access privacy: Challenges and directions," *IEEE Security & Privacy*, vol. 16, no. 4, pp. 38–45, 2018.
101. C. Cornelius, A. Kapadia, D. Kotz, D. Peebles, M. Shin, and N. Triandopoulos, "Anonysense: privacy-aware people-centric sensing," in *Proceedings of the 6th international conference on Mobile systems, applications, and services*, pp. 211–224, 2008.
102. S. Eberle, "Adaptive internet integration of field bus systems," *IEEE Transactions on Industrial Informatics*, vol. 3, no. 1, pp. 12–20, 2007.
103. M. Tubaishat and S. Madria, "Sensor networks: an overview," *IEEE potentials*, vol. 22, no. 2, pp. 20–23, 2003.
104. F. Heidtmann and D. Soffker, "Virtual sensors for diagnosis and prognosis purposes in the context of elastic mechanical structures," *IEEE Sensors Journal*, vol. 9, no. 11, pp. 1577–1588, 2009.
105. Y. Yan, T. Kuphal, and J. Bode, "Application of multiagent systems in project management," *International Journal of production economics*, vol. 68, no. 2, pp. 185–197, 2000.
106. L. Xu, C. Jiang, Y. Qian, J. Li, Y. Zhao, and Y. Ren, "Privacy-accuracy trade-off in differentially-private distributed classification: A game theoretical approach," *IEEE Transactions on Big Data*, vol. 7, no. 4, pp. 770–783, 2017.
107. H. Ning, H. Liu, J. Ma, L. T. Yang, and R. Huang, "Cybermatics: Cyber–physical–social–thinking hyperspace based science and technology," *Future generation computer systems*, vol. 56, pp. 504–522, 2016.
108. S. Chen, H. Xu, D. Liu, B. Hu, and H. Wang, "A vision of IoT: Applications, challenges, and opportunities with china perspective," *IEEE Internet of Things journal*, vol. 1, no. 4, pp. 349–359, 2014.
109. M. C. Shults, R. K. Rhodes, S. J. Updike, B. J. Gilligan, and W. N. Reining, "A telemetry-instrumentation system for monitoring multiple subcutaneously implanted glucose sensors," *IEEE Transactions on Biomedical Engineering*, vol. 41, no. 10, pp. 937–942, 1994.
110. W. Shen, Q. Hao, H. J. Yoon, and D. H. Norrie, "Applications of agent-based systems in intelligent manufacturing: An updated review," *Advanced engineering INFORMATICS*, vol. 20, no. 4, pp. 415–431, 2006.

111. G. Fortino, D. Parisi, V. Pirrone, and G. Di Fatta, "Bodycloud: A saas approach for community body sensor networks," *Future Generation Computer Systems*, vol. 35, pp. 62–79, 2014.
112. X. Cao, L. Liu, W. Shen, A. Laha, J. Tang, and Y. Cheng, "Real-time misbehavior detection and mitigation in cyber-physical systems over wlans," *IEEE Transactions on Industrial Informatics*, vol. 13, no. 1, pp. 186–197, 2015.

Chapter 6
Data-Driven Modeling and Simulation in Cyber Physical Systems

Abstract Data-driven modeling and simulation play a pivotal and transformative role in the realm of cyber physical systems (CPS), revolutionizing the way we analyze and understand system behavior within a virtual environment. By harnessing the power of machine learning algorithms, statistical techniques, and data mining methods, data-driven approaches bring forth invaluable insights, unveiling hidden patterns and relationships within vast datasets. This empowers CPS researchers and engineers with an unprecedented ability to comprehensively grasp the intricacies of complex system dynamics. With this newfound understanding, they can optimize system performance, predict potential anomalies, and significantly enhance overall system efficiency and reliability. The embrace of data-driven modeling and simulation signifies more than a mere technological advancement; rather, it represents a transformative shift in our perception and approach to CPS. Its profound impact resonates across various industries, including manufacturing, healthcare, transportation, and energy, fueling innovation and propelling humanity towards a future characterized by heightened intelligence, interconnection, and resilience.

The purpose of this chapter is to emphasize the critical importance of modeling and simulation in the domain of cyber physical systems (CPS). The chapter is presented to provide readers with a comprehensive understanding of the significance and benefits of complex system modeling and simulation in CPS.

This chapter sets the foundation by discussing the significance of modeling and simulation in CPS. The chapter aims to help readers recognize the essence of complex system modeling, which enables a deeper understanding of the intricate interplay between physical and computational elements. Additionally, it highlights the importance of complex system simulation, which allows for comprehensive analysis and evaluation of CPS performance across various scenarios. The subsequent chapters delve into data-driven modeling techniques, introducing readers to various data-driven models. The focus is on methods for model selection and validation to ensure the reliability and accuracy of these models. Real-world examples demonstrate the practical relevance of data-driven modeling, showcasing how it enhances CPS applications with improved prediction and decision-making capabilities. Furthermore, this chapter explores the realm of data-driven simulation, which leverages data insights to effectively simulate complex system behavior. By understanding different types

F. Li et al., *Data-Driven Cyber Physical Systems*,
https://doi.org/10.1007/978-981-96-8709-1_6

of data-driven simulation methods and techniques for model validation and uncertainty quantification, readers gain valuable insights into assessing complex system performance under different conditions. Finally, the inclusion of case studies offers real-life scenarios that showcase the practical application of data-driven models in simulating and testing CPS and illustrating the potential of data-driven modeling and simulation to revolutionize CPS domains, such as manufacturing, healthcare, transportation, and energy. By the end of this chapter, readers will have a comprehensive grasp of the importance and benefits of modeling and simulation in CPS. It equips researchers, engineers, and practitioners with the necessary knowledge and tools to tackle the unique challenges posed effectively. Embracing data-driven modeling and simulation techniques empowers CPS to optimize performance and resilience in the ever-evolving technological landscape.

6.1 Importance of Modeling and Simulation in Cyber Physical Systems

With the development of data acquisition systems, information technology, and network technology, the manufacturing industry has entered the era of digitalization [1, 2]. In the context of digitalization, the global manufacturing industry faces new challenges. In this background, advanced manufacturing strategies such as Industry 4.0 and Industrial Internet have been proposed successively by Germany [3], the United States, and China, with the common goal of achieving intelligent manufacturing. CPS, Internet of Things, cloud computing, big data, and other technologies occupy a central position in intelligent manufacturing [4–6]. Broadly understood, CPS refers to a controllable, trustworthy, and scalable networked physical device system that integrates computing, communication, and control capabilities based on environmental perception. CPS achieves deep integration and real-time interaction through feedback loops between computational and physical processes, thereby adding or extending new functions to monitor or control a physical entity in a secure, reliable, efficient, and real-time manner [7].

Undoubtedly, CPS are of great significance for realizing intelligent manufacturing and conducting interdisciplinary research. However, the current research on CPS still encounters significant challenges. The fundamental reasons that pose challenges for the field of computer science in CPS are the deep integration with the physical world and the scale (in terms of time and space) and complexity of the systems [8]. The first challenge lies in the lack of collaborative analysis and design [9]. Currently, the evaluation, modeling, and simulation of computer hardware and software, as well as computation and physical systems, are conducted in a disconnected manner, without sufficient co-design and co-analysis [10]. Furthermore, the interaction among various physical and computational components and systems in CPS presents a key challenge in system design, particularly in ensuring robustness, reliability, efficiency, and real-time performance. As the complexity of CPS increases, the required development time grows exponentially. Additionally, in the field of computer science, all higher-level abstractions currently conceal timing, making program execution unpredictable, especially within complex CPS environments. Due to the direct interaction between computation and the physical world in CPS, the abstraction of time cannot be separated [11, 12]. As the complexity of CPS systems increases, the behavior of real-time operating systems (RTOS) becomes imprecise and difficult to control [13]. This lack of precision in time behavior poses significant challenges in achieving reliable and predictable real-time performance.

Fortunately, with the rapid development of computer technology, new modeling and simulation techniques continue to emerge to address the complex problems and challenges of CPS. DDCPS is an emerging modeling approach in which actual data from information-physical systems are used to infer and construct system models. This modeling approach is based on the analysis and mining of data to reveal the behavioral patterns, correlations, and underlying patterns of the system. Data-driven modeling of information-physical systems enables better understanding and analysis of system behavior, improves system design and operation, and provides support and guidance for decision making [14]. Currently, research on modeling and simulation for data-driven cyber physical systems (DDCPS) can be broadly categorized into three main areas: modeling and simulation methods, modeling and simulation tools, and modeling and simulation environments. Through modeling and simulation, accurate models are established and the system is tested and evaluated in a simulated environment. This enables the discovery of potential issues and bottlenecks, optimization of system structure and parameters, and improvement of system efficiency and performance. Additionally, different work scenarios and application requirements can be simulated to evaluate the system's responsiveness, stability, and reliability, meeting the reliability requirements of CPS [15]. Modeling and simulation also provide support for decision-making. By establishing accurate models and testing different decision-making scenarios in a simulation environment, their impact on system performance, resource utilization, and cost-effectiveness can be evaluated [2]. This helps in making informed decisions and reducing uncertainty and risks during actual deployment. In conclusion,modeling and simulation in DDCPS can build an intelligent information system that integrates perception, interaction, and decision-making as shown in Fig. 6.1. They guide system design and optimization, predict and evaluate system performance, support decision-making, provide guidance for troubleshooting and optimization, and save costs and time. Through modeling and simulation, efficient and reliable CPS can be effectively developed and deployed [16].

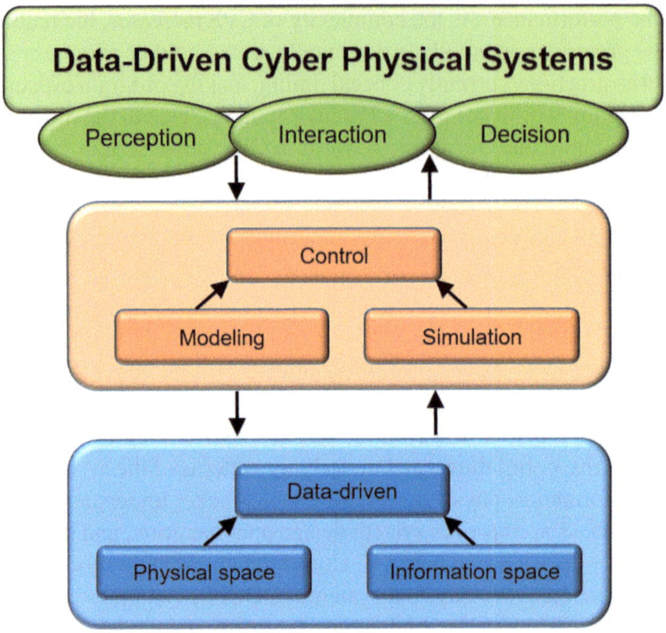

Fig. 6.1 Perception, interaction, and decision making integrated into CPS. The CPS combines seamlessly the ability to sense and interpret its surroundings, interacting with the environment and assisting people in making more informed choices, and data-driven to accentuate its role as an intelligent and adaptive system

6.1.1 Importance of Complex System Modeling for CPS

Within the realm of CPS, complex system modeling pertains to the process of modeling and describing the intricate behaviors and interactions among various components inherent to the networked physical systems. CPS establish a profound integration between the digital network and the physical world, enabling a seamless and tightly coupled operation. The primary drivers for developing and evolving such systems include reducing costs and development time while enhancing product quality. The essence of CPS lies in integrating computation and communication with physical processes to add new functionalities to physical systems. Achieved through the close integration of computation, communication, and control (3C), CPS provide real-time sensing, dynamic control, and information services to complex systems. Information systems serve as the core, enabling seamless integration and interaction among components [17]. Complex system modeling plays a pivotal role in gaining a deep understanding of CPS behavior by unraveling intricate interdependencies and interactions among components. This understanding is vital for comprehending the holistic operation, involving mutual influences between computational processes and physical phenomena. In the context of big data, data-driven enables CPS to

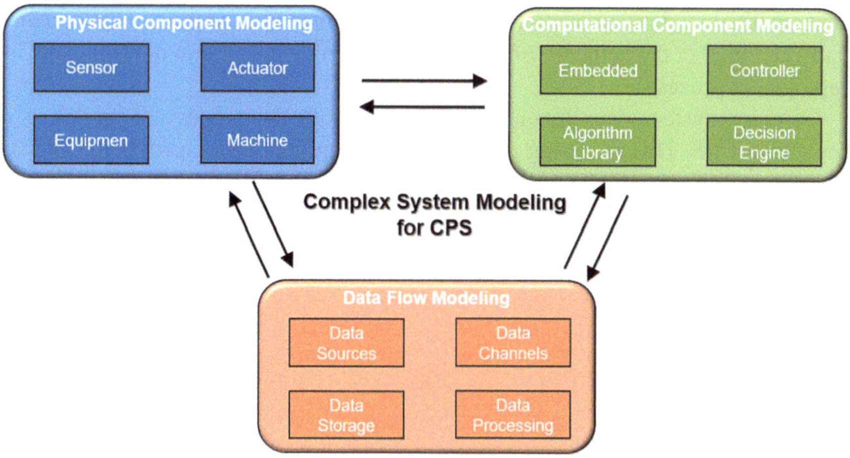

Fig. 6.2 Schematic diagram of complex system modeling for CPS. Physical component modeling, computational component modeling, and data flow modeling assist CPS by enhancing understanding, analysis, and control of complex system behavior

make real-time decisions, fault prediction and resource optimization, enhance intelligent self-adaptive capabilities, improve system efficiency and reliability, promote industrial innovation and energy conservation and emission reduction, and achieve smarter, more efficient and sustainable operation.

The modeling of CPS is shown in Fig. 6.2. It mainly includes the following aspects.

Physical component modeling: information physical systems contain physical components, such as sensors, actuators, machines, devices, etc. The modeling process requires a description of these physical components, including their characteristics, behavior, and modes of interaction. The above patterns can be done by representing and modeling the physical components using physical equations, kinematic models, circuit models, etc.

Computational component modeling: Information physical systems also contain computational components, such as embedded systems, controllers, algorithms, etc. The modeling process needs to describe the functions, input-output relationships, and execution logic of these computational components. This can be done by representing and modeling the computational components using state machines, logic models, algorithm descriptions, etc.

Data flow modeling is the key element in CPS, and the modeling process needs to consider the data generation, transmission, processing, and storage processes. This includes modeling aspects such as the format, transmission method, latency, and security of the data flow.

Modeling these aspects provides a better understanding of the behavior and performance of the information-physical system and provides the basis for system design, analysis, optimization, and control. Modeling can be achieved through the use of mathematical models, simulation tools, and experimental validation to better understand and predict the behavior of systems and to guide the system development and deployment process.

Complex system data-driven modeling refers to the process of modeling or describing CPS. This modeling process aims to capture the complex behaviors and interactions among various components within CPS. The primary motivations behind the advancement and progression of CPS include cost and time reduction in development, coupled with the improvement of the designed products. The rapid and substantial advancement of CPS has a profound impact on various facets of human existence, facilitating an expanded array of services and applications such as e-Health, intelligent residences, and e-Commerce, among others [1].

Firstly, complex system modeling allows engineers and researchers to capture the intricate interactions and dependencies within CPS. These systems involve numerous interconnected components, sensors, actuators, and communication networks. Modeling enables the representation of these components, their behaviors, and how they interact with eEach other. By understanding the complex dynamics of the system, it becomes possible to analyze the overall system performance and identify potential bottlenecks or areas for improvement. In the context of CPS, complex system modeling plays a crucial role in capturing the intricate behaviors and interactions among the components within a networked physical system [18]. The advent and proliferation of CPS have been driven by several key factors, including the reduction of development costs and time, as well as the enhancement of product quality. CPS leverages computation and communication to seamlessly integrate with physical processes, enabling the addition of new functionalities. One of the fundamental challenges in CPS lies in the irreversible nature of the physical operations carried out by most actuators. To address this challenge, decision information provided by information systems must not only exhibit reliability, security, and effectiveness but also possess real-time capabilities. CPS should be able to dynamically allocate resources to multiple real-time tasks, even under diverse and unpredictable circumstances, in order to meet the stringent timing requirements of each task.

Through the use of complex models, CPS can be simulated and analyzed under different conditions, enabling us to predict its responses and performance in diverse scenarios. Furthermore, complex system modeling enables the evaluation and optimization of CPS design and control strategies. By constructing accurate models and analyzing simulation results, we can identify and address potential bottlenecks and limiting factors within the system. This optimization process encompasses resource allocation, scheduling, and overall system efficiency enhancement, ultimately leading to improved performance [19]. Additionally, complex system modeling facilitates the analysis and assessment of potential risks and vulnerabilities within CPS. By simulating system behavior in various scenarios, we can identify possible security threats, performance bottlenecks, and failure points. These insights enable us to implement appropriate measures to mitigate risks, enhance system reliability, and ensure a secure operational environment [20]. In summary, complex system modeling plays a pivotal role in CPS, offering benefits such as improved accuracy and enhanced system performance. It enables a comprehensive understanding of system behavior, prediction of responses, risk identification, and optimization of system design and control strategies. By harnessing the power of complex models, CPS development and application can be propelled forward, ushering in the era of intelligent manufacturing. In manufacturing, complex system modeling is essential for optimizing production processes. By modeling the manufacturing line, including machines, robots, sensors, and material flow, engineers can identify potential areas of inefficiency or potential failures. Through simulations, they can evaluate different scenarios and make informed decisions regarding process redesign, resource allocation, and scheduling, ultimately enhancing productivity and reducing costs.

6.1.2 Importance of Complex System Simulation for CPS

Simulation is a vital tool in the realm of CPS, encompassing various essential aspects that collectively contribute to a comprehensive understanding of system behavior and performance. In this context, the simulation of CPS entails diverse facets, including physical simulation, environmental simulation, data flow simulation, and performance evaluation simulation [21]. Each aspect plays a crucial role in validating and optimizing CPS operations. By employing advanced numerical methods, mathematical models, and simulation platforms, researchers can delve into the behavior and interactions of physical components, environmental factors, data flow processes, and system performance metrics. This multifaceted simulation approach empowers CPS researchers to fine-tune system responses, ensure adaptability under diverse conditions, and enhance overall efficiency and reliability in this dynamically evolving domain. The simulation of CPS includes the following main aspects.

Physical simulation: Physical simulation is an important part of information physical system simulation. It involves the simulation of the behavior and interactions of physical components to obtain the physical properties and response of the system. By building physical models, the motion, mechanical properties, and energy transfer of the physical components can be simulated. Physical simulations can be performed using numerical methods (e.g., finite element method, multi-body dynamics, etc.) or mathematical models based on physical laws.

Environmental simulation: The environment has a significant impact on the behavior and performance of information physical systems [22]. Environmental simulation involves the simulation of external environmental factors to verify the performance and adaptability of the system under different environmental conditions. Environmental simulation can be performed using simulation platforms, sensor simulators, etc.

Data flow simulation: The data flow in an information-physical system plays a key role in the behavior and performance of the system. Data flow simulation involves the simulation of data generation, transmission, processing, and storage processes to verify the accuracy, timing, and reliability of the data flow. Data flow simulation can be performed using network simulation tools, timing analysis tools, data flow modeling tools, etc.

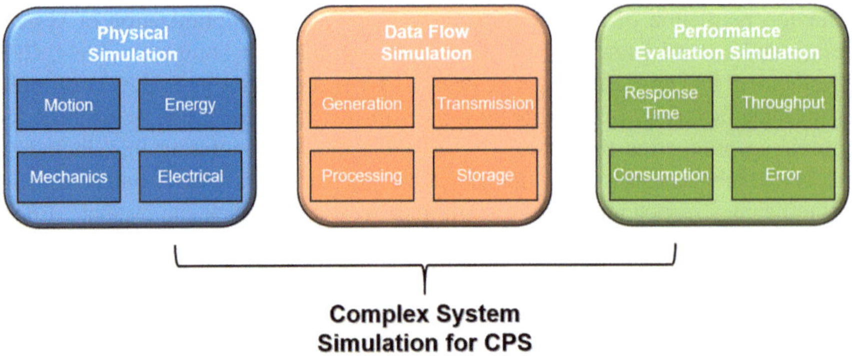

Fig. 6.3 Schematic diagram of complex system simulation for CPS. Physical component simulation, data flow simulation, and performance evaluation simulation aid CPS by enabling accurate modeling, analysis, and optimization of system behavior and efficiency

Performance evaluation simulation: Performance evaluation is one of the important goals of information physical system simulation. Performance evaluation simulation involves the simulation and evaluation of system performance metrics such as response time, throughput, and energy consumption. Through performance evaluation simulation, the system design can be optimized and the parameter configuration can be adjusted to meet the performance requirements. The above four simulation components together make up the simulation involved in CPS, as shown in Fig. 6.3.

Simulation tool software is particularly important in the study of CPS simulation techniques. Ptolemy II, developed by UC Berkeley, is an open-source modeling and simulation tool that serves as a pioneer in constructing heterogeneous systems [23]. It is capable of designing, modeling, and simulating real-time embedded systems. The execution semantics of Ptolemy II rely on the indicators within the Actor Model called Directors. We consider Ptolemy II to be an excellent tool that has made significant contributions to heterogeneous modeling [24]. However, the execution of models is constrained by the computational models (Models of Computation, MoC) provided by the tool itself. When the model semantics cannot be adequately represented by the available computational models, system designers have to modify the core components of the tool to accommodate their requirements. In the context of modeling complex CPS systems, the supported computational models of Ptolemy II are still insufficient.

Furthermore, as an important component of MATLAB, Simulink is commonly used for modeling and simulating dynamic systems. It supports event-based modeling as well as physical modeling and provides visual simulation and testing capabilities for systems with discrete and continuous characteristics. Simulink is widely applied in various domains such as control systems, signal processing, wireless communications, computer vision, and more [25]. Although existing simulation tools and techniques can to some extent accomplish the modeling and simulation tasks in CPS, we believe that when constructing complex and heterogeneous CPS, it is important not to limit ourselves to a single tool, not to confine our modeling to the existing elements of the tool, and not to adhere to fixed execution patterns dictated by the tool. In order to enhance the efficiency of CPS system development and facilitate communication between domain experts and system designers/developers, the development of CPS modeling languages suitable for experts and design/development personnel has attracted the attention of researchers. For example, Badreddin [26] proposed a textual language called fSysML specifically designed for CPS modeling. It can be used to model modules, requirements, use cases, or perform compositional modeling. The language was exemplified with an adaptive system in the healthcare domain, demonstrating its applicability and effectiveness.

Traditional simulation methods in the CPS domain mainly focus on simulating physical systems or communication networks and often fail to meet the demands of complex CPS under realistic conditions. Therefore, there is a need for research on collaborative simulation techniques specifically tailored for CPS. Currently, CPS simulation techniques primarily employ two approaches: functional extension and co-simulation. For instance, TrueTime developed by Lund University extends the functionality of MATLAB by incorporating network protocol models and event-triggered functions, enabling the simulation of distributed real-time control systems [27]. However, excessively simplified models can lead to a lack of accurate representation of dynamic processes within the system. Co-simulation, on the other hand, involves separately simulating the physical and communication systems using specialized simulation tools, providing higher accuracy. However, achieving synchronization and data exchange between these systems is a key challenge. Non-real-time co-simulation does not require high-precision synchronization and achieves data exchange by associating the physical and information subsystems within the same time domain through a synchronization module. Real-time hybrid simulation, on the other hand, establishes parallel simulation tools to synchronize the timelines. The emergence of large-scale parallel high-speed simulators has effectively addressed the synchronization and coordination challenges in parallel simulation, significantly advancing the development of real-time hybrid simulation techniques in CPS [28]. These advancements in CPS simulation techniques open up new possibilities for accurately modeling and simulating complex CPS systems, facilitating better understanding, analysis, and optimization of their behavior and performance.

Simulation can provide numerical reflections of system operations, which is crucial for information transmission in complex CPS [29]. Through simulation, the performance of CPS can be evaluated, including metrics such as response time, throughput, and energy consumption. By adjusting system parameters and algorithms, CPS performance can be optimized to enhance efficiency and reliability. Simulation of complex systems helps identify potential issues and bottlenecks, as well as diagnose faults at a system level. Additionally, simulation can be employed to optimize system structure and parameters, improving system reliability and stability. Researchers like Jha, Amitkumar V, through analysis and validation of simulation results, can identify bottlenecks and fault causes, and propose corresponding optimization strategies [30]. For example, they can adjust the transmission capacity of a power network, optimize resource allocation strategies, and enhance communication network performance to improve reliability and stability in smart grid systems. Simulation of complex systems can simulate different working scenarios and decision-making strategies, evaluating their impact on CPS performance and resource utilization [31]. This aids in making informed decisions and reducing uncertainty and risks during actual deployment. Through simulation, problems can be identified before actual development and deployment, reducing costs associated with trial and error and repetitive design. Furthermore, simulation can expedite the system development cycle and improve deployment efficiency.

6.1.3 Benefits of Modeling and Simulation in CPS

Since their conceptualization by Helen Gill of the National Science Foundation in 2006, CPS have garnered growing attention from academic and industrial circles. CPS have found wide-ranging applications in intelligent domains that are closely intertwined with people's lives, including smart grids, smart buildings, intelligent transportation, and healthcare. These applications have not only greatly improved people's quality of life but also exerted profound influences on societal development. Specifically, CPS can effectively map the design concepts, prototyping, processes, and testing involved in product development into virtual space for simulation, iteration, optimization, and formal verification. This enables friendly information sharing and reuse among different stages and developers throughout the entire product development cycle. Furthermore, the deep integration of CPS with industrial manufacturing breaks down information silos between components during production [32]. By establishing a CPS from the underlying hardware devices to the upper-level flexible management platform, real-time optimization control and flexible organizational configuration of processes can be achieved. CPS provide intelligent services, rational management, and scheduling of various production resources, thus enabling the transition from "manufacturing" to "smart manufacturing". The main drivers of CPS in industry is shown in Fig. 6.4.

Compared to traditional embedded systems designed for specific applications, CPS possess an important characteristic: they are interdisciplinary and span multiple fields [33]. In addition to knowledge in computer science, CPS involve expertise from various domains such as environmental science, mechanical engineering, electronics, biology, chemistry, aerospace, medicine, and transportation. This interdisciplinary nature of CPS design and implementation often leads to complexity, posing a challenging problem of effectively integrating knowledge from different domains. When seeking a comprehensive understanding of complex phenomena, modeling has proven to be the optimal choice and a fundamental mechanism for dealing with real-world complexity. It assists scientists and engineers in describing and predicting systems. For example, physicists employ the "Big Bang Theory" model to explain the expansion of the universe, while economists utilize economic models to forecast market trends. Unlike natural models, engineering models serve as blueprints for undeveloped systems, representing an exploration of system solutions during

Fig. 6.4 The main drivers of CPS in industry. The control, computation, and interaction shape the dynamics of cyber-physical systems, enabling intelligent, data-driven actions and interconnected experiences within the information-physical realm

the development process. Modeling reduces system complexity through abstraction, enabling a condensed representation of knowledge. In essence, models are simulations of real systems, providing us with a means to approximate and approach the idealized representation of a system.

The various characteristics of CPS make modeling and simulation of their dynamic behavior particularly important. CPS modeling falls under engineering modeling, where system designers not only need to consider models from various domains but also clearly depict the interactions between these models and the dynamic behavior they exhibit. When modeling CPS, three key characteristics of CPS dynamic behavior deserve special attention. Firstly, CPS exhibit hybridity, as they rely on discrete behavior from computational systems and continuous behavior from the physical environment such as continuous time variations, temperature changes, and non-abrupt mechanical changes, among others [34]. The heterogeneous nature of CPS results in its behavior being hybrid in nature. Secondly, CPS behavior often incorporates randomness due to operating in open environments affected by uncertain factors such as network delays and weather fluctuations as well as user behavior. These factors contribute to the stochastic nature of CPS behavior. Thirdly, CPS are frequently applied in safety-critical systems, such as smart healthcare systems, train control systems, and intelligent driving systems. This necessitates ensuring the safety and trustworthiness of CPS behavior. By utilizing relevant simulation tools such as Simulink and Ptolemy II, models can be executed, and each execution of a model is referred to as a simulation [35]. Through simulation, system designers can better observe system variables, understand system behavior, rectify errors in system models, and improve system models to ultimately achieve an optimal system design.

6.2 Data-Driven Modeling Techniques

Data-driven modeling techniques are a method of inferring relationships between system variables using data. This modeling approach does not require a deep understanding of domain-specific knowledge beforehand but relies on statistical analysis and machine learning algorithms applied to large datasets. Data-driven modeling techniques have applications in various fields such as engineering, science, finance, and more, helping us understanding and predicting the behavior of complex systems [36, 37]. As shown in Fig. 6.5, the fields of data-driven CPS and artificial intelligence are intersecting fields, where artificial intelligence empowers machine perception and provides data-driven modeling processes through machine learning or more advanced deep learning.

The main advantage of data-driven modeling is its ability to uncover hidden patterns and correlations that exist within the data. By analyzing the data, we can discover non-linear relationships between variables, uncover latent causal relationships, and identify other complex structures. These models can be used to predict future trends, evaluate the impact of different factors on system performance, and support decision-making and problem-solving processes.

The development of data-driven modeling techniques has been facilitated by the era of big data, where we have the capability to collect and store large-scale, high-dimensional datasets, as well as leverage powerful computational resources and machine learning algorithms for data analysis [38]. Common data-driven modeling methods include regression analysis, neural networks, decision trees, support vector machines and more. These methods can be chosen and adjusted based on the specific problem and characteristics of the data to obtain accurate models and predictive results. In summary, data-driven modeling techniques are powerful tools that aid in understanding the behavior of complex systems, predicting future trends,

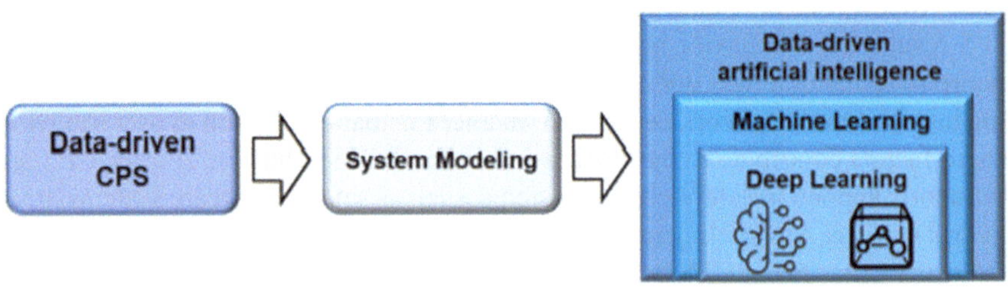

Fig. 6.5 Artificial intelligence empowers modeling patterns for CPS

and supporting decision-making processes. With the continuous advancements in data science and machine learning, data-driven modeling techniques will play an increasingly important role across various fields.

In the field of CPS modeling, there are various techniques available depending on the type of modeling. These techniques include component-oriented, multi-agent-based, actor-oriented, event-based, as well as structural or behavioral based approaches. The choice of technique depends on factors such as domain-specificity or multi-domain applicability and whether the model is timed or un-timed or a hybrid combination. Furthermore, these modeling techniques can be applied for different purposes specification, analysis, verification, and validation. Modeling involves the application of systematic techniques to design and develop CPS. The majority of current modeling methods are derived from in-depth analysis conducted by experts the field, known as first-principle based approaches. Numerous tools and resources are available for undertaking first-principle based modeling of CPS. There exist several modeling languages for CPS, including Architecture Analysis and Design Language (AADL), Unified Modeling Language (UML), Planning Domain Definition Language (PDDL), CyPhyML, ESMol, etc. Moreover, there are various simulation tools such as Modelica, Simulink, LabView, Ptolemy, etc. [39].

The approach based on first principles offers us the CPS model, which possesses formal and provable deterministic characteristics. This model facilitates simulation and analysis for identifying design flaws. However, it does have limitations arising from its inability to effectively represent non-deterministic or stochastic properties of CPS. These properties emerge during the physical realization of the model due to factors such as environmental variations, platform noise, component failures, network delays, and uncontrollable scheduling.

These constraints can be mitigated through the implementation of data-driven modeling techniques. Data-driven modeling involves utilizing specific system data to estimate or infer the correlation between variables within the system, such as input and output variables, without necessitating domain expertise understanding of the system's behavior. While the utilization of data-driven modeling necessitates a substantial amount of data, the integration of sensors in CPS has led to an abundance of operational data that can be leveraged for developing models for future generations of CPS. The primary benefit of employing a data-driven approach in modeling is its ability to leverage the side-channels, as explained in the following section, for inferring diverse system variables across different domains. Furthermore, the methodology of data-driven modeling takes advantage of advancements in artificial intelligence, machine learning, and data mining to construct models that accurately estimate relationships between system variables paragraph delves into various methodologies for data-driven modeling aimed at capturing the interaction between variables in both cyber and physical domains.

6.2.1 Introduction to Data-Driven Modeling

The rapid increase in computational power over the past few decades has led to significant advancements in statistical and machine learning techniques. This has resulted in the development of a wide range of algorithms and data mining methods, which serve as the fundamental mathematical framework for artificial intelligence () agents. While AI has a rich history in scientific exploration, modern computers now have the capability to efficiently process and incorporate algorithms on a large scale [40–42]. This remarkable progress has been made possible primarily due to the significant reduction in expenses associated with sensors, computational capabilities, and data storage technologies. Undoubtedly, these extensive volumes of data provide us with novel prospects for uncovering insights through data-centric exploration, commonly known as the fourth scientific paradigm.

In various domains of engineering, physical and biological sciences, physical models are typically represented by a collection of governing constitutive relations, spatio-temporal relations, and/or dynamical systems. Data-driven exploration in these domains is deliberately designed to uncover fundamental connections and distinct governing equations across diverse disciplines, encompassing principles of conservation or the transmission of nonlinear waves, sporal patterns, advancement of prognostic methodologies for nanoscale flow dynamics, and health monitoring. In complex domains of life sciences, the utilization of data-driven approaches holds significant potential in uncovering previously unidentified biological mechanisms. Consequently, there is a growing emphasis on securing funding for the development of novel methodologies, software applications, and Big Data sector.

Data-driven modeling is a powerful approach that utilizes data to build models and make predictions without relying on explicit knowledge of the underlying system dynamics. This modeling technique leverages large datasets and advanced computational algorithms to extract patterns, relationships, and insights from the data. The key idea behind data-driven modeling is to learn from observed data and use it to create models that can accurately capture the behavior of complex systems. Instead of relying on predefined equations or assumptions about the system, data-driven models derive their structure and parameters directly from the available data. This enables the modeling of highly nonlinear and intricate relationships that may be challenging to capture using traditional analytical methods. The process of data-driven modeling typically involves several steps. First, relevant data is collected and preprocessed to ensure its quality and suitability for analysis. Then, various statistical and machine learning techniques are applied to the data to identify patterns, correlations, and trends. These techniques may include regression analysis, decision trees, neural networks, support vector machines, or other advanced algorithms. Data-driven conventional modeling model, as shown in Fig. 6.6.

The availability of measurement data for physical systems has experienced a recent and significant surge, leading to emergence of numerous data-driven techniques for modeling and forecasting dynamics. Deep neural networks (DNNs) have emerged as one of the most prominent approaches in this field. DNNs not only demonstrate

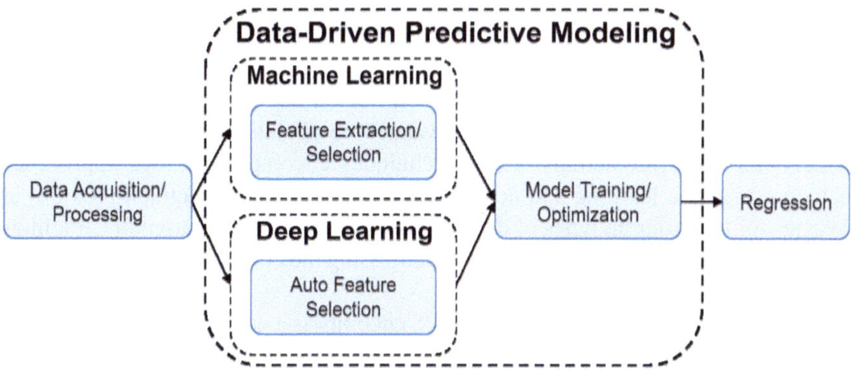

Fig. 6.6 Data-driven modeling schematic. Deep learning and machine learning: data input, model training, evaluation, and prediction—enabling intelligent systems and insights

exceptional performance in tasks image classification, but they have also proven to be effective in predicting the future state of dynamic systems. However, a significant drawback of DNNs and similar data-driven approaches is their limited interpretability: their focus lies solely on prediction without offering explicit governing equations or easily interpretable models based on the original variables. "In complex domains of life sciences, the utilization of data-driven approaches holds significant potential in uncovering previously unidentified biological mechanisms". Consequently, there is a growing emphasis on securing funding for the development of novel methodologies, software applications, and Big Data sector.

However, the process of uncovering new information can be rephrased using sparse regression, offering a computationally feasible alternative. This allows us to harness the capabilities of symbolic regression while maintaining computational feasibility. These distinct approaches demonstrate the range of strategies available for extracting significant physical insights from data. They also emphasize the potential of machine learning and artificial intelligence algorithms to acquire knowledge of physics principles and limitations. By utilizing contemporary sparse regression architectures and neural networks, various essential tasks can be accomplished solely through data analysis: (i) the discovery of first principles models, (ii) the identification of physical constraints and conservation laws, and (iii) improved models using known physics. A variety of structures enable the development of black box and gray box1 modeling strategies for intricate systems with limited knowledge of physics. Moreover, these structures not only facilitate the imposition of physics constraints or embedding physics but also enable the identification of additional physical constraints that should be embedded, i.e., one can restrict learning and learn restrictions. As a result of these strategies, it is possible to discover physical models that are both parsimonious and interpretable. Additionally, important insights such as conservation laws and physical constraints may also be uncovered. These advancements have the potential to generate models that can be adapted for multi-scale physics, noisy systems, and limited data, making them more widely applicable.

6.2.2 Types of Data-Driven Models Used in CPS

Data-driven modeling techniques have become essential in the field of CPS for their ability to leverage large datasets and advanced computational algorithms to build models and make predictions. These techniques encompass various approaches, including machine learning and deep learning models. This paragraph provides an overview of the different types of data-driven models commonly used in CPS, highlighting their applications and benefits in addressing the complexities and challenges of CPS domains.

Data-driven modeling techniques in CPS encompass a wide range of methods that leverage data to construct models and gain insights into the behavior and dynamics of cyber-physical systems. These techniques have transformed traditional modeling approaches by enabling the automatic extraction of patterns and relationships from data, thereby reducing reliance on manual model construction and expert domain knowledge. Among the various data-driven modeling techniques, machine learning and deep learning models have emerged as particularly powerful tools in CPS applications.

These ML algorithms help to solve different business problems like linear regression: Linear regression models are used to establish a linear relationship between the input features and the output, fitting the data by minimizing the squared error between the predicted and true values [43–45]; Logistic regression: Logistic regression models are used in dichotomous problems to predict the probability of a sample belonging to a category by fitting a logistic function; Principal component analysis (PCA): Principal component analysis is a dimensionality reduction technique that maps high-dimensional data to low-dimensional space by linear transformation to retain the main information in the data; Decision tree: Decision tree models use a tree structure for classification or regression tasks, dividing the data set into different subsets through a series of splitting rules; Random forest: Random forest is an integrated learning method that builds multiple decision trees and uses voting or averaging to make predictions; Support vector machine (SVM): Support vector machine models perform classification tasks by constructing a hyperplane to separate different classes of samples; K-nearest neighbors (KNN): The K-nearest neighbors model classifies samples based on the distance between them and predicts the label of a new sample by finding the K closest samples; Naive Bayes: The Naive Bayes model performs classification by computing posterior probabilities based on Bayes' theorem and the assumption of conditional independence between features; Ensemble learning: Ensemble learning methods improve prediction performance by combining multiple base models. Common integration methods include Bagging, Boosting and Stacking. Detailed information can be found in Fig. 6.7.

(1) Linear regression: Linear regression is a basic statistical analysis method used to model the linear relationship between the input and output variables, the linear regression schematic is shown in Fig. 6.8. The model assumes a linear functional relationship between the input and output variables and performs parameter estimation by minimizing the difference between the predicted and

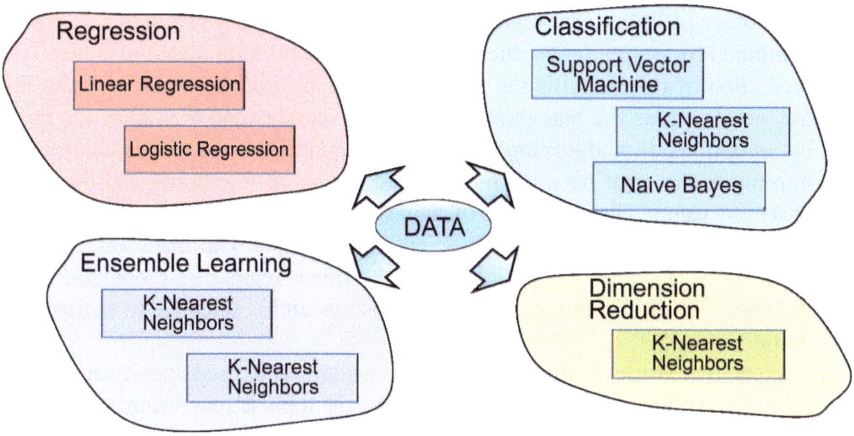

Fig. 6.7 Types of machine learning in DDCPS. Machine learning encompasses regression for predicting values, classification for categorizing, ensemble learning for combining models, and dimensionality reduction for simplifying data representation

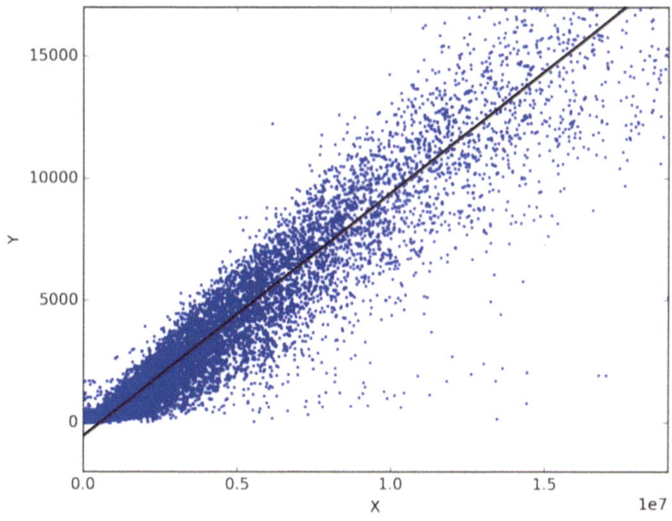

Fig. 6.8 Schematic of linear regression. A straight line fits the data points (blue), modeling the relationship linearly

true values [46]. In linear regression, the input and output variables are linearly combined by weighting coefficients with a constant term added as a bias. The prediction of the model is the sum of the products of the input variables and the weights, plus the bias term. The optimal weights and bias term are found by an optimization algorithm such as least squares or gradient descent, which minimizes the error between the predicted value of the model and the actual observed value. The advantage of linear regression model is that it is simple to understand, computationally efficient, and fits data with linear relationships better [47]. However, the limitations of the linear regression model are that it fits poorly to data with non-linear relationships and is sensitive to outliers and outliers.

(2) Logistic regression: Logistic regression is a commonly used classification algorithm for predicting dichotomous problems, the logistic regression schematic is shown in Fig. 6.9. Unlike linear regression, the output of logistic regression is a probability value processed by an activation function, indicating the probability that a sample belongs to a certain category. Logistic regression is based on a linear regression model that converts continuous predictions into probability values by mapping the output of linear regression to a range of Sigmoid functions (also called logistic functions) [48]. Such a transformation allows the

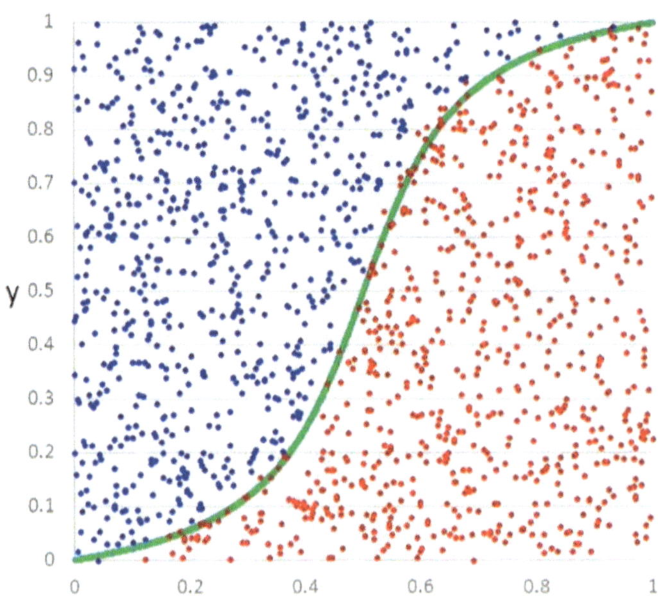

Fig. 6.9 Schematic of logistic regression. Logistic regression result: S-shaped curve classifies data points, illustrating the probability of binary outcomes

logistic regression model to output probability values between 0 and 1, indicating the probability that a sample belongs to a certain category. In logistic regression, optimization algorithms such as maximum likelihood estimation or gradient descent are used to make the model's predictions as close as possible to the actual labels by adjusting the weights and biases of the model. Based on the predicted probability values, a threshold can be used to determine the categorical labels of the samples. The logistic regression model has the advantages of simplicity, ease of interpretation and high computational efficiency. However, logistic regression also has some limitations, such as poor fit for data with nonlinear relationships and sensitivity to outliers and outliers.

(3) Principal component analysis: PCA is a commonly used dimensionality reduction technique and data pre-processing method, the PCA is shown in Fig. 6.10. It can transform high-dimensional data into low-dimensional data, while retaining the most important information in the data [49]. The objective of PCA is to transform the original data into a different coordinate system using a linear transformation, aiming to maximize the variance of the data in this new coordinate system. In particular, PCA computes the covariance matrix of the initial data and subsequently acquires the eigenvalues and their corresponding eigenvectors through an eigenvalue decomposition performed on said covariance matrix. The eigenvectors are the principal components, and they represent the orientation of the data under the new coordinate system. In PCA, the dimensionality reduction of the data can be achieved by selecting the principal components with the largest eigenvalues. By selecting the first k principal components, the original data can be projected into the k-dimensional space, and the dimensionality reduction of the data can be achieved. The dimensionality reduction retains the most important structure and change trends in the original data, while reducing

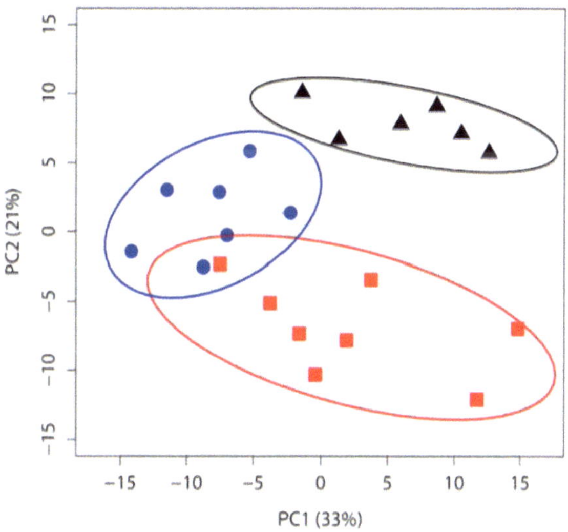

Fig. 6.10 Schematic of principal component analysis. Data points are transformed into a new coordinate system, emphasizing variance along the principal components. This technique simplifies data representation, identifies clusters, and reduces dimensionality for a clearer understanding of the data structure

the dimensionality of the data and improving the computational efficiency. PCA has a wide range of applications in data analysis and feature extraction. It can be used for dimensionality reduction pre-processing to reduce the dimensionality of data sets, remove redundant information, and improve the training effect of subsequent models. In addition, PCA can be used for data visualization to help us understanding the distribution and relationships of data by mapping high-dimensional data into two- or three-dimensional space. Note that PCA assumes that the data obeys a linear distribution and is sensitive to the shape and variance distribution of the data. Before applying PCA, it may be necessary to pre-process the data to satisfy these assumptions.

(4) Decision tree: Decision tree is a common machine learning algorithm that is widely used in classification and regression problems, the decision tree is shown in Fig. 6.11. It constructs a tree structure by partitioning a data set to make predictions and decisions about the data. The process of building a decision tree is a top-down recursive process [50]. It starts from the root node, selects an optimal feature, and divides the dataset into different subsets based on that feature. Then, the same partitioning process is applied recursively to each subset until some termination condition is met, such as reaching the maximum depth of a leaf node or the number of samples in the subset no longer changes. Each leaf node represents the result of a classification or regression. One of the advantages of decision trees is that they are highly interpretable, and the tree structure allows visualization of the decision process. It can automatically select important features and classify or predict them based on their split points, and is applicable to both discrete and continuous features. In addition, the decision tree is robust to outliers and missing values. However, decision trees also have some limitations. It is prone to overfitting, especially when dealing with high-dimensional data and complex tasks. To solve this problem, the model can be regularized by using pruning techniques, limiting the maximum depth of the tree or the minimum

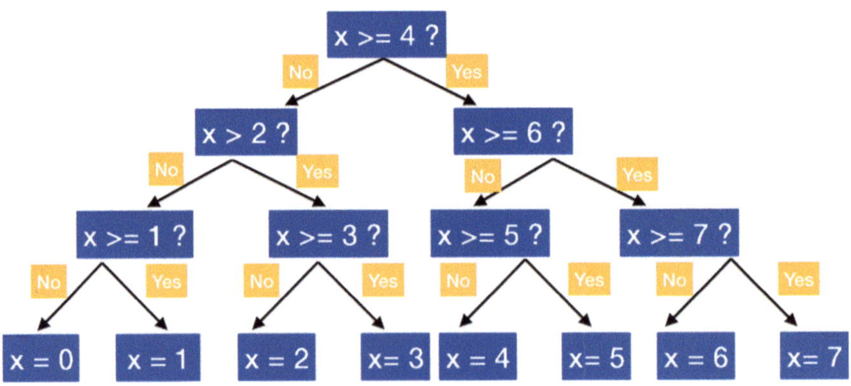

Fig. 6.11 Schematic of decision tree. A tree-like model divides data based on attributes, providing a visual way to make decisions and predict outcomes

number of samples of the nodes. In addition, decision trees are sensitive to noise and irrelevant features in the data, which may lead to unstable prediction results. To improve the performance of decision trees, some extended algorithms such as Random Forest and Gradient Boosting Decision Tree have also emerged. They improve the accuracy and robustness of the model by integrating the prediction results of multiple decision trees.

(5) Random forest: Random forest is a powerful integrated learning algorithm that performs classification or regression tasks by constructing multiple decision trees and combining their prediction results, the random forest is shown in Fig. 6.12. It combines the simplicity of decision trees and the advantages of integrated learning, with good accuracy, robustness and interpretability [51]. The process of constructing a random forest consists of two main steps: random sampling and random feature selection. First, a portion of samples from the original training set is randomly selected for sampling with put-back to form multiple subsets. Then, for each subset, a random selection of features from all features is used for decision tree construction. This randomness is introduced to help reducing overfitting and increasing the diversity of the model. In random forests, each decision tree is trained independently and the final prediction is obtained by voting or averaging. This integrated approach makes the random forest highly accurate and robust. In addition, random forests can evaluate the importance of features and determine the extent of their contribution to the prediction by analyzing the use of each feature in the decision tree. Random forest is suitable for various types of datasets and is also good for handling high-dimensional data and large-scale datasets. It is robust to noise and missing values, and can handle

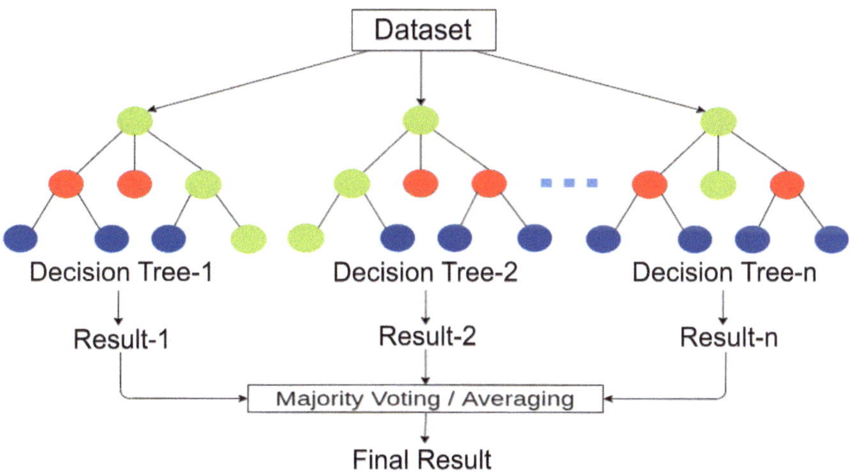

Fig. 6.12 Schematic of random forest. Multiple decision trees collaborate to make predictions, with each tree's output contributing to the final result, offering robust and accurate insights (see the color figure before)

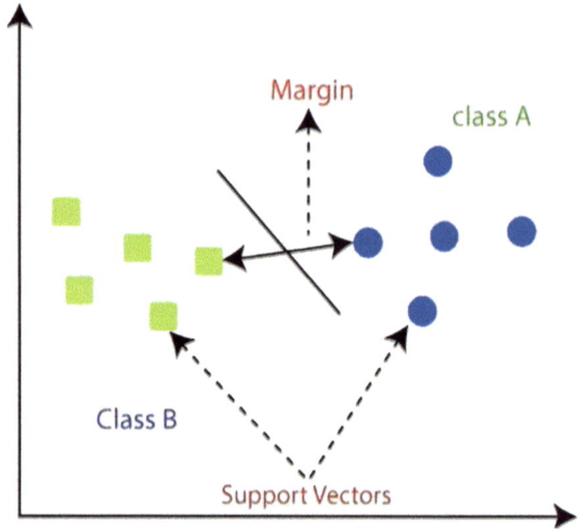

Fig. 6.13 Schematic of Support vector machine. A hyperplane optimally separates data points into distinct classes with a maximal margin, aiding classification tasks

nonlinear relationships and interaction features. In addition, random forest can perform parallel computation to accelerate the training process of models.

(6) Support vector machine: Support vector machine (SVM) is a commonly used supervised learning algorithm, mainly for classification and regression tasks. The goal of SVM is to find an optimal hyperplane in the feature space that can maximize the interval between different classes, thus achieving effective classification, the SVM is shown in Fig. 6.13. The core idea of SVM is to map samples to a high-dimensional feature space, so that samples that are indistinguishable in the low-dimensional space become distinguishable in the high-dimensional space. By finding the optimal hyperplane, i.e., the hyperplane that maximizes the interval between two different classes, SVM is able to classify the new samples accurately. For the linearly divisible case, SVM uses a linear hyperplane for classification, while for the linearly indistinguishable case, SVM maps the samples to a high-dimensional space by a kernel function, which enables nonlinear classification. In the training process of SVM, the optimal hyperplane parameters are solved by an optimization problem. The goal of the optimization problem is to minimize the regularized loss function and make the sample points as far away from the classification hyperplane as possible to form support vectors. These support vectors are the key points of the decision boundary and play a decisive role in the classification results [52]. SVMs have several advantages, including high adaptability to high-dimensional feature spaces and large-scale data sets, the ability to handle linear and nonlinear classification problems, and good generalization capabilities. In addition, SVMs can use different kernel functions to adapt to different data structures, such as linear kernels, polynomial kernels, and Gaussian kernels. However, SVMs also have some limitations, such as slower training speed for large-scale data sets and more sensitive in handling noise and

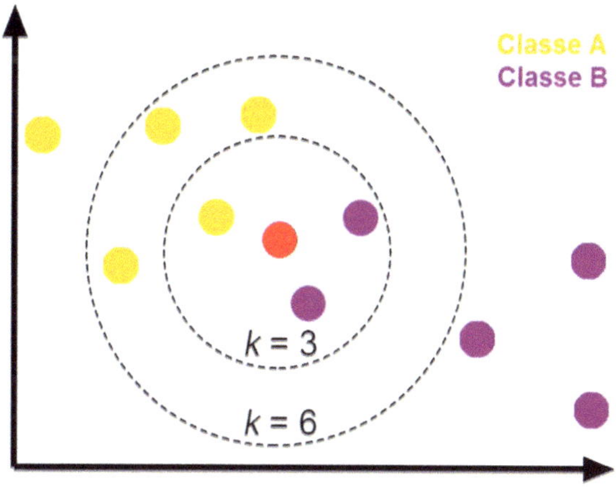

Fig. 6.14 Schematic of K-nearest neighbors. Data points are assigned labels based on the most prevalent class among their k nearest neighbors, depicting clustering and classification patterns (see the color figure before)

outliers. In addition, the interpretability of SVM is low, which makes it difficult to explain the decision process of the model.

(7) K-nearest neighbors: K-nearest neighbors (KNN) is a basic supervised learning algorithm commonly used for classification and regression tasks, the KNN is shown in Fig. 6.14. The principle of KNN is simple, by calculating the distance between samples to determine the class of new samples [53]. KNN works on the principle of instance-based learning, which labels all training samples as data points in the feature space. When a new sample needs to be classified, the KNN algorithm finds the K nearest neighbors to that sample, and then votes on the categories of those neighbors and selects the category with the majority as the category of the new sample. In the regression problem, KNN will predict the value of the new sample based on the average of the K nearest neighbors. The key parameter of KNN is the value of K, which indicates the number of neighbors selected. A smaller K value makes the model more noise sensitive, while a larger K value makes the model smoother but prone to category balancing problems. Therefore, it is necessary to choose the appropriate K value for the specific problem when using KNN. The advantages of the KNN algorithm include its simplicity, ease of implementation, insensitivity to outliers, and its applicability to multi-category classification and regression problems. In addition, KNN can be applied to nonlinear data and unbalanced data sets. However, KNN algorithms also have some limitations. First, the computational complexity of the KNN algorithm is high when the dataset is large because it needs to calculate the distance of each sample from all training samples. Second, the KNN algorithm is more sensitive to the selection and scaling of features, and different scaling of features may lead to different distance calculation results.

(8) Naive Bayes: Naive Bayes is a simple and effective classification algorithm based on Bayes' theorem, the Naive Bayes is shown in Fig. 6.15. It assumes that the features are independent of each other, hence the term "plain". Naive

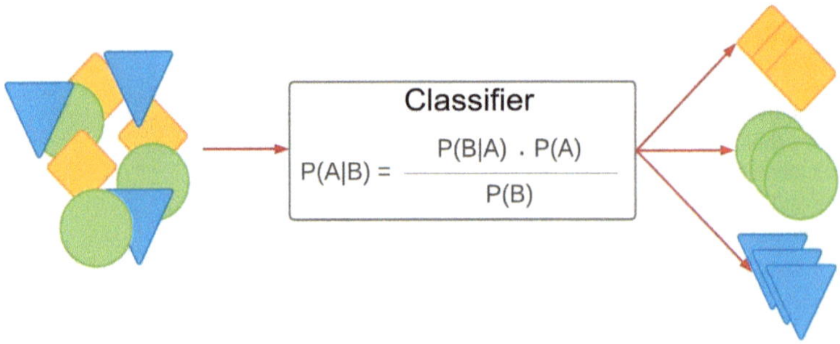

Fig. 6.15 Schematic of Naive Bayes. Probabilistic method that predicts class labels by calculating conditional probabilities based on independent features, making it suitable for classification tasks

Bayes algorithm is commonly used in text classification, spam filtering, sentiment analysis, etc. [54]. The core idea of plain Bayesian algorithm is to use Bayes' theorem to calculate the probability of samples belonging to different categories and make classification decisions based on the probability. In the training phase, the algorithm estimates the probability distribution of each feature under each category by counting the sample data. In the classification phase, for a new sample, the plain Bayesian algorithm calculates the probability that the sample belongs to each category and selects the category with the highest probability as the predicted outcome. The advantages of the plain Bayesian algorithm are its simplicity and efficiency. Since the features are assumed to be independent of each other, the computational speed of the plain Bayesian algorithm is fast, which is especially suitable for handling large-scale data sets. In addition, the plain Bayesian algorithm is robust to missing data and noise, and performs well for high-dimensional data sets. However, the plain Bayesian algorithm also has some limitations. Since features are assumed to be independent of each other, the complex relationships between features cannot be captured. When there is a strong correlation between features, the plain Bayesian algorithm may produce inaccurate results. In addition, the plain Bayesian algorithm makes more stringent assumptions about the distribution of the input data and may not perform well for non-linear and non-Gaussian distributed data.

Deep learning models, on the other hand, have gained significant attention in recent years due to their ability to handle large-scale, high-dimensional data. Deep learning models, including neural networks, convolutional neural networks (CNNs), and recurrent neural networks (RNNs), are capable of capturing intricate patterns and dependencies in data. They have shown remarkable performance in tasks such as image and speech recognition, natural language processing, and time series analysis, making them well-suited for various CPS applications [55].

Some commonly used deep learning algorithms include convolutional neural networks (CNNs): CNNs are primarily used for image and video analysis tasks. They utilize convolutional layers to automatically learn and extract features from input images. Recurrent neural networks (RNNs): RNNs are designed to process sequential data, such as text or time series data. They use recurrent connections to retain and utilize information from previous time steps; Generative adversarial networks (GANs): GANs consist of two neural networks, a generator and a discriminator, which are trained in an adversarial manner. GANs are commonly used for generating realistic synthetic data, such as images or text; Deep reinforcement learning (DRL): DRL combines deep learning with reinforcement learning principles. It enables an agent to learn optimal actions in an environment by trial and error, typically used in tasks like game playing or robotics; Transformer networks: Transformer networks have gained popularity in natural language processing tasks, such as machine translation and text summarization. They utilize self-attention mechanisms to capture global dependencies in the input data. Detailed information can be found in Fig. 6.16.

(1) CNNs: CNNs is a deep learning model specialized for processing tasks with grid-structured data, especially image and vision related tasks. CNN extracts and learns features of the input data through components such as convolutional layer, pooling layer and fully connected layer, the CNN structure is shown in Fig. 6.17. In the convolutional layer, CNN uses convolutional operations to extract features from the input data [56]. The convolution operation uses a set of learnable filters

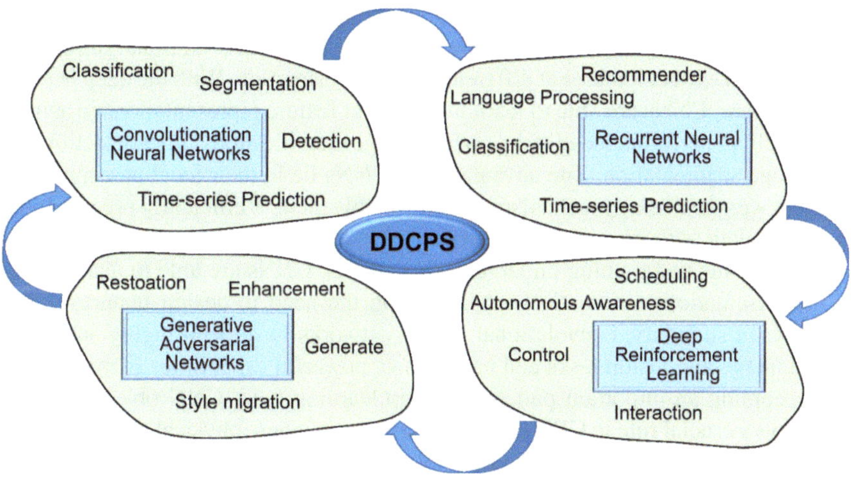

Fig. 6.16 Types of deep learning in DDCPS. Convolutional networks, recurrent neural networks, generative adversarial networks, and deep reinforcement learning form the core of DDCPS's deep modeling

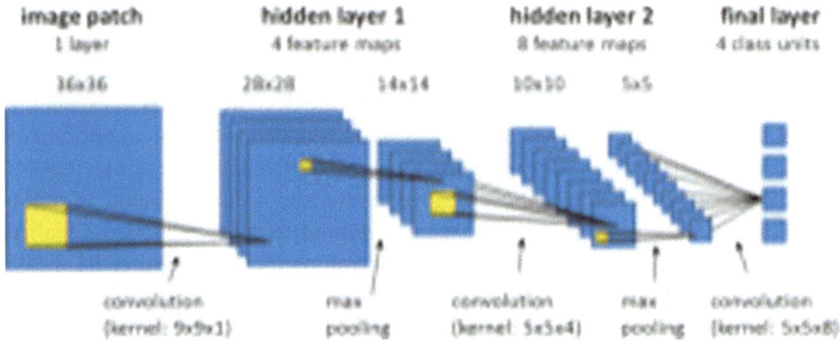

Fig. 6.17 Schematic of CNNs. Hierarchical layers with convolution, pooling, and fully connected layers, specialized for image analysis and feature extraction

(or convolution kernels) to perform element-by-element multiplication and summation with the input to generate feature maps. These feature maps can capture local features such as edges, textures and shapes at different locations. Pooling layers are used to downsample the feature maps, reducing model complexity and preserving important features. The most common pooling operation is maximum pooling, which selects the maximum value in each pooling window as the output. Finally, the fully connected layer converts the pooled feature maps into the final predictions. The neurons in the fully connected layer are connected to all neurons in the previous layer to generate the output by learning the weights. Convolutional neural networks have multiple stacks of convolutional and pooling layers to extract features at different levels of abstraction. With the deep network structure, CNNs are able to learn higher-level feature representations to achieve excellent performance in tasks such as image classification, target detection, and image segmentation. The advantages of CNNs lie in their local perceptiveness and weight sharing mechanisms, which enable them to efficiently process large-scale data with translational invariance. In addition, through back-propagation algorithms and training on large-scale datasets, CNNs are able to learn feature representations automatically, eliminating the need to design features manually. In summary, convolutional neural networks have been highly successful in image and vision tasks and have shown powerful capabilities in many areas, becoming an important part of the deep learning field. CNN object detection plays a crucial role in CPS. CPS refer to systems where physical entities interact with computational systems, encompassing domains such as autonomous driving, intelligent transportation, and smart manufacturing. By employing CNN object detection, accurate detection and recognition of objects within CPS can be achieved, enabling more intelligent and secure applications. CNN object detection algorithms, through network architectures comprising convolutional layers,

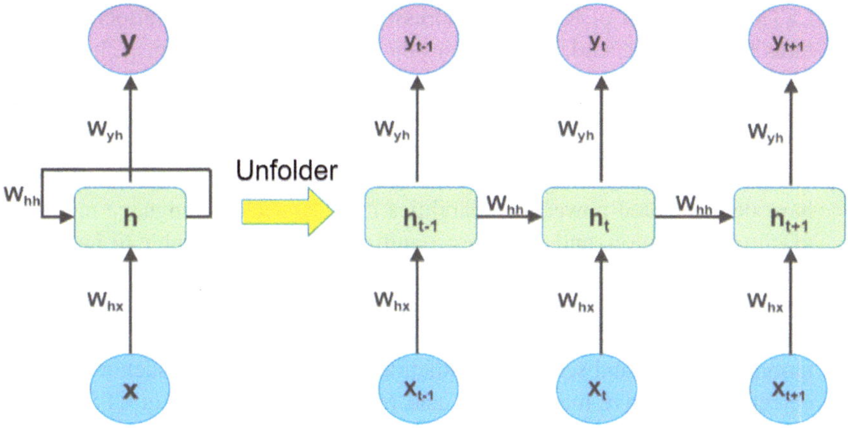

Fig. 6.18 Schematic of RNNs. Looped connections allow information retention, making it suitable for sequence and time-series data analysis

pooling layers, and fully connected layers, automatically learn and extract features from images, allowing for the classification and localization of objects across different classes.

(2) RNNs: RNNs are a type of deep learning model specifically designed for handling sequential data by maintaining contextual relevance through recurrent connections, the RNN structure is shown in Fig. 6.18. Among them, Long short-term memory (LSTM) is a variant of RNN that addresses gradient problems and effectively captures long-term dependencies by introducing gating mechanisms [57]. The key idea of RNNs is to pass previous information to future states by cycling through time, using the output of the previous moment as the input of the current moment. This cyclic connection allows RNNs to model inputs of arbitrary length in a sequence and shows superior performance in dealing with long-term dependency problems. This cyclic connection allows RNNs to model inputs of arbitrary length in a sequence and shows superior performance in dealing with long-term dependency problems. The basic units of RNNs are recurrent units, most commonly LSTM and gated recurrent unit (GRU). These units have internal gating mechanisms to selectively remember, forget and output information, thus better handling long-term dependencies. Recurrent neural networks have many advantages in modeling sequential data. It is able to capture temporal dynamics and semantic correlations in sequences and improve the predictive power of the model by learning contextual information. In addition, RNNs can be trained end-to-end by back-propagation algorithms to automatically learn feature representations without the need to design features manually. However, traditional RNN suffer from gradient disappearance and gradient explosion, making it difficult to capture long-term dependencies. To address these issues, several improved RNN structures have emerged, such as LSTM with gating mechanism and GRU.

LSTM is a variant of RNN that addresses gradient issues and effectively captures long-term dependencies by introducing gating mechanisms. The gates in LSTM allow for the selective input, forget, and output of information, enabling the model to control the flow of information and remember important contextual information. This makes LSTM a powerful tool for modeling sequences, especially when dealing with data with long-term dependencies. RNN and LSTM have demonstrated powerful capabilities in fields such as language modeling, machine translation, and speech recognition [58]. They are able to handle the dynamic properties and contextual relevance of sequence data, making them a significant research direction in deep learning. RNN are used for sequence prediction in CPS. CPS involve the interaction of physical entities with computational systems, and sequence prediction is crucial for tasks such as anomaly detection, predictive maintenance, and resource optimization in CPS applications. RNNs are well-suited for handling sequential data and capturing temporal dependencies. They can effectively model and predict sequences by leveraging their recurrent connections, which allow information to flow through time and maintain contextual relevance. In CPS, RNN can be trained on historical sensor data to predict future system states, anticipate failures, or optimize resource allocation. The ability of RNN to capture complex patterns and dynamics in sequential data makes them a valuable tool for sequence prediction tasks in CPS.

(3) GANs: GANs are deep learning models that consist of a generator and a discriminator network. The goal of GANs is to learn to generate sample data with high fidelity by adversarial training, the GAN structure is shown in Fig. 6.19. The generator network aims to generate samples that resemble real data by receiving

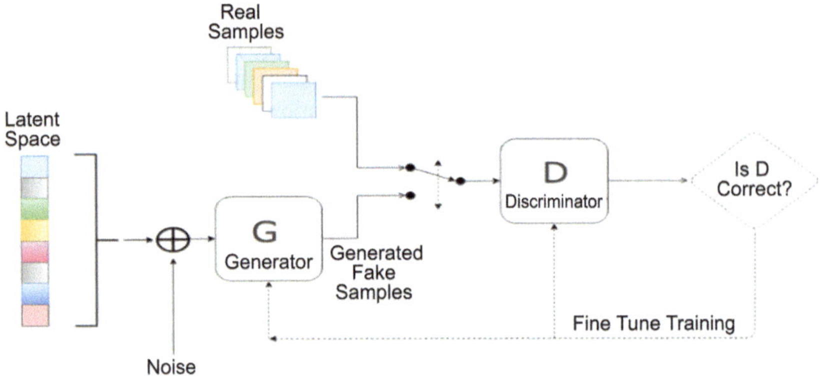

Fig. 6.19 Schematic of GANs. Comprising a generator and discriminator, GANs engage in a competitive process to generate and evaluate data quality

random noise as input and generating synthetic data through a series of transformations and mapping operations [59]. The discriminator network, on the other hand, is responsible for distinguishing the generator-generated samples from the real data, and it does so by learning to distinguish the generated samples from the real samples. The core idea of GANs is to pit the generator network and the discriminator network against each other to improve the performance of the model through continuous adversarial training. The generator continuously adjusts the generated samples to be closer to the real samples, while the discriminator strives to improve the ability to distinguish the generated samples from the real samples. This adversarial training process iterates until the generator is able to generate samples that are realistic enough that the discriminator cannot distinguish the generated samples from the real samples. Generative adversarial networks have made significant breakthroughs in the areas of image generation, image restoration, style migration, and text generation. It can generate image samples with realism and diversity, and even perform image synthesis without real samples. In addition, GANs have been applied to tasks such as data enhancement, anomaly detection, and model compression. In summary, generative adversarial network is a deep learning model that confronts each other through generator and discriminator networks. It is capable of generating highly realistic samples and demonstrates powerful generative capabilities and potential creativity in a variety of tasks. Generative adversarial networks (GANs) are utilized in CPS to generate synthetic data and enable small-sample learning. CPS involves the interaction between physical and computational components. GANs provide a powerful framework for generating realistic synthetic data that mimics the distribution of real CPS data. This is particularly useful when collecting large amounts of labeled data is challenging. By training a GAN on limited real data, it can learn the underlying patterns and generate additional synthetic samples. These synthetic samples augment the training set, increasing its size and diversity. This benefits machine learning models in CPS, especially in small-sample learning scenarios. GANs also contribute to data augmentation, anomaly detection, and data privacy in CPS. Synthetic data generated by GANs closely resemble real CPS data, allowing for extensive testing and simulation without the need for additional physical resources. GANs can detect anomalies by comparing real and synthetic samples. Additionally, GANs preserve data privacy by generating synthetic data that masks sensitive information.

(4) DRL: DRL is a method that combines deep learning with reinforcement learning and aims to allow machine agents to learn optimal decision strategies by interacting with the environment, the DRL structure is shown in Fig. 6.20. In deep reinforcement learning, the machine agent takes actions and interacts with the environment by observing the state of the environment, and guides the learning process through reward signals [60]. Deep neural networks are used as function approximators that map states to probability distributions of actions. By continuously interacting with the environment and adjusting the network weights, the goal of deep reinforcement learning is to learn an optimal policy that maximizes the cumulative reward. Deep reinforcement learning has achieved important

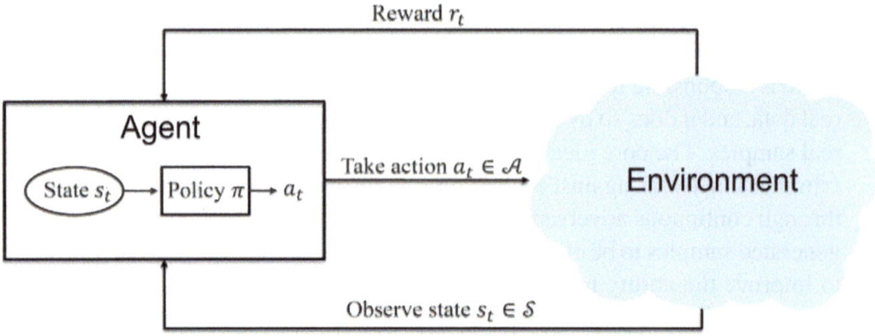

Fig. 6.20 Schematic of DRL. An agent interacts with an environment, learning optimal actions through deep neural networks and reward-based feedback

results in many fields, such as AlphaGo and AlphaZero, as well as practical applications such as robot control and autonomous driving. Its strength lies in its ability to handle high-dimensional state and action spaces and to extract useful features from raw inputs through end-to-end learning. However, deep reinforcement learning also faces some challenges, such as low sample efficiency, training instability and sensitivity to hyperparameters. In recent years, several improved algorithms and techniques have been proposed, such as deep Q-nets (DQN), policy gradient methods, and Actor-Critic models, to improve the stability and efficiency of deep reinforcement learning. In summary, deep reinforcement learning is a method that combines deep learning and reinforcement learning to learn optimal policies by interacting with the environment [61]. It has potential for solving complex tasks and decision problems and has demonstrated a wide range of applications in several fields. CPS involves the interaction between physical entities and computational systems. DRL, a combination of deep learning and reinforcement learning, offers a powerful approach for training intelligent agents to make decisions and take actions in CPS environments. DRL is leveraged in CPS for adaptive control, where agents learn to adapt their control policies based on changing environmental conditions. By continuously interacting with the CPS environment and receiving feedback, the agents can adjust their control actions in real-time, ensuring optimal performance and adaptability. This adaptive control mechanism enables CPS to operate efficiently under dynamic and uncertain conditions. DRL in CPS facilitates adversarial training to enhance system resilience and enables adaptive control, allowing CPS to dynamically respond to changing circumstances and optimize their performance. By leveraging the power of deep learning and reinforcement learning, DRL contributes to the advancement of intelligent and adaptive CPS applications.

Machine learning finds various applications in CPS, spanning across different domains and tasks. It is commonly utilized in anomaly detection, where models are developed to learn normal behavior patterns from sensor data, enabling the identification of abnormal system behavior or faults. CPS resource optimization benefits from machine learning techniques by predicting energy demands and optimizing resource allocation, such as energy generation, transmission, and storage, based on these predictions [62]. Machine learning also plays a crucial role in prediction and decision-making within CPS, enabling the forecasting of future events and behaviors to make informed decisions. Additionally, adaptive control systems employ machine learning to dynamically adjust control strategies based on real-time feedback and sensor data, enhancing system flexibility and adaptability to varying environmental conditions. Wireless communication optimization in CPS leverages machine learning to improve communication protocols and resource allocation, optimizing factors like throughput, latency, and energy efficiency by learning network data traffic patterns and communication requirements. Lastly, machine learning aids in security and privacy protection within DDCPS by learning patterns of normal and malicious behavior, enhancing threat identification and response capabilities.

In recent years, both traditional machine learning methods and emerging deep learning techniques have played pivotal roles in various aspects of CPS, including process control, automation, and decision-making. Traditional machine learning approaches, exemplified by the work of Li et al. [63, 64]. In anomaly detection, have contributed significantly to CPS applications. These methods have been effective in tasks such as predictive modeling, resource optimization, and fault detection. They offer interpretability and generalization capabilities, making them suitable for handling structured and labeled data in CPS domains. Biocides, the advent of deep learning has brought about a paradigm shift in CPS applications. Deep learning methods, characterized by multi-layer neural networks, have shown exceptional performance in handling unstructured and high-dimensional data, such as images, videos, and sensor streams. Deep learning models can automatically learn hierarchical representations from data, capturing intricate patterns and correlations. This capability has enabled significant advancements in areas such as image recognition, natural language processing, and time series analysis, which are vital for CPS.

By combining traditional machine learning methods and deep learning methods, CPS can benefit from the strengths of both methodologies. This integration fosters comprehensive data analysis, adaptive control, and informed decision-making in real-time scenarios. As CPS continue to evolve, the synergy between machine learning and deep learning will undoubtedly drive further advancements in optimizing CPS performance, enhancing system resilience, and enabling intelligent and autonomous operation.

6.2.3 *Methods for Model Selection and Validation*

Selecting a suitable model is an essential part of developing an effective predictive or explanatory model for understanding mechanism behind data generation [65]. While numerous research papers have been published on the topic of model selection, a crucial and largely unresolved query remains: How can we effectively determine the most suitable modeling procedure that encompasses both model selection and parameter estimation? In practical scenarios, it is often challenging to ascertain which procedure aligns best with the given data. Rather than rigidly adhering to personal preferences for a specific procedure, it would be more prudent to consider an adaptive approach when selecting a modeling procedure. In this paragraph, our main objective is to utilize cross validation for the purpose of selecting a modeling procedure in regression scenarios. This becomes particularly relevant when there is uncertainty regarding whether the true model can be classified as finite or infinite dimensional within a classical framework. Additionally, it also applies when determining if the true regression function exhibits sparsity in either a linear or additive form within high dimensional settings.

One aspect of our research involves the exploration of optimal for a specific machine learning technique [66]. Hyperparameters refer to the parameters that need to be specified before fitting the model are derived from the fitting process. In the context of a logistic regression model, it is necessary to specify the regularization strength (and potentially the type of regularization) as hyperparameters before fitting. On the other hand, the coefficients obtained from are considered model parameters. The identification of appropriate hyperparameters influences the performance of the model on specific data.

Additionally, it may be necessary to choose the most suitable learning approach (along with their corresponding "optimal" hyperparameters) from a pool of eligible machine learning methods. This process can be referred to as algorithm selection. When dealing with a classification problem, one might question whether the given task would benefit more from utilizing a logistic regression model or a random forest classifier in order to achieve optimal classification performance.

Before delving into the specifics of various approaches to model selection and when they should be utilized, it is crucial to address an additional aspect: model assessment. The purpose of model assessment is to estimate the generalization error of the chosen model, which refers to how effectively the selected model performs on unfamiliar data. It goes without saying that a proficient machine learning model not only excels with training data (otherwise it could simply memorize it), but also demonstrates proficiency with unseen data. Consequently, prior to deploying a use, it is imperative to have reasonable confidence in its ability to maintain performance levels when presented with new data.

Model validation involves assessing the accuracy of a model in representing system behavior. It is important to evaluate the validity of the model both operationally, by comparing its output with observed data, and conceptually, by examining the justifiability of the underlying theory and assumptions. Models can undergo validation

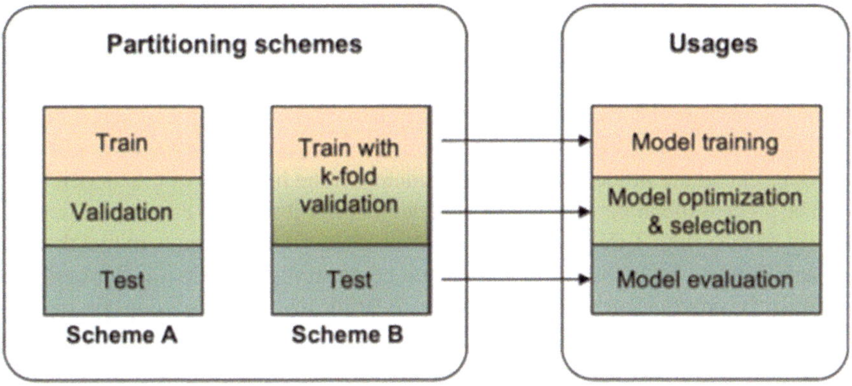

Fig. 6.21 Data-driven model evaluation. CPS model are trained through iterative processes involving forward and backward passes, with evaluation based on performance metrics and validation datasets

by comparing their output to external field or experimental datasets that correspond with the simulated scenario. Nevertheless it is crucial to take into account the data's quality (e.g., the extent of measurement error), its true representation of the system, and whether it serves as the most optimal evaluation for the model. It may be challenging to conduct operational validation of the model with independent data in situations where the simulated scenario goes beyond observed conditions (e.g., predicting responses to future climate change) or when utilizing probabilistic forecasts that account for uncertainty in system processes. In the latter scenario, choosing between a deterministic or probabilistic framework involves a trade-off between precision and accuracy. Generally, deterministic models exhibit greater precision but lower accuracy compared to those that integrate uncertainty. Nevertheless, irrespective of the simulation type, conceptual validation remains achievable. The data-driven model validation process is shown in Fig. 6.21.

Cross-validation is a widely utilized technique for assessing the predictive capabilities of a model, whether it is predetermined or developed through modeling procedures. Essentially, this method involves dividing the data into segments, where one portion is used to train each competing model and the remaining data is employed to evaluate their predictive performance by measuring validation errors. Ultimately, the model exhibiting superior overall performance is chosen. Due to these reasons, cross-validation has gained significant popularity in data mining as an effective means for selecting models or modeling procedures [65].

Conducting sensitivity analyses constitutes an essential component of the model validation procedure. The objective behind performing a sensitivity analysis is to assess the impact of parameters, conditions, and alternative assumptions on the output of the model. The process involves an iterative approach, allowing for feedback to enhance the model. A sensitivity analysis is conducted to compare response variables across multiple model iterations. In each comparison run, all parameters remain constant except for the specific parameter under examination. When a model parameter is found to have an excessive impact on the simulation's output, which deviates from reality, it becomes necessary to reassess the characterization of the model. The simulation process necessitates conducting comprehensive sensitivity analyses in order to comprehend the influence of each parameter on the behavior of the model.

The process of validating a model enhances its credibility and ensures the dependability of its results. Effective simulation necessitates constructing a model that reasonably reflects the biological phenomenon being studied. While it is impossible to definitively prove accuracy of a model, validation involves assessing the dependability and credibility of its performance. Given the ever-changing nature of natural systems, continuous model validation should be conducted, particularly when fresh data becomes accessible from the physical system.

6.2.4 Examples of Data-Driven Modeling in CPS Applications

The combination of physics and data driven has been applied in some fields. And the classic application of data-driven models in CPS is shown in Fig. 6.22. In the field of biomedicine, the integration of physics and data-driven models has proven to be a valuable approach in enhancing the analysis of medical images [67], leading to improved treatment protocol development. In the field of geosciences and climate, the integration physics and data-driven approaches has played a significant role in predicting climate fluctuations and mitigating potential abrupt disasters [68]. In the domain of additive manufacturing, combining physics with data-driven methods offered outstanding benefits, such as low training time, data-driven decision logic, real-time responsiveness, and uncertainty quantification. In the field of intelligent construction, it is possible to collaboratively model the requirements related to different construction scenarios. Recently, various terms have been suggested to approaches that integrate physics with data-driven models, such as "theory-informed" "informed by physics" "based on physical principles" "constrained by physics" "assisted by simulations", and "guided by physics". Physics enhanced the coherence of data-driven approaches. Theory-informed data science incorporated models based on theory into various algorithms, output samples, and observational data to guarantee the alignment between data-driven methods and physics principles, referred to as physical coherence [9]. Physics-informed machine learning integrates physical constraints into the training data, model structures, and inference algorithms of data-driven models to obtain solutions that are consistent with principles of physics. Physics-based machine learning employs reduced-dimensional representations of high-fidelity simulation results as the training dataset for machine learning [62, 69, 70]. Physics-constrained machine learning utilizes partial differential equations to restrict the optimization of data-driven models and enhance convergence rates. Additionally, numerous studies have highlighted the integration of

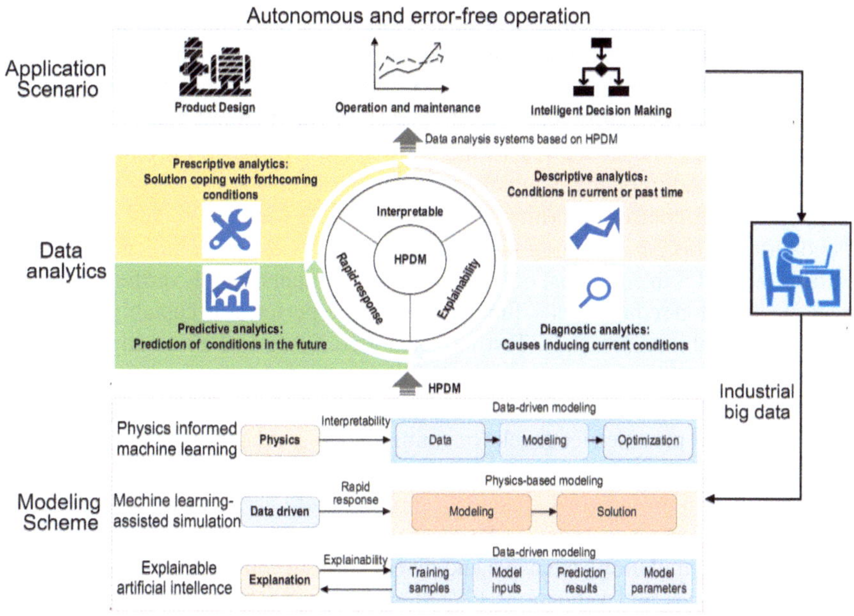

Fig. 6.22 A holistic operational paradigm based on data-driven modeling in CPS. The CPS data-driven modeling thrives on a three-pronged approach: modeling, analysis, and application, delivering a robust operational paradigm

physics-based models in guiding the development of data-driven models. Simulation-assisted machine learning incorporates simulation models into various aspects such as training samples, hypothesis set, algorithm, and final hypothesis to provide guidance for constructing data-driven models. The integration of physics-based models with machine learning techniques was employed in the preprocessing, model design, and regularization stages of data-driven models to steer the development of hybrid models and guarantee their interpretability.

6.3 Simulation and Testing of Cyber Physical Systems Using Data-Driven Models

In the development of modern CPS, simulation and testing play a crucial role. CPS involve the integration of computer systems with physical entities and are applied in critical domains such as intelligent transportation, smart manufacturing, and energy systems. Given the complexity and real-time requirements of CPS, simulation and testing have become indispensable in the development process [71].

Simulation involves constructing models and systems to mimic the behavior and interactions of CPS in a virtual environment. It helps engineers simulating and evaluating the performance, functionality, and feasibility of CPS systems, as well as identifying potential issues and risks [33, 72]. Through simulation, design flaws can be identified and addressed in the early stages, reducing development costs and time [73]. Additionally, simulation can be used for optimizing system parameters, validating control algorithms, assessing system robustness, and providing a foundation for decision-making [74]. Testing involves the practical validation and evaluation of CPS in either real or simulated environments [75]. It verifies if the CPS meet specification requirements, functions correctly, performs acceptably, and meets safety and reliability demands. Testing helps identifying and eliminating potential faults and defects, improving system reliability and robustness. It also assesses the performance of CPS under different operating conditions, identifies performance bottlenecks, and provides feedback for continuous system improvement [76].

Simulation and testing are interrelated and mutually supportive in CPS development. Simulation provides a virtual environment where various aspects of CPS can be modeled, system behavior can be predicted, and design choices can be evaluated. Through simulation, extensive and diverse testing can be performed, covering different scenarios and validations [77]. Testing, on the other hand, validates the simulation results in real or simulated environments to confirm system performance and functionality. Simulation and testing complement each other, working in synergy to ensure the correctness, reliability, and robustness of CPS. Simulation and testing play pivotal roles in CPS development, ensuring the design, evaluation, and optimization of CPS systems. They help predicting and resolving issues before actual deployment, reducing development risks and costs. As the CPS field continues to advance and innovate, simulation and testing will evolve and improve to adapt to the increasing complexity and intelligence of CPS systems. In the future, with the introduction of more advanced simulation and testing techniques, we can expect higher performance, improved safety, and greater reliability in CPS.

Data-driven models have emerged as powerful tools for enhancing simulation and testing processes in the field of CPS. By leveraging large amounts of data collected from CPS systems, data-driven models can provide valuable insights into system behavior, improve accuracy in simulation, and enhance the effectiveness of testing.

Data-driven models offer a valuable means of enhancing simulation by incorporating real-world data into the modeling process [8]. Traditional simulation models often rely on simplified assumptions and predefined parameters, which may not accurately capture the complex dynamics and uncertainties present in CPS. Data-driven models, on the other hand, leverage historical data and machine learning techniques to learn from the observed behavior of the system. By training on real-world data, these models can capture the intricate relationships and patterns within the CPS, leading to more accurate and realistic simulation results. Data-driven models also play a crucial role in improving the effectiveness of testing in CPS development. In traditional testing approaches, predefined test cases are designed to assess the system's behavior under specific conditions. However, these test cases may not encompass the full range of scenarios that a CPS system may encounter in real-world operation. Data-driven models can be used to analyze historical data, identify patterns, and generate additional test cases that cover a broader spectrum of system behavior. By incorporating these diverse test cases, data-driven models enable more comprehensive testing, enhancing the system's reliability and robustness [78].

Data-driven models have proven particularly effective in predictive maintenance and fault detection in CPS [79]. By analyzing sensor data and historical maintenance records, these models can identify patterns indicative of potential system failures or performance degradation. By continuously monitoring the system's condition and comparing it with established models, data-driven approaches can predict and detect anomalies, enabling proactive maintenance actions [80]. This helps prevent catastrophic failures, minimize downtime, and optimize system performance.

While data-driven models offer significant potential in enhancing simulation and testing in CPS, there are challenges that need to be addressed. Data quality, data privacy, and the interpretability of complex models are among the key concerns. Additionally, as CPS systems become more interconnected and dynamic, the scalability and real-time requirements of data-driven models pose further challenges. Future research should focus on developing scalable algorithms, addressing data privacy concerns, and integrating real-time data streams into data-driven models. By overcoming these challenges, data-driven models have the potential to revolutionize CPS simulation and testing, enabling more accurate predictions, improved system performance, and enhanced reliability [78].

6.3.1 Introduction to Data-Driven Simulation

Data-driven simulation refers to a simulation approach that utilizes large amounts of data to improve the accuracy and realism of simulated systems [81]. Traditional simulation methods often rely on mathematical models and predefined assumptions, which may not fully capture the complexities and uncertainties present in real-world systems [82]. Data-driven simulation, on the other hand, leverages historical data and advanced data analysis techniques, such as machine learning and statistical modeling, to inform and enhance the simulation process.

The availability of vast amounts of data from various sources, such as sensors, logs, and historical records, provides an opportunity to extract valuable insights about system behavior and dynamics. Data-driven simulation involves training models on this data to learn patterns, relationships, and dependencies within the system. These models can then be used to simulate the behavior of the system under different scenarios, enabling more accurate predictions and better understanding of system performance.

The use of data-driven simulation offers several advantages [83, 84]. Firstly, it allows for the inclusion of real-world factors and variability that traditional simulation models may overlook. By incorporating actual data, the simulation can better capture the complexities and uncertainties present in the system, leading to more reliable results. Secondly, data-driven simulation can provide a more realistic representation of system behavior by learning from observed data, enabling simulations to closely match real-world outcomes [73]. This can be particularly valuable in domains where accurate predictions and understanding of system dynamics are crucial, such as transportation, healthcare, and environmental modeling. Moreover, data-driven simulation can be used to optimize system parameters, evaluate different strategies or policies, and identify potential issues or risks before implementing them in the real world. It enables engineers and researchers to test hypotheses, assess the impact of different factors, and make informed decisions based on the simulation outcomes. Additionally, data-driven simulation can facilitate the exploration of alternative scenarios and what-if analyses, allowing for a better understanding of system behavior under different conditions.

CPS systems that are connected to the cyber space not only enhance system control flexibility but also increase the likelihood of infrastructure being susceptible to attacks. To enhance the security of CPS systems and mitigate losses caused by hacker attacks or misconfigurations, continuous monitoring of the system is essential [22]. This monitoring provides a global perspective on the current operating state of the system and enables the timely detection of faults. Moreover, it is crucial to conduct hypothetical analysis on critical operations that could potentially lead to system failures, such as system reconfigurations and the addition or removal of devices. By selecting the optimal operational approach, the likelihood of failures resulting from these operations can be minimized. Digital twin technology offers a feasible solution for real-time system monitoring and hypothetical analysis. It represents a digital model that continuously interacts with its physical counterpart. Compared to traditional simulation techniques, digital twins achieve synchronization through information exchange with physical entities. They can evolve into a continuously improving system when combined with machine learning and deep learning, making them advanced simulation models. Digital twins enable the monitoring of real-time system behavior and provide valuable insights into potential issues and vulnerabilities [85]. By simulating and analyzing various scenarios, they facilitate proactive identification of faults and the development of effective mitigation strategies. The integration of digital twin technology with CPS systems enhances their resilience, security, and operational efficiency.

Overall, data-driven simulation offers a powerful approach to enhance the accuracy, realism, and predictive capabilities of simulations. By leveraging historical data and advanced data analysis techniques, it provides a valuable tool for researchers, engineers, and decision-makers to gain insights, improve system designs, and make informed decisions in various domains. As the field of data science and machine learning continues to advance, the potential applications and benefits of data-driven simulation are expected to grow, further enabling the development and optimization of complex systems in a wide range of industries.

6.3.2 Types of Data-Driven Simulation Used in CPS

Currently, research on modeling and simulation for CPS can be broadly categorized into three main areas: simulation methods, simulation tools and simulation enviroment. Detailed information can be found in Fig. 6.23.

Research on Simulation Methods: Literature suggests that due to the unique characteristics of CPS, its capabilities cannot be achieved through individual physical or information entities alone. Based on existing simulation tools and practical experiences, it can be argued that the current tools may not fully meet the simulation requirements of CPS. Therefore, there is a need for new technologies and perspectives to address these limitations.

Simulation can be classified based on several principles. One classification criterion is the type of model used. Physical simulation involves creating physical prototypes or replicas to simulate real-world systems. Computer simulation, also known as mathematical simulation, utilizes mathematical models and algorithms to simulate systems on a computer without physical components [86, 87]. Hybrid simulation combines physical and mathematical models to simulate systems that involve both physical and computational aspects, such as CPS. Another criterion is the type of computer used in the simulation. Analog simulation utilizes analog computers to simulate continuous systems by directly modeling their physical behavior. Digital simulation, on the other hand, employs digital computers to simulate discrete systems that operate in distinct time steps or events. Hybrid simulation combines analog and digital computers to simulate systems with both continuous and discrete components. Simulation can also be classified based on the nature of the signal flow within the simulation [88]. Continuous system simulation focuses on systems with continuous signals or variables that change smoothly over time. Discrete system simulation, on the other hand, deals with systems where signals or variables change at distinct time points or events. The relationship between simulation time and real-time is another criterion for classification. Real-time simulation progresses in sync with real-world

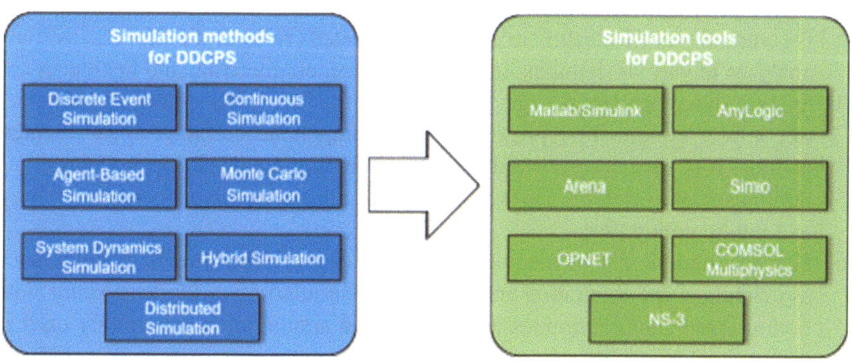

Fig. 6.23 The relationship between simulation methods and simulation tools in CPS

time, allowing for simulations that reflect real-time constraints and interactions. Fast-time simulation compresses longer time periods into shorter simulation durations, while slow-time simulation expands shorter time periods into longer simulation durations. Lastly, simulation can be categorized based on the nature of the objects being simulated. Aerospace simulation focuses on simulating aerospace systems, such as spacecraft or aircraft, to study their behavior and performance [89]. Chemical process simulation involves modeling and simulating chemical processes, including reactions and flows, to analyze their dynamics and optimize operations [90]. Economic system simulation simulates economic systems, such as markets or financial networks, to study their behavior, predict outcomes, and test policy scenarios. These classifications provide different perspectives for understanding and categorizing simulation methods, enabling researchers and practitioners to choose the most suitable approach for their specific simulation needs and objectives.

Simulation methods can be classified into various categories based on different criteria. Here are some common classifications of simulation methods:

(1) Discrete event simulation: This method focuses on modeling and simulating systems where events occur at discrete points in time. It is widely used for modeling systems with a dynamic sequence of events and processes that evolve over time, such as queuing systems, manufacturing processes, and supply chain [91].
(2) Continuous simulation: This method models and simulates systems where variables change continuously over time. It is commonly used in modeling physical systems governed by differential equations, such as fluid dynamics, electrical circuits, and mechanical systems [92].
(3) Agent-based simulation: This method models individual entities or agents within a system, each with its own behavior, decision-making rules, and interactions with other agents. Agent-based simulation is suitable for studying complex systems with emergent behaviors, such as social networks, traffic systems, and ecological systems.
(4) Monte carlo simulation: This method involves running simulations based on randomly generated inputs to analyze the behavior and outcomes of a system [93]. It is commonly used to estimate probabilities, assess risks, and perform sensitivity analyses in various domains, including finance, engineering, and healthcare.
(5) System dynamics simulation: This method focuses on modeling the behavior of complex systems by representing their components and their dynamic interactions. It is particularly useful for understanding feedback loops, delays, and nonlinearities in systems and is often applied in areas such as economics, ecology, and public policy [94].
(6) Hybrid simulation: This method combines different simulation techniques, such as discrete event simulation and continuous simulation, to model systems with both discrete and continuous aspects. Hybrid simulation is commonly used in CPS, where physical processes and computational components interact [95].
(7) Distributed simulation: This method involves distributing the simulation process across multiple computing resources or platforms to improve scalability and

performance. It is commonly used for large-scale simulations, such as network simulations, military simulations, and virtual reality applications [96].

Research on Simulation Tools:The current research on simulation tools encompasses various aspects, and some of the major simulation tools include:

(1) Matlab/Simulink: Matlab is a widely used numerical computing and simulation environment, while Simulink is an extension tool of Matlab used for modeling, simulation, and analysis of dynamic systems across multiple domains.
(2) AnyLogic: AnyLogic is a multi-method simulation software that supports continuous, discrete, and hybrid simulations. It provides tools for modeling and analyzing complex systems in various domains such as logistics, supply chain, transportation, manufacturing, etc.
(3) Arena: Arena is a discrete event simulation software used for modeling and analyzing various business and industrial systems, including production lines, transportation networks, hospitals, etc.
(4) Simio: Simio is an object-oriented simulation software that combines discrete event simulation and system dynamics simulation. It is used for modeling and analyzing complex dynamic systems.
(5) OPNET: OPNET is a network simulation software primarily used for modeling and simulating various communication networks, such as computer networks, wireless sensor networks, etc.
(6) COMSOL Multiphysics: COMSOL Multiphysics is a multiphysics simulation software used for modeling and simulating systems involving multiple physical phenomena, such as electromagnetic fields, fluid dynamics, structural mechanics, etc.
(7) NS-3: NS-3 is an open-source network simulator used for research and evaluation of communication network protocols and algorithms. It supports a wide range of network technologies and application domains.

Simulation enviroment: In the context of CPS, a simulation environment specifically tailored for CPS is a software platform or framework that enables the modeling, simulation, and analysis of complex interactions between cyber and physical components.

6.3.3 Model Validation and Uncertainty Quantification

Model validation and uncertainty quantification are crucial steps in the simulation and testing of CPS. These processes help ensuring the reliability, accuracy, and predictive capability of simulation models used in CPS development. Model validation involves assessing the degree to which a simulation model accurately represents the real-world system it intends to simulate [97].

The first step in model validation is to compare the model's behavior and outputs with real-world data or observations. This can involve collecting empirical data from physical experiments, field measurements, or historical records. By comparing the model's predictions with actual observations, researchers can identify any discrepancies or differences. Statistical analysis techniques are often used to quantify the level of agreement between the model and the real system. Uncertainty quantification is another important aspect of model validation. It involves characterizing and quantifying the uncertainties inherent in the simulation model and its inputs. Uncertainties can arise from various sources, including measurement errors, parameter estimation, simplifications in the model, and incomplete knowledge of system behavior. Uncertainty quantification techniques help estimating the range of possible outcomes and the confidence level associated with the model predictions. To validate a simulation model, researchers employ various techniques, such as sensitivity analysis, calibration, and verification. Sensitivity analysis helps identifying influential model parameters and their impact on the model's outputs. Calibration involves adjusting the model parameters to improve the agreement between the model predictions and real-world data. Verification involves checking whether the simulation model accurately represents the mathematical equations and algorithms it is based on.

In the field of verification, significant research efforts have been made to explore the validation of CPS. As highlighted by K.G. Larsen in, timed automata have become a useful form of modeling real-time embedded systems. UPPAAL, based on the theory of timed automata and symbolic state space exploration, is a commonly used modeling and verification tool for real-time systems and has emerged as a leader in the field [98]. The use of Statistical Model Checking (SMC) techniques in the verification of CPS models has also gained wide recognition. SMC leverages simulation traces and statistical techniques to obtain probability intervals for model properties. Currently, UPPAALSMC supports SMC algorithms. InJ. Suryadevara proposed a method to transform MARTE/CCSL pattern specifications into timed automata to facilitate model verification using the UPPAAL tool. These research efforts demonstrate the ongoing exploration and development of effective verification techniques for CPS. By utilizing tools such as UPPAAL and techniques like SMC and timed automata, researchers can enhance the validation process and ensure the correctness and reliability of CPS models. These advancements contribute to the overall improvement of CPS development and the assurance of system properties and behavior [99].

6.3.4 Case Studies of Simulation and Testing Using Data-Driven Models in CPS Applications

In the research of CPS simulation technology, considering the influence of information systems, traditional simulation software that focuses only on physical systems or communication networks is inadequate. Therefore, research on collaborative simulation technology suitable for CPS is necessary. Currently, CPS simulation technology primarily adopts two approaches: functional extension and co-simulation. For example, TrueTime developed by Lund University extends the functionality of MATLAB by adding network protocol models and event-triggered functions, enabling the simulation of distributed real-time control systems. However, excessive simplification of models can lead to the loss of accurate simulation of dynamic processes [100]. The application pattern of DDCPS simulation in industry is shown in Fig. 6.24.

Co-simulation, on the other hand, involves using specialized simulation tools to separately simulate the physical system and communication system, providing higher accuracy. However, the key challenge lies in achieving synchronization and data exchange between the two. Non-real-time co-simulation does not require high-precision synchronization and achieves data interaction through a synchronization module that associates the physical subsystem and information subsystem in the same time domain. Real-time hybrid simulation, on the other hand, involves establishing parallel simulation tools or multiple sets of tools to synchronize the time axes. The emergence of large-scale parallel high-speed simulators has effectively solved the synchronization and coordination challenges in parallel simulation, greatly advancing the development of real-time hybrid simulation technology in CPS.

These case studies demonstrate the effectiveness of data-driven approaches in improving system performance, optimizing resource allocation, and enhancing decision-making processes. By showcasing real-world examples, we aim to provide insights into the successful integration of data-driven simulation and testing methodologies in CPS applications.

Case study 1: Smart grid optimization. The first case study is the application of a data-driven model for smart grid system optimization. By utilizing historical power consumption data, weather data, and network information, a data-driven model is developed to simulate and optimize power distribution, load balancing, and demand response strategies. The data-driven simulation of DDCPS effectively identifies optimal grid configurations in smart grid environments, improves energy efficiency and reduces operational costs [101].

Case study 2: Healthcare system simulation. In this case study, DDCPS also plays a role in data-driven integrated simulation and diagnostics in healthcare systems. By leveraging patient data, medical records, and treatment outcomes, a data-driven simulation model is developed to evaluate different healthcare policies, resource allocation strategies, and patient flow optimization [73]. The data-driven simulation-based model effectively helps healthcare administrators to make informed decisions, improve patient care, and optimize resource utilization in complex healthcare environments.

Case study 3: Traffic management and control. The third case study focuses on the application of data-driven simulation in traffic management and control systems. By integrating traffic data from sensors, historical traffic patterns, and real-time information, a data-driven model is developed to simulate traffic flow, evaluate congestion scenarios, and optimize traffic signal timings. The data-driven simulation-based modeling helps to reduce traffic congestion, enhance traffic safety, and improve overall traffic efficiency [102].

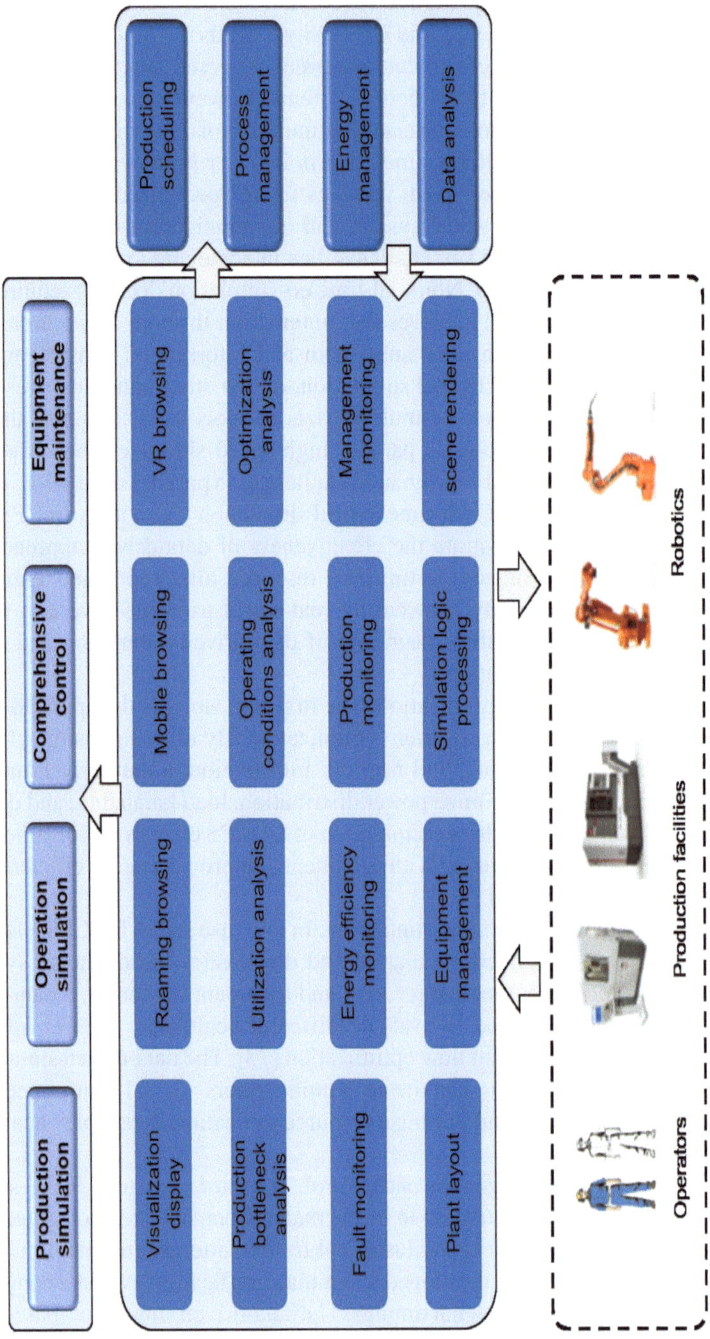

Fig. 6.24 Application patterns of DDCPS simulation in industry

Case study 4: Industrial process optimization. In this case study, we delve into the application of data-driven models for simulating and optimizing industrial processes. By utilizing sensor data, historical process parameters, and maintenance records, a data-driven simulation model is developed to identify process bottlenecks, optimize resource allocation, and predict equipment failures. Data-driven simulation process helps operatirs to realize predictive maintenance, increased productivity, and reduced downtime in industrial environments through data-driven simulation [103].

Through these case studies, we have illustrated the successful integration of data-driven simulation and testing in diverse CPS applications. The application of data-driven models has shown significant potential in optimizing system performance, improving decision-making processes, and enhancing resource allocation strategies. These case studies highlight the importance of leveraging real-world data to develop accurate and reliable simulation models, leading to more effective CPS design, operation, and management.

6.4 Summary

This chapter explores the significance of modeling and simulation in the context of DDCPS. This section provides a comprehensive overview of the key components, emphasizing the importance of complex system modeling and simulation, data-driven modeling techniques, and the simulation and testing of CPS using data-driven models.

The first section underscores the critical role of modeling and simulation in CPS. It highlights the understanding of complex system modeling for CPS, emphasizing the need for accurate representations of the intricate dynamics inherent in these systems. Furthermore, it emphasizes the importance of complex system simulation, which allows for the analysis, evaluation, and optimization of CPS performance under different scenarios.

The subsequent section focuses on data-driven modeling techniques. It introduces the concept of data-driven modeling and explores various types of data-driven models used in CPS. The chapter also delves into methods for model selection and validation, ensuring the reliability and accuracy of the models employed. It further provides real-world examples of data-driven modeling in CPS applications, demonstrating the practical relevance and effectiveness of these approaches.

The final section addresses the simulation and testing of cyber physical systems using data-driven models. It introduces data-driven simulation, which leverages the insights gained from data analysis to simulate CPS behavior. The chapter discusses different types of data-driven simulation techniques employed in CPS and emphasizes the importance of model validation and uncertainty quantification. It concludes with case studies showcasing the application of data-driven models in simulating and testing CPS, providing valuable insights for system analysis, optimization, and performance evaluation.

Overall, this chapter emphasizes the significance of modeling and simulation in DDCPS. By leveraging data-driven approaches, researchers and practitioners can effectively model complex CPS, optimize system performance, predict system behavior, and enhance system availability. The book offers valuable insights and methodologies that can be applied across various industries, including manufacturing, healthcare, transportation, and energy, leading to improved productivity, energy management, and predictive maintenance.

Competing Interests

1. The research is supported by the National Key Research and Development Project under Grant 2023YFB3307300, the National Natural Science Foundation of China under Grants 62373014, 62422301, 62125301, and 92467205.
2. The authors have no conflicts of interest to declare that are relevant to the content of this chapter.

References

1. E. A. Lee, "The past, present and future of cyber-physical systems: A focus on models," *Sensors*, vol. 15, no. 3, pp. 4837–4869, 2015.
2. F. Tao, Q. Qi, L. Wang, and A. Nee, "Digital twins and cyber–physical systems toward smart manufacturing and industry 4.0: Correlation and comparison," *Engineering*, vol. 5, no. 4, pp. 653–661, 2019.
3. V. Alcácer and V. Cruz-Machado, "Scanning the industry 4.0: A literature review on technologies for manufacturing systems," *Engineering science and technology, an international journal*, vol. 22, no. 3, pp. 899–919, 2019.
4. J. Dang, Y. Liang, L. Huang, L. Zhu, Y. Tang, Q. Wang, and B. Lv, "Accurate modeling method of power system load based on online measurement under cps environment," in *2017 IEEE 7th Annual International Conference on CYBER Technology in Automation, Control, and Intelligent Systems (CYBER)*, pp. 1361–1366, IEEE, 2017.
5. G. Karsai and J. Sztipanovits, "Model-integrated development of cyber-physical systems," in *Software Technologies for Embedded and Ubiquitous Systems: 6th IFIP WG 10.2 International Workshop, SEUS 2008, Anacarpi, Capri Island, Italy, October 1–3, 2008 Proceedings 6*, pp. 46–54, Springer, 2008.
6. R. Drechsler and U. Kühne, "Formal modeling and verification of cyber-physical systems," *1st International Summer School on Methods and Tools for the Design of Digital Systems, Bremen, Germany*, 2015.
7. R. Baheti and H. Gill, "Cyber-physical systems," *The impact of control technology*, vol. 12, no. 1, pp. 161–166, 2011.
8. J.-P. A. Yaacoub, O. Salman, H. N. Noura, N. Kaaniche, A. Chehab, and M. Malli, "Cyber-physical systems security: Limitations, issues and future trends," *Microprocessors and microsystems*, vol. 77, p. 103201, 2020.
9. R. Alguliyev, Y. Imamverdiyev, and L. Sukhostat, "Cyber-physical systems and their security issues," *Computers in Industry*, vol. 100, pp. 212–223, 2018.
10. P. Derler, E. A. Lee, and A. S. Vincentelli, "Modeling cyber–physical systems," *Proceedings of the IEEE*, vol. 100, no. 1, pp. 13–28, 2011.
11. H. Chen, "Applications of cyber-physical system: a literature review," *Journal of Industrial Integration and Management*, vol. 2, no. 03, p. 1750012, 2017.
12. F. Barat and R. Lauwereins, "Reconfigurable instruction set processors: a survey," in *Proceedings 11th International Workshop on Rapid System Prototyping. RSP 2000. Shortening the Path from Specification to Prototype (Cat. No. PR00668)*, pp. 168–173, IEEE, 2000.
13. S. W. Golomb, "Mathematical models: Uses and limitations," *IEEE Transactions on Reliability*, vol. 20, no. 3, pp. 130–131, 1971.

14. Y. Liu, X. Ma, L. Shu, G. P. Hancke, and A. M. Abu-Mahfouz, "From industry 4.0 to agriculture 4.0: Current status, enabling technologies, and research challenges," *IEEE Transactions on Industrial Informatics*, vol. 17, no. 6, pp. 4322–4334, 2020.
15. T. D. Oesterreich and F. Teuteberg, "Understanding the implications of digitisation and automation in the context of industry 4.0: A triangulation approach and elements of a research agenda for the construction industry," *Computers in industry*, vol. 83, pp. 121–139, 2016.
16. M. Thiebaut de Schotten and S. J. Forkel, "The emergent properties of the connected brain," *Science*, vol. 378, pp. 505–510, 2022.
17. P. Dallasega, E. Rauch, and C. Linder, "Industry 4.0 as an enabler of proximity for construction supply chains: A systematic literature review," *Computers in industry*, vol. 99, pp. 205–225, 2018.
18. Y. Yuan, X. Tang, W. Zhou, W. Pan, X. Li, H.-T. Zhang, H. Ding, and J. Goncalves, "Data driven discovery of cyber physical systems," *Nature communications*, vol. 10, no. 1, p. 4894, 2019.
19. D. Sonntag, S. Zillner, P. van der Smagt, and A. Lörincz, "Overview of the cps for smart factories project: Deep learning, knowledge acquisition, anomaly detection and intelligent user interfaces," *Industrial Internet of Things: Cybermanufacturing Systems*, pp. 487–504, 2017.
20. N. Boccara and N. Boccara, *Modeling complex systems*, vol. 1. Springer, 2010.
21. I. Priyadarshini, R. Kumar, L. M. Tuan, L. H. Son, H. V. Long, R. Sharma, and S. Rai, "A new enhanced cyber security framework for medical cyber physical systems," *SICS Software-Intensive Cyber-Physical Systems*, pp. 1–25, 2021.
22. E. A. Lee, "Cps foundations," in *Proceedings of the 47th design automation conference*, pp. 737–742, 2010.
23. J. Buck, S. Ha, E. A. Lee, and D. G. Messerschmitt, "Ptolemy: A framework for simulating and prototyping heterogeneous systems," in *Readings in hardware/software co-design*, pp. 527–543, 2001.
24. C. Ptolemaeus, *System design, modeling, and simulation: using Ptolemy II*, vol. 1. Ptolemy. org Berkeley, 2014.
25. H.-L. Tsai, C.-S. Tu, Y.-J. Su, *et al.*, "Development of generalized photovoltaic model using matlab/simulink," in *Proceedings of the world congress on Engineering and computer science*, vol. 2008, pp. 1–6, San Francisco, USA, 2008.
26. O. Badreddin, V. Abdelzad, T. Lethbridge, and M. Elaasar, "fsysml: Foundational executable sysml for cyber-physical system modeling.," in *GEMOC@ MoDELS*, pp. 38–51, 2016.
27. D. Henriksson, A. Cervin, and K.-E. Årzén, "Truetime: Real-time control system simulation with matlab/simulink," in *Proceedings of the Nordic MATLAB Conference*, Copenhagen, Denmark, 2003.
28. J. E. Carrion, B. Spencer Jr, and B. M. Phillips, "Real-time hybrid simulation for structural control performance assessment," *Earthquake Engineering and Engineering Vibration*, vol. 8, no. 4, pp. 481–492, 2009.
29. M. Krishna, S. M. B. Chowdary, P. Nancy, and V. Arulkumar, "A Survey on Multimedia Analytics in Security Systems of Cyber Physical Systems and IoT," in *2021 2nd International Conference on Smart Electronics and Communication (ICOSEC)*, pp. 1–7, Oct. 2021.
30. A. Mishra, A. V. Jha, B. Appasani, A. K. Ray, D. K. Gupta, and A. N. Ghazali, "Emerging technologies and design aspects of next generation cyber physical system with a smart city application perspective," *International Journal of System Assurance Engineering and Management*, pp. 1–23, 2022.
31. Y. Gil, D. Garijo, D. Khider, C. A. Knoblock, V. Ratnakar, M. Osorio, H. Vargas, M. Pham, J. Pujara, B. Shbita, *et al.*, "Artificial intelligence for modeling complex systems: taming the complexity of expert models to improve decision making," *ACM Transactions on Interactive Intelligent Systems*, vol. 11, no. 2, pp. 1–49, 2021.
32. F. Dressler, "Cyber physical social systems: Towards deeply integrated hybridized systems," in *2018 International Conference on Computing, Networking and Communications (ICNC)*, pp. 420–424, IEEE, 2018.
33. N. Jazdi, "Cyber physical systems in the context of industry 4.0," in *2014 IEEE international conference on automation, quality and testing, robotics*, pp. 1–4, IEEE, 2014.

34. J. Lee, B. Bagheri, and H.-A. Kao, "A cyber-physical systems architecture for industry 4.0-based manufacturing systems," *Manufacturing letters*, vol. 3, pp. 18–23, 2015.
35. N. Pandiarajan and R. Muthu, "Mathematical modeling of photovoltaic module with simulink," in *2011 1st International Conference on Electrical Energy Systems*, pp. 258–263, IEEE, 2011.
36. A. Majeed, Y. Zhang, S. Ren, J. Lv, T. Peng, S. Waqar, and E. Yin, "A big data-driven framework for sustainable and smart additive manufacturing," *Robotics and Computer-Integrated Manufacturing*, vol. 67, p. 102026, 2021.
37. Z. Ge, "Review on data-driven modeling and monitoring for plant-wide industrial processes," *Chemometrics and Intelligent Laboratory Systems*, vol. 171, pp. 16–25, 2017.
38. P. O'Donovan, K. Leahy, K. Bruton, and D. T. O'Sullivan, "An industrial big data pipeline for data-driven analytics maintenance applications in large-scale smart manufacturing facilities," *Journal of Big Data*, vol. 2, no. 1, pp. 1–26, 2015.
39. S. Li and P. Zhao, "Big data driven vehicle battery management method: A novel cyber-physical system perspective," *Journal of Energy Storage*, vol. 33, p. 102064, 2021.
40. J. N. Kutz, *Data-driven modeling & scientific computation: methods for complex systems & big data*. Oxford University Press, 2013.
41. N. Zhang, "A cloud-based platform for big data-driven cps modeling of robots," *IEEE Access*, vol. 9, pp. 34667–34680, 2021.
42. J. N. Kutz, S. L. Brunton, B. W. Brunton, and J. L. Proctor, *Dynamic mode decomposition: data-driven modeling of complex systems*. SIAM, 2016.
43. Z.-H. Zhou, *Machine learning*. Springer Nature, 2021.
44. M. I. Jordan and T. M. Mitchell, "Machine learning: Trends, perspectives, and prospects," *Science*, vol. 349, no. 6245, pp. 255–260, 2015.
45. H. Wang, Z. Lei, X. Zhang, B. Zhou, and J. Peng, "Machine learning basics," *Deep learning*, pp. 98–164, 2016.
46. D. C. Montgomery, E. A. Peck, and G. G. Vining, *Introduction to linear regression analysis*. John Wiley & Sons, 2021.
47. X. Su, X. Yan, and C.-L. Tsai, "Linear regression," *Wiley Interdisciplinary Reviews: Computational Statistics*, vol. 4, no. 3, pp. 275–294, 2012.
48. D. G. Kleinbaum, K. Dietz, M. Gail, M. Klein, and M. Klein, *Logistic regression*. Springer, 2002.
49. H. Abdi and L. J. Williams, "Principal component analysis," *Wiley interdisciplinary reviews: computational statistics*, vol. 2, no. 4, pp. 433–459, 2010.
50. S. R. Safavian and D. Landgrebe, "A survey of decision tree classifier methodology," *IEEE transactions on systems, man, and cybernetics*, vol. 21, no. 3, pp. 660–674, 1991.
51. L. Breiman, "Random forests," *Machine learning*, vol. 45, pp. 5–32, 2001.
52. I. Steinwart and A. Christmann, *Support vector machines*. Springer Science & Business Media, 2008.
53. G. Guo, H. Wang, D. Bell, Y. Bi, and K. Greer, "Knn model-based approach in classification," in *On The Move to Meaningful Internet Systems 2003: CoopIS, DOA, and ODBASE: OTM Confederated International Conferences, CoopIS, DOA, and ODBASE 2003, Catania, Sicily, Italy, November 3-7, 2003. Proceedings*, pp. 986–996, Springer, 2003.
54. I. Rish *et al.*, "An empirical study of the naive bayes classifier," in *IJCAI 2001 workshop on empirical methods in artificial intelligence*, vol. 3, pp. 41–46, 2001.
55. Y. LeCun, Y. Bengio, and G. Hinton, "Deep learning," *nature*, vol. 521, no. 7553, pp. 436–444, 2015.
56. N. Kalchbrenner, E. Grefenstette, and P. Blunsom, "A convolutional neural network for modelling sentences," arXiv:1404.2188, 2014.
57. M. Schuster and K. K. Paliwal, "Bidirectional recurrent neural networks," *IEEE transactions on Signal Processing*, vol. 45, no. 11, pp. 2673–2681, 1997.
58. S. Hochreiter and J. Schmidhuber, "Long short-term memory," *Neural computation*, vol. 9, no. 8, pp. 1735–1780, 1997.
59. I. Goodfellow, J. Pouget-Abadie, M. Mirza, B. Xu, D. Warde-Farley, S. Ozair, A. Courville, and Y. Bengio, "Generative adversarial networks," *Communications of the ACM*, vol. 63, no. 11, pp. 139–144, 2020.

60. V. Mnih, K. Kavukcuoglu, D. Silver, A. Graves, I. Antonoglou, D. Wierstra, and M. Riedmiller, "Playing atari with deep reinforcement learning," arXiv:1312.5602, 2013.
61. T. P. Lillicrap, J. J. Hunt, A. Pritzel, N. Heess, T. Erez, Y. Tassa, D. Silver, and D. Wierstra, "Continuous control with deep reinforcement learning," arXiv:1509.02971, 2015.
62. R. Rai and C. K. Sahu, "Driven by data or derived through physics? a review of hybrid physics guided machine learning techniques with cyber-physical system (cps) focus," *IEEE Access*, vol. 8, pp. 71050–71073, 2020.
63. E. A. Lee, "Cyber physical systems: Design challenges," in *2008 11th IEEE international symposium on object and component-oriented real-time distributed computing (ISORC)*, pp. 363–369, IEEE, 2008.
64. A. A. Jamal, A.-A. M. Majid, A. Konev, T. Kosachenko, and A. Shelupanov, "A review on security analysis of cyber physical systems using machine learning," *Materials Today: Proceedings*, vol. 80, pp. 2302–2306, 2023.
65. S. Arlot and A. Celisse, "A survey of cross-validation procedures for model selection," 2010.
66. P. Probst, M. N. Wright, and A.-L. Boulesteix, "Hyperparameters and tuning strategies for random forest," *Wiley Interdisciplinary Reviews: data mining and knowledge discovery*, vol. 9, no. 3, p. e1301, 2019.
67. Q. Liu, L. Yu, L. Luo, Q. Dou, and P. A. Heng, "Semi-supervised medical image classification with relation-driven self-ensembling model," *IEEE transactions on medical imaging*, vol. 39, no. 11, pp. 3429–3440, 2020.
68. J. Wang, Y. Li, R. X. Gao, and F. Zhang, "Hybrid physics-based and data-driven models for smart manufacturing: Modelling, simulation, and explainability," *Journal of Manufacturing Systems*, vol. 63, pp. 381–391, 2022.
69. M. A. Chao, C. Kulkarni, K. Goebel, and O. Fink, "Fusing physics-based and deep learning models for prognostics," *Reliability Engineering & System Safety*, vol. 217, p. 107961, 2022.
70. A. Agrawal, C. M. Ahmed, and E.-C. Chang, "Poster: Physics-based attack detection for an insider threat model in a cyber-physical system," in *Proceedings of the 2018 on Asia Conference on Computer and Communications Security*, pp. 821–823, 2018.
71. R. Alur, *Principles of cyber-physical systems*. MIT press, 2015.
72. J. Baudrillard, *Simulacra and simulation*. University of Michigan press, 1994.
73. E. Negri, L. Fumagalli, and M. Macchi, "A review of the roles of digital twin in cps-based production systems," *Procedia manufacturing*, vol. 11, pp. 939–948, 2017.
74. S. Weyer, T. Meyer, M. Ohmer, D. Gorecky, and D. Zühlke, "Future modeling and simulation of cps-based factories: an example from the automotive industry," *Ifac-Papersonline*, vol. 49, no. 31, pp. 97–102, 2016.
75. I. A. Oyewumi, A. A. Jillepalli, P. Richardson, M. Ashrafuzzaman, B. K. Johnson, Y. Chakhchoukh, M. A. Haney, F. T. Sheldon, and D. C. de Leon, "Isaac: The idaho cps smart grid cybersecurity testbed," in *2019 IEEE Texas Power and Energy Conference (TPEC)*, pp. 1–6, IEEE, 2019.
76. S. Mittal, A. Tolk, A. Pyles, N. Van Balen, and K. Bergollo, "Digital twin modeling, co-simulation and cyber use-case inclusion methodology for iot systems," in *2019 Winter Simulation Conference (WSC)*, pp. 2653–2664, IEEE, 2019.
77. Y. Jiang, H. Song, Y. Yang, H. Liu, M. Gu, Y. Guan, J. Sun, and L. Sha, "Dependable model-driven development of cps: From stateflow simulation to verified implementation," *ACM Transactions on Cyber-Physical Systems*, vol. 3, no. 1, pp. 1–31, 2018.
78. M. Schmidt and C. Åhlund, "Smart buildings as cyber-physical systems: Data-driven predictive control strategies for energy efficiency," *Renewable and Sustainable Energy Reviews*, vol. 90, pp. 742–756, 2018.
79. O. Niggemann, G. Biswas, J. S. Kinnebrew, H. Khorasgani, S. Volgmann, and A. Bunte, "Data-driven monitoring of cyber-physical systems leveraging on big data and the internet-of-things for diagnosis and control.," in *DX*, pp. 185–192, 2015.
80. B. B. Cassoli, N. Jourdan, P. H. Nguyen, S. Sen, E. Garcia-Ceja, and J. Metternich, "Frameworks for data-driven quality management in cyber-physical systems for manufacturing: A systematic review," *Procedia CIRP*, vol. 112, pp. 567–572, 2022.

81. G. Hüllen, J. Zhai, S. H. Kim, A. Sinha, M. J. Realff, and F. Boukouvala, "Managing uncertainty in data-driven simulation-based optimization," *Computers & Chemical Engineering*, vol. 136, p. 106519, 2020.
82. J. McKinney, D. A. Cook, D. Wood, and R. Hatala, "Simulation-based training for cardiac auscultation skills: systematic review and meta-analysis," *Journal of general internal medicine*, vol. 28, pp. 283–291, 2013.
83. F. Müller-Plathe, "A simple nonequilibrium molecular dynamics method for calculating the thermal conductivity," *The Journal of chemical physics*, vol. 106, no. 14, pp. 6082–6085, 1997.
84. J. De Winter, P. M. van Leeuwen, R. Happee, et al., "Advantages and disadvantages of driving simulators: A discussion," in *Proceedings of measuring behavior*, vol. 2012, pp. 28–31, 2012.
85. F. Tao and Q. Qi, "Make more digital twins," *Nature*, vol. 573, no. 7775, pp. 490–491, 2019.
86. W. F. Van Gunsteren and H. J. Berendsen, "Computer simulation of molecular dynamics: methodology, applications, and perspectives in chemistry," *Angewandte Chemie International Edition in English*, vol. 29, no. 9, pp. 992–1023, 1990.
87. N. H. Malik, "A review of the charge simulation method and its applications," *IEEE Transactions on electrical insulation*, vol. 24, no. 1, pp. 3–20, 1989.
88. R. Sinha, C. J. Paredis, V.-C. Liang, and P. K. Khosla, "Modeling and simulation methods for design of engineering systems," *J. Comput. Inf. Sci. Eng.*, vol. 1, no. 1, pp. 84–91, 2001.
89. A. K. Noor, S. L. Venneri, D. B. Paul, and M. A. Hopkins, "Structures technology for future aerospace systems," *Computers & Structures*, vol. 74, no. 5, pp. 507–519, 2000.
90. D. J. Higham, "Modeling and simulating chemical reactions," *SIAM review*, vol. 50, no. 2, pp. 347–368, 2008.
91. A. Varga, "Discrete event simulation system," in *Proc. of the European Simulation Multiconference (ESM'2001)*, pp. 1–7, 2001.
92. F. E. Cellier and E. Kofman, *Continuous system simulation*. Springer Science & Business Media, 2006.
93. S. F. Railsback, S. L. Lytinen, and S. K. Jackson, "Agent-based simulation platforms: Review and development recommendations," *Simulation*, vol. 82, no. 9, pp. 609–623, 2006.
94. B. K. Bala, F. M. Arshad, K. M. Noh, et al., "System dynamics," *Modelling and Simulation*, vol. 274, 2017.
95. G. Figueira and B. Almada-Lobo, "Hybrid simulation–optimization methods: A taxonomy and discussion," *Simulation Modelling Practice and Theory*, vol. 46, pp. 118–134, 2014.
96. R. M. Czekster, C. Morisset, J. A. Clark, S. Soudjani, C. Patsios, and P. Davison, "Systematic review of features for co-simulating security incidents in cyber-physical systems," *Security and Privacy*, vol. 4, no. 3, p. e150, 2021.
97. X. Zheng, C. Julien, M. Kim, and S. Khurshid, "Perceptions on the state of the art in verification and validation in cyber-physical systems," *IEEE Systems Journal*, vol. 11, no. 4, pp. 2614–2627, 2015.
98. K. Wan, K. Man, and D. Hughes, "Specification, analyzing challenges and approaches for cyber-physical systems (cps).," *Engineering Letters*, vol. 18, no. 3, 2010.
99. M. A. Mohamed, G. Kardas, and M. Challenger, "Model-driven engineering tools and languages for cyber-physical systems–a systematic literature review," *IEEE Access*, vol. 9, pp. 48605–48630, 2021.
100. V. Gunes, S. Peter, T. Givargis, and F. Vahid, "A survey on concepts, applications, and challenges in cyber-physical systems.," *KSII Trans. Internet Inf. Syst.*, vol. 8, no. 12, pp. 4242–4268, 2014.
101. K. M. Tan, V. K. Ramachandaramurthy, and J. Y. Yong, "Integration of electric vehicles in smart grid: A review on vehicle to grid technologies and optimization techniques," *Renewable and Sustainable Energy Reviews*, vol. 53, pp. 720–732, 2016.
102. A. Agneessens, F. Buemi, S. Delucchi, M. Massa, G. Agosta, A. Barenghi, C. Brandolese, W. Fornaciari, G. Pelosi, E. Ferrari, et al., "Safe cooperative cps: A v2i traffic management scenario in the safecop project," in *2016 International Conference on Embedded Computer Systems: Architectures, Modeling and Simulation (SAMOS)*, pp. 320–327, IEEE, 2016.
103. E. Marilungo, A. Papetti, M. Germani, and M. Peruzzini, "From pss to cps design: a real industrial use case toward industry 4.0," *Procedia Cirp*, vol. 64, pp. 357–362, 2017.

Chapter 7
Fault Detection and Predictive Maintenance in Cyber Physical Systems

Abstract This chapter explores data-driven fault detection and predictive maintenance techniques in the context of cyber physical systems (CPS). We present the development history of fault detection and predictive maintenance technologies in CPS, and provide a detailed overview of the emergence and significance of data-driven approaches. In addition, we present application background and case studies that demonstrate the practical application and effectiveness of data-driven techniques in fault detection and predictive maintenance within the CPS fields.

Cyber physical systems (CPS) integrates sensors, computers, communication, and control technologies, enabling a deep fusion of the physical and digital worlds. The development of fault detection and predictive maintenance technology in CPS provides assurance for the safe and reliable operation of these systems. However, traditional fault detection and maintenance methods often rely on human experience and rules, which lack intelligence and efficiency, making them insufficient to address the challenges posed by the complexity of CPS. As a result, data-driven fault detection and predictive maintenance technology have emerged as a promising direction to tackle CPS failures. Through case studies, we demonstrate the practical effectiveness of data-driven techniques in fault detection and predictive maintenance, ensuring the smooth operation of CPS.

7.1 An Overview of Fault Detection and Maintenance

In an era of rapid technological advancement, CPS have propelled the development of various fields, including manufacturing, transportation, and healthcare. However, with increasing complexity and interconnectedness, the reliable operation of CPS has become challenging. In the CPS environment, faults can lead to system failures and compromise security. Therefore, the investigation of effective fault detection and predictive maintenance strategies is a worthy endeavor. Fault detection involves real-time monitoring of system conditions in CPS to identify potential faults or anomalies. Predictive maintenance relies on data collection and analysis to proactively take maintenance measures, preventing faults from occurring or escalating, and

F. Li et al., *Data-Driven Cyber Physical Systems*,
https://doi.org/10.1007/978-981-96-8709-1_7

maximizing the reliability and stability of CPS. Our aim is to explore and develop innovative technologies that leverage data analytics, machine learning (ML), and advanced sensing techniques to seamlessly integrate fault detection and predictive maintenance. By detecting faults in real-time and implementing proactive maintenance, we can significantly enhance the reliability, availability, and lifespan of CPS, optimizing system performance, and reducing downtime. The ultimate goal is to prevent system failures caused by faults and mitigate their potential consequences, ensuring the safe and smooth operation of CPS across various domains.

7.1.1 The Development of CPS Fault Detection

Fault detection refers to the process of identifying and diagnosing anomalies or failures that can occur in the physical or cyber components of the system [1]. The objective of fault detection is to ensure the reliability and safety of the system by detecting and mitigating faults before they cause significant damage [2]. Fault detection involves various techniques such as monitoring, modeling, and analyzing system behavior, and comparing it to expected or normal behavior [3]. Fault detection ensures that CPS operate optimally and mitigate potential hazards or accidents [4].

The real-time fault detection method is introduced in detail. Considering diverse fault detection sources, including equipment signals, process variables, and semantic data, and the essential requirement for extracting figurative feature data independently [5, 6], we classify current real-time fault detection methods into three categories: independent feature extraction, end-to-end neural networks, and qualitative knowledge reasoning. The classification diagram of methods is shown in Fig. 7.1.

Independent feature extraction: The fault detection methods based on feature extraction [7] are widely used. The concept behind methods focusing on autonomous feature extraction is to derive valuable feature information from monitoring data, enabling the establishment of a connection between these features and system states for fault detection through feature classification. The primary step of this approach involves the extraction and classification of characteristic information that distinguishes between normal and faulty states. In instances where noise reduction or simplification of computational demands is necessary, techniques like principal component analysis (PCA) [8], linear discriminant analysis (LDA) [9], can be used to compress features and eliminate redundant information. Reduced-dimensional feature components play a pivotal role in enhancing the performance of models. Presently, independent feature extraction-based methods are widely applied in the field of fault detection, particularly in the analysis of vibration and acoustic signals from mechanical equipment.

End-to-End neural networks: The methods based on machine learning [10] can meet the need of fault detection. With the powerful feature learning and pattern recognition capabilities, neural networks perform the entire abstract feature learning and classification process in an end-to-end manner without needing to perform independent feature extraction. The principle of end-to-end neural network-based fault detection is as follows: first, the network automatically learns deep abstract features

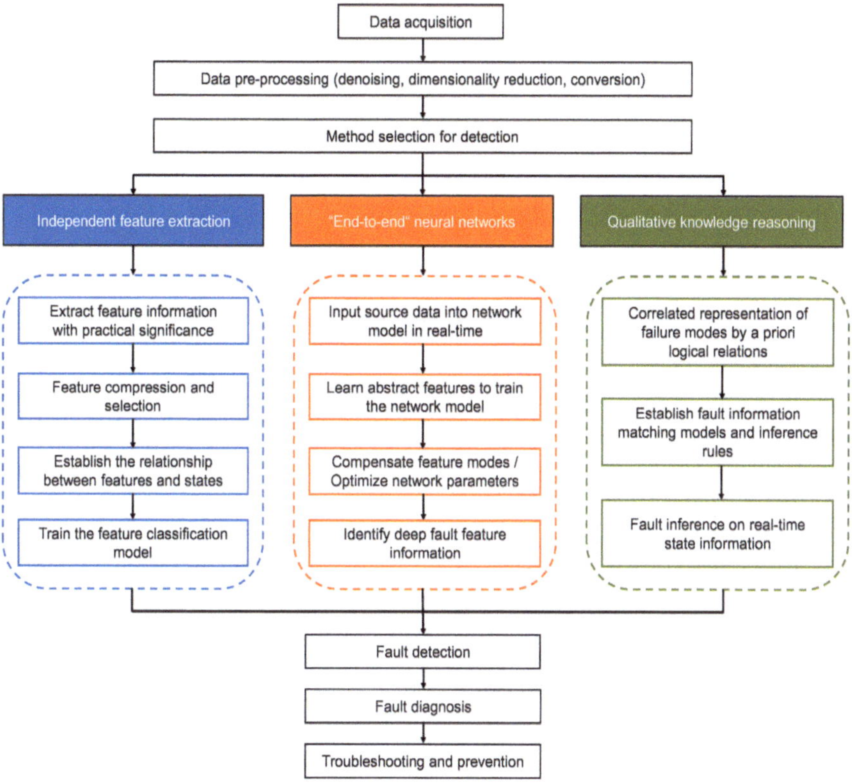

Fig. 7.1 Classification of fault detection methods. Methods are divided into three categories, which are: independent feature extraction, end-to-end neural networks, and qualitative knowledge reasoning

of a large amount of input data layer by layer, then the network is trained according to the mapping of feature attributes to fault patterns. Finally, the trained network is used for fault detection of new monitoring data. Some deep hierarchical networks are even capable of taking source data directly as input and monitoring it in real time. In addition, researchers can further enhance the fault detection performance of the model by introducing optimization algorithms to optimize the network structure and parameters.

Qualitative knowledge reasoning: The methods based on qualitative knowledge [11] is a common fault detection technique. The methods usually use a combination of qualitative and quantitative analysis, establish inference rules with product knowledge, and use data as an auxiliary basis to achieve real-time reasoning about fault information. The brief process of this kind of method is as follows: first, the relevant representation of failure modes is made using a priori qualitative knowledge or data. Then, the fault information matching model and inference rules are established to perform realtime fault inference on the system state.

Significant progress has been made in the field of CPS fault detection [12, 13]. By monitoring and analyzing system operation data in real time, fault detection technology can detect potential problems early and effectively prevent system failures or performance degradation. Complex CPS structure and massive data processing require efficient and accurate fault detection methods. Future researches will continue to address existing challenges and further drive innovation and application of CPS fault detection technology.

7.1.2 The Development of CPS Maintenance

Maintenance in CPS ensures the stable and efficient operation of the system [14]. Effective maintenance minimizes downtime, reduces operational costs, and ensures continuous and stable operation of the system under various challenges and workloads. Fault maintenance is not only applicable in industrial manufacturing and transportation sectors but also plays a crucial role in various fields such as healthcare and smart cities [15, 16]. CPS collect data from various sources, including sensors, machine logs, and historical maintenance records. By analyzing the data, potential issues can be detected, and faults can be identified and promptly addressed. Continuous improvement and innovative fault maintenance methods make CPS systems safer, more efficient, and more reliable, providing people with a better experience in their production and daily lives.

Maintenance techniques can be divided into three types: condition-based [17], time-based [18], and predictive analytics-based [19]. Condition-based maintenance uses sensors and other monitoring devices to measure the condition of a physical component and predict when it will fail. Time-based maintenance involves replacing a component after a set amount of time, regardless of its condition. Predictive analytics-based maintenance uses data analytics and ML to analyze large volumes of data and predict when maintenance is needed.

Condition-based maintenance (CBM): Condition-based maintenance uses sensors and other monitoring devices to measure the condition of a physical component and predict when it will fail. CBM techniques can detect changes in temperature, vibration, sound, and other factors to identify potential issues. By monitoring the health of physical components in real-time, CBM techniques can help organizations take proactive maintenance measures, reducing the risk of unplanned downtime and increasing the longevity of their systems.

Time-based maintenance (TBM): Time-based maintenance involves replacing a component after a set amount of time, regardless of its condition. This approach is often used for components that have a predictable wear-and-tear pattern or that are not critical to system function. TBM is less effective than CBM for identifying potential issues, but it is still a useful approach for ensuring that components are replaced before they fail.

Predictive analytics-based maintenance (PABM): Predictive analytics-based maintenance uses data analytics and ML to analyze large volumes of data and predict when maintenance is needed. PABM identifies trends and patterns that may indicate potential issues, such as unusual behavior or deviations from normal operating conditions, which helps organizations identify potential problems earlier, allowing them to take action before they become major issues. In addition, PABM considers a variety of factors, such as usage patterns, weather conditions, and historical performance data, to provide a more accurate prediction of when maintenance is needed.

Predictive maintenance techniques can be used to monitor the health and performance of physical systems in real time, providing organizations with the ability to take proactive measures to address potential issues before they result in costly downtime or system failure [20]. By continuously monitoring physical components using sensors and other monitoring devices, predictive maintenance can detect changes in various factors such as temperature, vibration, and sound, allowing for early identification of potential problems [21]. Real-time monitoring enables timely action, mitigating risks and preventing system failures. By adopting a proactive approach to maintenance, organizations can reduce maintenance costs, increase equipment lifespan, and improve overall system reliability and performance. CPS are complex systems that integrate physical and digital components, and any disruption in the functioning of these components can have serious consequences [22]. Predictive maintenance enables organizations to monitor the health and performance of their physical systems in real-time and take proactive maintenance measures, reducing the need for reactive maintenance that can be time-consuming and expensive.

7.1.3 Future Trends of Fault Detection and Predictive Maintenance

In CPS, fault detection and predictive maintenance enable early identification and diagnosis of faults, which not only prevents accidents and injuries, but also reduces maintenance costs and improves system efficiency [23].

Rule-driven methods utilize predefined rules and logic for fault detection and predictive maintenance [24]. The rules can be based on domain experts' knowledge and experience or specific rule sets. Expert system-driven methods combine knowledge representation and inference techniques to simulate the decision-making process of domain experts for fault detection and predictive maintenance. Expert system-driven methods typically employ domain experts' rules and logic for inference and decision-making to achieve automated fault detection and maintenance decision-making.

Model-driven methods in CPS offer several advantages in the domain of fault detection and predictive maintenance [25]. First, the model-driven methods utilize the knowledge of system dynamics and behavior captured in the model. By comparing the actual system response with the expected model output, deviations can be detected in time and accurate fault detection can be achieved. Second, the model-driven methods promote proactive maintenance by predicting potential failures based on the model's analysis of system conditions and performance, which allows organizations to schedule maintenance activities in advance, minimizing unexpected downtime and maximizing operational efficiency. Additionally, model-driven methods support decision-making processes by simulating different scenarios and evaluating the impact of maintenance actions on system performance, which helps optimize resource allocation and reduces maintenance costs [26]. Lastly, the use of models allows for continuous monitoring and real-time analysis of system health, enabling early detection of anomalies and timely intervention to prevent catastrophic failures. The model-driven methods empower organizations to enhance reliability, safety, and productivity in CPS by leveraging accurate predictions and informed maintenance strategies.

Model-driven fault detection and predictive maintenance have some limitations in CPS. One of the main drawbacks is the need for accurate and up-to-date models, which can be challenging in dynamic and complex systems. The models may not accurately capture all possible system states and behaviors, leading to false alarms or missed detections. Additionally, the models may become outdated or invalid as the system evolves or new components are introduced, requiring significant effort and resources to update or retrain. Another limitation is the computational cost and complexity of implementing model-based approaches, which may not be feasible in real-time or resource-constrained systems.

Data-driven fault detection and predictive maintenance leverage a vast amount of real-time data for analysis and processing, enabling the real-time monitoring of system status and early detection of potential faults or anomalies, thereby helping to prevent system failures or further escalation. Moreover, data-driven methods automate the process of fault detection and predictive maintenance, reducing manual intervention, and enhancing efficiency and scalability. Data-driven fault detection and predictive maintenance represent a crucial direction for future development.

7.2 Data-Driven Approaches for Fault Detection and Predictive Maintenance

In CPS, ensuring safety and stability is of utmost importance [27]. With the integration of physical and computational components, CPS operate in dynamic and complex environments where any failure or instability can have severe consequences. The safety of individuals, the integrity of the system, and the protection of valuable assets are critical concerns in CPS applications such as autonomous vehicles, smart grids, and industrial automation. To ensure safety and stability, fault detection and predictive maintenance techniques are employed [28]. By continuously monitoring the system's performance and analyzing data patterns, potential faults and anomalies can be detected in advance. The proactive methods enable timely maintenance actions to be taken, preventing critical failures and minimizing disruptions to the systems.

In the field of CPS, model-driven fault detection and predictive maintenance heavily rely on accurate mathematical models of the physical systems under consideration. However, constructing such models can be a complex and time-consuming task, requiring in-depth domain knowledge and expertise [29]. Additionally, model-driven approaches may struggle to capture the intricate dynamics and complexities of real-world systems, leading to suboptimal fault detection performance. In contrast, data-driven approaches leverage large amounts of sensor data to automatically learn patterns and correlations, enabling the detection and prediction of anomalies or faults without explicitly relying on precise models. By using data-driven techniques, CPS can overcome the limitations of model-driven approaches and achieve more robust and accurate fault detection capabilities.

Data-driven fault detection and predictive maintenance in CPS offer several advantages [30]. First, data-driven approaches can handle the complex and dynamic nature of CPS, as they rely on the analysis of real-time sensor data to identify anomalies and patterns. The flexibility allows for more accurate and timely detection of faults, leading to proactive maintenance and reduced downtime. Second, data-driven methods can adapt to evolving system behavior by continuously learning from new data, enabling the detection of previously unknown or changing fault patterns. Additionally, data-driven techniques can leverage ML to uncover hidden correlations and complex relationships within the data, enhancing the accuracy of fault detection and prediction. Moreover, data-driven approaches eliminate the need for explicit mathematical models, reducing the burden of model development and maintenance.

7.2.1 Data-Driven Fault Detection Approaches

In recent years, advancements in data analytics and ML have paved the way for leveraging data-driven approaches in various domains [31]. We provide an overview of the principles and benefits associated with data-driven fault detection. By harnessing the power of data analysis and modeling [32], the methods offer the potential to enhance fault detection accuracy and improve overall systems reliability. We explore the fundamental concepts and concrete methods of data-driven fault detection, highlighting its significance in addressing the challenges of fault detection in complex systems.

With the advent of Industry 4.0, current industrial processes are transforming into smart ones [33]. In particular, many modernized industrial processes are equipped with sensors to gather process-related data for discovering faults existing or arising in processes as well as monitoring the process status, which provides favorable conditions for realizing data-driven fault detection. The basic functions of fault detection can be summed up into two parts , namely (1) monitoring the behavior of process (variables) and (2) revealing the fault presence. To maintain high process yield and throughput in industrial operations, it is crucial to swiftly, accurately, and effectively detect and address potential process or equipment faults that could undermine the overall system performance [34]. Fault detection has played an important role in

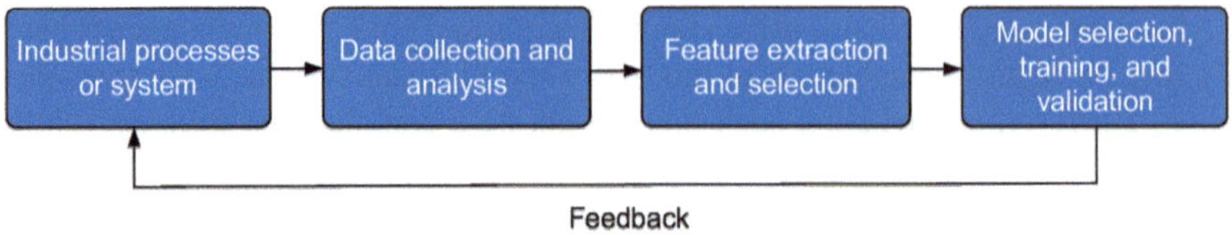

Fig. 7.2 Implementation procedure of a conventional fault detection method. First, collect the original data and carry out necessary pre-processing, such as denoising or dimension reduction, to ensure data quality and accuracy. Then, based on specific requirements and fault categories, select appropriate real-time fault detection methods for fault diagnosis. The final step is to troubleshoot and take preventive measures

various industrial engineering processes, for example, semiconductor manufacturing, chemical engineering, and software engineering [35]. For the economical and optimal operation of processes, effective detection of suspicious faults is increasingly demanded to prevent any deterioration in the process, which could eventually lead to a decline in product yield or process throughput. Generally, as a fundamental method for supervision of the processes, the fault detection task is performed based on several process and equipment data measured by instruments (i.e., sensors). The implementation process of a conventional fault detection method is shown in Fig. 7.2.

In data-driven fault detection methods, fault detection methods based on PCA and partial least squares (PLS) are considered powerful tools [36]. Among various statistical fault detection approaches, PCA-based approaches have been successfully implemented to highly complex industrial processes for detecting unknown faults during the operations. With the general PCA, several different types of PCA techniques, for example, dynamic principal component analysis (DPCA), and also data-driven reconstruction-based contribution approaches have been frequently adopted for fault detection in many industrial processes. To address the detection of hidden disturbances in chemical processes, an extended version of general PCA models was proposed. This extension, incorporating two key properties for statistical process monitoring known as static and dynamic, aimed to improve the effectiveness of the monitoring system. By integrating these properties into the PCA models, the detection of hidden disturbances in chemical processes becomes more accurate and robust [37]. The method used a method of 'time lag shift' for describing dynamic behaviors in general PCA, namely DPCA.

To capture important features of semiconductor production processes, a fault detection approach based on a pattern recognition technique was proposed. This method enables effective pattern recognition and fault detection, facilitating the improvement of process quality and reliability in semiconductor manufacturing [38]. The contribution plot technique which does not require a priori knowledge has been used for many fault isolation problems. However, due to the smearing effect, this approach may provide misleading conclusions on the faulty variables. To address the given problem, a contribution plot approach based on the maximization of reduction of combined index using the missing data approach was presented. The approach aimed to resolve the problem by identifying the contributions of different variables or factors to the combined index reduction. By utilizing the missing data approach, the proposed method enables an effective analysis of the factors impacting the problem, providing valuable insights for problem resolution [39].

In the process monitoring and fault detection of nonlinear processes, a simple fault identification method called local kernel principal component analysis (LKPCA) is proposed based on the local structure of data, emphasizing the importance of considering local features of data when dealing with nonlinear processes [40]. The proposed LKPCA method provides a direct method for fault identification in such processes by utilizing the principles of kernel principal component analysis (KPCA). Specifically, the optimization problem is first formulated by incorporating structural analysis into KPCA, and then transformed into a generalized eigenvalue decomposition problem using general kernel techniques. Based on this method, two monitoring statistics (Hotelling T2 and Q) were proposed and confidence limits were calculated using kernel density estimation method. Finally, sensitivity analysis is used to establish contribution graphs for two monitoring statistical data.

The fault detection method based on PCA improves the fault detection ability when the process model is missing. Use generalized likelihood ratio (GLR) testing for fault detection, and then establish a model based on principal component analysis technology. The effectiveness of the method in improving fault detection capability without a process model was verified on simulated data of a continuous stirred tank reactor [41]. A framework was presented for diagnosing faults arising in dynamic industrial processes [42]. The developed on-line framework incorporates multiscale PCA (MSPCA) for extracting key process features and utilizes an adaptive neuro-fuzzy inference system (ANFIS) to establish fault-signal correlations through the process data. Specifically, the features extracted from the use of MSPCA were partitioned into two spaces (i.e., score and residual spaces), and an attempt was made to classify different kinds of process faults using multiple ANFIS. A fault detection technique used a multi-level PCA (MLPCA) technique which is an extended version of a multi-block consensus PCA (CPCA), and then, based on the proposed MLPCA method, they presented a decentralized fault detection method [43].

To solve the autocorrelation problem in traditional Hotelling's T2 and Q statistics, a multivariate statistical database based on DPCA and de-correlated residuals is proposed. The problem of estimation and detection of small and unpredictable disturbances caused by early failures is studied. The Kullback-Leibler divergence (KLD) theory is used as a means of detecting early faults, which tend to cause subtle distortions in the process. By utilizing KLD theory to effectively identify and resolve these early faults, otherwise the use of traditional fault detection methods may be overlooked [44]. Based on divergence, an early estimation of fault amplitude was derived, followed by an assessment of divergence performance based on detection error probabilities. Simultaneously, the fault-to-noise ratio was considered as a standard for comparing fault levels and noise. In the dynamic process monitoring method, the Gaussian and non-Gaussian data were separated using the Jarque-Bera test. DPCA and dynamic independent component analysis (DICA) methods were applied to monitor dynamic subspaces for Gaussian and non-Gaussian blocks, respectively. The process of dynamic fault detection and analysis is shown in Fig. 7.3.

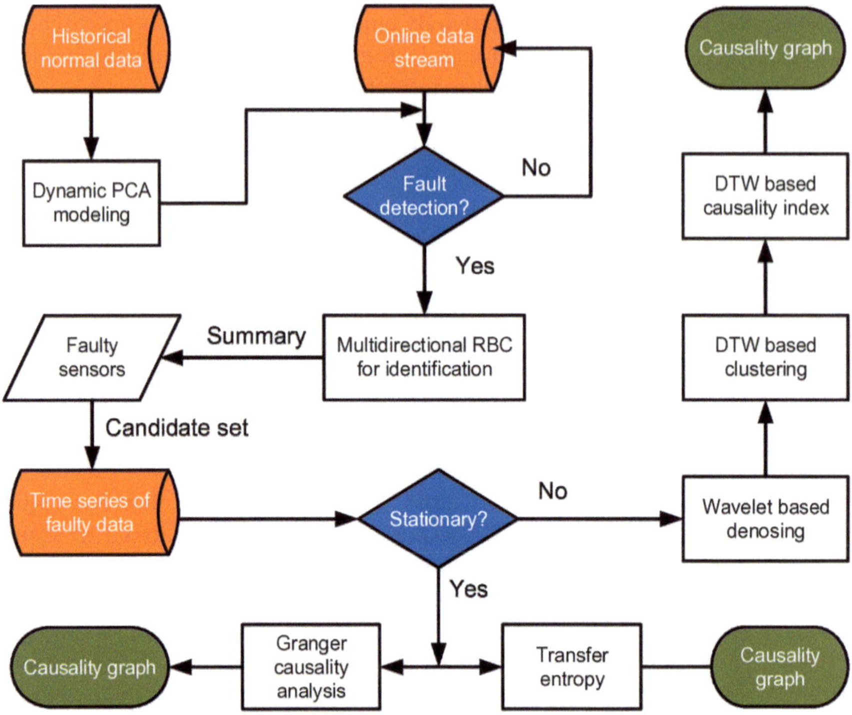

Fig. 7.3 The process of dynamic fault detection and analysis. First, real-time data is obtained from various sensors and sources to capture the dynamic behavior of the system. This data is then processed and analyzed using advanced techniques such as statistical analysis, machine learning, or deep learning algorithms

ML approaches have recently been applied for detecting and diagnosing faults. Signal processing and feature are most commonly performed using these methods, which entail intelligent signal processing and data collection. Multi-fault classification of IMs using vibration and current signals has been achieved using a variety of classifiers such as support vector machines (SVM), artificial neural networks (ANN), and k-nearest neighbors (kNN). The random forest (RF) method enables the diagnosis of machine faults with satisfactory accuracy in a short period of time [45]. The method outperforms other classifiers like the kNN, SVM, and decision trees in terms of performance. The extreme gradient boosting and RF classifiers are very promising for bearing fault detection using machine learning models. However, choosing the fault features that are most indicative of induction machine faults remains a challenge. Compared with conventional methods for extracting and selecting features, deep learning is more effective at extracting and learning the representative patterns of the signals automatically. CNN, for instance, show outstanding performance for pattern recognition and have also proven useful as fault detection methods. A CNN model that used vibration spectrum features to detect rotary machinery faults was developed [46]. However, due to computational time, deep learning is not suitable for online condition monitoring. In order to make the model more accurate and expedient, an appropriate learning approach must be explored.

The kNN-based fault detection methods can effectively capture unique characteristics in many real industrial processes, for example, multimode batch trajectories and non-linearity. To account for the correlations among numerous process variables in semiconductor manufacturing, novel methods have been developed, for example, multivariate control charts and statistics. However, the traditional multivariate-based techniques may not work well when the variables are non-Gaussian distributions. In the fault detection problem in semiconductor manufacturing, the use of adaptive Markov distance based false alarm detection method can effectively detect faults.

In many industrial processes, it is very common that the number of abnormal (or faulty) instances may be smaller than that of normal (stable) ones and also it is moderately difficult task to label each fault according to the fault types. A semi-supervised form of the fault detection model was presented [47], which considered both labeled and unlabeled data for fault classification. Fault detection in scenarios with imbalanced class distributions, where the distribution of fault categories is unequal, was considered. The aim was to improve the accuracy and reliability of fault detection in real-world applications [48]. An incremental clustering method was used to construct an online fault detection method, and the unbalance of classes was considered. Simulation data were collected from the actual semiconductor plasma etching process, and the effectiveness of the method was verified by testing.

For the detection of bearing faults, motor signal vibration analysis has been widely employed [49]. For simulations and practical applications, a faster Fourier transform was introduced for vibration signal analysis. In addition, other signal processing approaches based on time-frequency analysis such as wavelet transform (WT), short-time fourier transform (STFT), and Hilbert-Huang transform (HHT) are successfully employed to diagnose motor bearing faults [50]. The WT has gained increasing popularity as a fault detection method because it can enhance the resolution in both the time and frequency domains, replacing the STFT's disadvantage. HHT is also a powerful technique for analyzing motor vibration signals, along with Fourier transforms and WTs. Detection of transient amplitude and nonlinear frequency content reveals the presence of motor faults. It is possible to filter vibration signals contaminated by wavelet packet decomposition (WPD), and the energy contained in the WPD coefficients might be applied to identify rolling bearing failures. With the help of time–frequency analysis and the WPD, bearing fault features were successfully extracted.

For the detection of process faults, many intelligent system-based methods have been developed. In particular, ANN-based approaches have been successfully implemented for detecting incipient machine faults. A data-driven hybrid fault detection framework was presented for a power transmission-line network [51]. A CNN model-based fault detection method was presented [52]. The convolutional layers of the traditional CNN model were modified to capture the structural features of the data. Testing was conducted on semiconductor CVD process data, validating the effectiveness of the approach. Based on the time series data, a wind turbine fault detection model was established, and the early fault detection ability of the model and the prediction error of the secondary fault were verified. In addition, the full signal reconstruction methods and autoregressive methods were used to detect gearbox bearing faults and state temperature anomalies.

7.2.2 Data-Driven Predictive Maintenance Approaches

The prognostic and health management concept has become the inevitable tendency in the context of smart manufacturing and industrial big data [53], and it provides a reliable solution for managing the health status of equipment. In addition, the predictive maintenance strategy for industrial equipment can accurately perceive performance degradation since it was designed to achieve near-zero failures, near-zero hidden dangers, near-zero accidents, and near-zero pollution throughout the entire manufacturing process. In particular, the prediction of machine health can, not only significantly reduce the unexploited downtime and expensive labor costs but also ensure safe operation and optimize the maintenance plan.

In recent years, industrial wireless sensor networks and industrial CPS have become an emerging data acquisition technology in the complex industrial environment, and mechanical data can be collected using various types of sensors with high reliability and in a real-time manner. Obviously, because of the continuous improvement of data acquisition ability, as well as the exponential growth of data volume, data-driven methods for health monitoring have achieved great success and received widespread attention regarding the predictive maintenance of industrial equipment. Artificial intelligence (AI) algorithms have achieved important progress during the past five years [54]. The flowchart of the data-driven method for predictive maintenance is shown in Fig. 7.4.

DL solutions bring a unique application dimension and utilize data that differs somewhat from other methods. For instance, DL methods often involve the use of generated acoustic emissions and images. For example, consider a proposed "cyber-physical ball screw prognostics system framework" that employs data acquired from accelerometers, thermometers, and acoustic emission sensors. It utilizes the RBM algorithm for predictive detection of various failure modes within the component. Another method, which employs acoustic emission data and DL techniques, is designed for fault detection in bearings. This approach involves preprocessing sensor signals using the short-time Fourier transform. It constructs a simplified spectrum matrix obtained through this transform and utilizes a DL architecture optimized for efficient large memory storage retrieval, thereby enabling the diagnosis of faults.

A methodology was introduced to process and transform the acquired vibration measurement data obtained from a simulated motor-driven vibration system. This process aimed to create a dataset suitable for training and testing an artificial neural network (ANN) model. The goal of this ANN model is to predict the future state of the equipment, specifically, to forecast when a failure might occur [37]. The approach entails utilizing both frequency and amplitude data for the purpose of

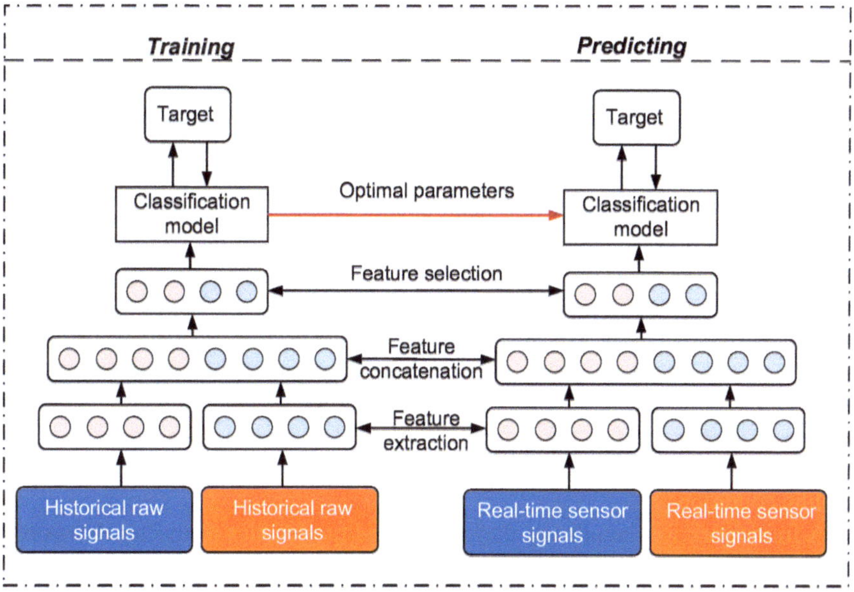

Fig. 7.4 Flowchart of the data-driven method for predictive maintenance. A data-driven predictive maintenance system comprises two primary phases. First, there is a learning phase, which involves training a model using historical raw sensor signals. Then, in the second phase, the trained model is deployed to predict target outcomes and make informed decisions

dataset classification and establishing a method for determining the failure time of the vibrating system. To accomplish the prediction task, a multilayer perceptron (MLP) methodology was employed, primarily for its straightforward implementation and its notable capability for generalization [55]. The proposed ANN model was subsequently assessed for its efficiency. It was compared to other machine learning techniques, specifically regression tree (RT), RF, and SVM, using the root mean square error performance index. The comparative analysis and training results were satisfactory, demonstrating that the ANN outperformed the other methods. When it came to medium-term and long-term predictions, the ANN excelled in comparison. However, in terms of short-term predictions, the ANN, RF, and RT exhibited similar levels of generalization.

The application of ANN and SVM algorithms is employed to create predictive models for gauge degradation measurements in two different types of rail tracks, encompassing both straight and curved segments [38]. The performance evaluation of the proposed models employs mean squared error and the coefficient of determination. Based on the study's results, satisfactory and somewhat similar outcomes are provided by both the ANN and SVM models. However, in predicting gauge deviation in straight segments, a slight advantage is shown by the ANN models over the SVM models. Additionally, a test rig was designed to simulate the operational conditions of a real wind turbine. The goal is for the time-domain vibration characteristics of several key components within the wind turbine to be studied and their status to be monitored using this information. By capturing the vibration characteristics associated with the health and fault states of these key components, an artificial neural network model was developed and applied for classifying their respective fault and health states. A study was reported on the monitoring of starter degradation in the auxiliary power unit using both a physics-based model and a neural network model [39]. A generic modeling technique was employed to address the limitations arising from the absence of component characteristics. A comparative analysis between back-propagating and forward-feeding neural network models was conducted, with training and testing being performed. Both models were applied under normal and deteriorated conditions, and their capabilities were verified. Based on the collected data, it was determined that the physics-based approach yielded more consistent results in cases involving degraded starters. In contrast, the neural network model exhibited superior performance when dealing with starters in a healthy condition.

A framework for data-driven diagnostics and prognostics of machines was proposed to enhance efficiency and lower maintenance costs [40]. Furthermore, a precise data labeling methodology was developed for supervised learning by comparing the serial numbers of target components on adjacent dates. The study utilized real data from vending machines to validate the proposed framework with three distinct classifiers: SVM, RF, and gradient boosting machines (GBM) [56]. Additionally, two models were created for predictive maintenance: one for diagnostics and the other for two-stage prognostics. Cross-validated simulations produced results demonstrating that the diagnostics model can attain an accuracy rate exceeding 80%. Therefore, the developed Support vector machine (SVM) model can be effectively employed for the detection and prognostic monitoring of complex vending machines. Furthermore, the prognostics model surpasses the performance of one-stage conventional prediction models.

A case study involving the application of predictive maintenance to an actual machining process was conducted [41]. The aim of the study was to enhance the tool life of the machine by employing machine learning methods for predicting RUL. Real-time data was collected from the computer, followed by data approximation during the analysis. The study applied linear and quadratic regression models to estimate RUL in the design application. Ultimately, the study produced precise results for RUL prediction, facilitating meaningful comparisons. In addition, a predictive maintenance approach driven by a digital twin was implemented for machine tools to prevent faults and associated issues [42]. A hybrid approach was employed to compute RUL results, which indicate the prediction error ratio, i.e., the disparity between the actual value and the predicted value.

The growing abundance of industrial data has facilitated the emergence and implementation of data-driven predictive maintenance. This approach leverages advanced computational techniques to extract valuable insights from the expanding pool of operational data, offering crucial information about equipment status. Data-driven predictive maintenance has found extensive application in industrial manufacturing, making use of ML and DL algorithms [57]. Traditional ML algorithms, (e.g., logistic regression (LR) [58], SVM [59], typically demand the collection of substantial data encompassing various health conditions and failure scenarios for model training. Subsequently, feature engineering is carried out, involving time, frequency, and time-frequency domain analysis. This helps in extracting relevant features, which are then used for learning and representing the device's health condition [60]. In contrast, DL, which encompasses various neural network (NN) models, eliminates the need for intricate feature engineering. It achieves this by employing an end-to-end learning approach, accomplished by incorporating multiple deep layers between the raw data and the prediction outcome. Consequently, these deep models can be seen as a "black box" that directly produces prediction results from the input data. This fundamental distinction sets DL apart from traditional ML [61]. Given these factors, both ML and DL have found extensive use in the domain of predictive maintenance. Currently, the proportion of maintenance costs in an enterprise's operational expenses is on the rise, primarily due to inefficient maintenance practices, which can involve either insufficient maintenance or excessive maintenance. Therefore, there is a pressing need to develop a well-balanced maintenance strategy. Such a strategy should not only mitigate the inefficiencies and waste associated with over-maintenance but also guarantee that equipment receives the necessary maintenance to remain in optimal working condition.

As shown in Fig. 7.5, the implementation process of data-driven predictive maintenance is outlined in detail and aligns closely with a structured design methodology. Overall, this process can be broken down into four key stages: (1) Operational assessment: This initial stage involves a thorough assessment of the operational environment to identify key factors and requirements. (2) Data acquisition: In this phase, data is collected and acquired from relevant sources. (3) Feature engineering: The data is pre-processed and engineered to extract relevant features. It's important to note that feature engineering becomes less crucial when employing DL algorithms. (4) Modeling: At this stage, predictive models are built and trained. Model training typically necessitates substantial computing power and storage capacity, often conducted on cloud or server platforms. However, the deployment of the trained model for real-time data flow analysis can be accomplished on edge nodes like edge servers or edge gateways. This approach optimizes the use of transmission and computing resources, particularly considering limitations in computing power, storage, and time delays.

Predictive maintenance plays a crucial role in the context of CPS by addressing the imperative need for proactive and efficient maintenance strategies. With the integration of sensors, real-time data collection, and advanced analytics, CPS enable the prediction of potential failures before they occur, allowing for timely interventions [62]. The approach offers several benefits, including increased equipment reliability, minimized downtime, optimized maintenance schedules, and cost savings. By leveraging historical data, ML algorithms, and predictive models, CPS can detect early warning signs of impending faults, enabling maintenance teams to

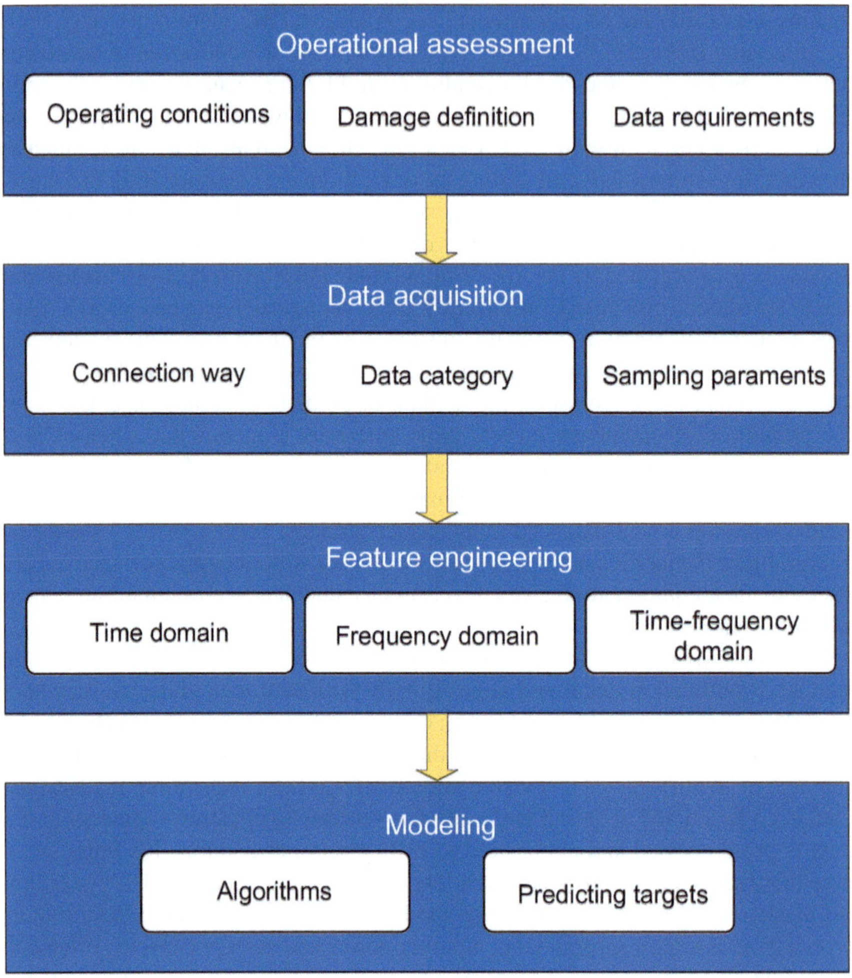

Fig. 7.5 Implementation process of the data-driven predictive maintenance. The process can be divided into four stages: (1) operational assessment; (2) data acquisition; (3) feature engineering; and (4) modeling

take preventive actions, such as repairing or replacing components, at the right time. This not only reduces the risk of catastrophic failures but also maximizes operational efficiency and extends the lifespan of critical assets. Ultimately, predictive maintenance empowers organizations to achieve higher levels of productivity, safety, and customer satisfaction in CPS environments.

7.2.3 Discussion of Fault Detection and Predictive Maintenance

In the realm of CPS, the discussion surrounding fault detection and predictive maintenance is of paramount importance. With CPS being characterized by the seamless integration of physical components and data-driven computational systems, the application of data-driven approaches in fault detection and predictive maintenance becomes crucial. These methods leverage advanced data analytics techniques, including ML, statistical modeling, and pattern recognition, to uncover hidden patterns, anomalies, and potential failures in CPS. By harnessing the power of data, CPS can achieve real-time monitoring, early detection of faults, and accurate prediction of maintenance needs. The data-driven nature of these approaches enables organizations to move away from traditional reactive maintenance practices towards proactive strategies, optimizing resource allocation, reducing downtime, and minimizing costs. Moreover, data-driven fault detection and predictive maintenance enhance the overall reliability, safety, and efficiency of CPS, facilitating continuous operation and improved performance of critical systems.

With the widespread application of CPS in everyday life and various industries, the issue of fault detection in CPS has become profoundly significant. However, there are numerous challenges in achieving effective fault detection in CPS. The complexity of CPS poses a major challenge. CPS integrate various technologies such as computing, networking, and sensors. Its system architecture and operational mechanisms are diverse and intricate, leading to the complexity and diversity in fault detection. Addressing this necessitates comprehensive consideration of knowledge across different levels and domains to design fault detection methods suitable for specific CPS. The diversity and scale of data present a significant challenge. CPS generate diverse types of data, including structured, unstructured, real-time data, etc. These data exhibit high dimensionality and complexity, placing high demands on the design and performance of fault detection algorithms. Moreover, CPS typically generate a vast amount of data, requiring efficient processing and analytical methods to achieve real-time and accurate fault detection.

Furthermore, the real-time and high-reliability requirements of CPS also add to the complexity of the fault detection. Many CPS applications have stringent requirements for real-time response, such as intelligent transportation systems and industrial automation. Hence, fault detection algorithms need to be not only accurate but also possess rapid response capabilities to ensure normal operation and safety of the system. Safety and privacy preservation are critical considerations in CPS fault detection. CPS span various domains, including transportation, healthcare, energy, etc., which concern individual privacy and societal safety. Therefore, in designing fault detection algorithms, due consideration must be given to data security and privacy protection, avoiding potential data breaches and security risks.

In summary, fault detection in CPS encounters challenges from diverse system architectures, diverse and large-scale data, real-time requirements, high-reliability demands, and safety and privacy concerns. Addressing and overcoming these challenges are essential to propel the development of fault detection techniques in CPS to higher levels, ensuring the stable, secure, and efficient operation of CPS.

To reach a mature application level, there are still many challenges to overcome in predictive maintenance, as illustrated in Fig. 7.6. Operational assessment plays a pivotal role and essentially serves as the litmus test for the success or failure of predictive maintenance. In essence, it sets the tone for the entire implementation process. In general, factors such as time and economic costs should be given initial consideration. Striking the right balance between the expense of introducing a predictive maintenance system and the losses incurred from equipment failures is crucial. Additionally, safety concerns are of paramount importance, and the specific implementation is subject to various restrictions, including the placement of sensors, which must conform to safety regulations and considerations.

Crucially, data-driven approaches integrate data acquisition as a fundamental component of the comprehensive design for predictive maintenance, furnishing valuable insights tailored to specific tasks. In academic research, many groups opt to procure

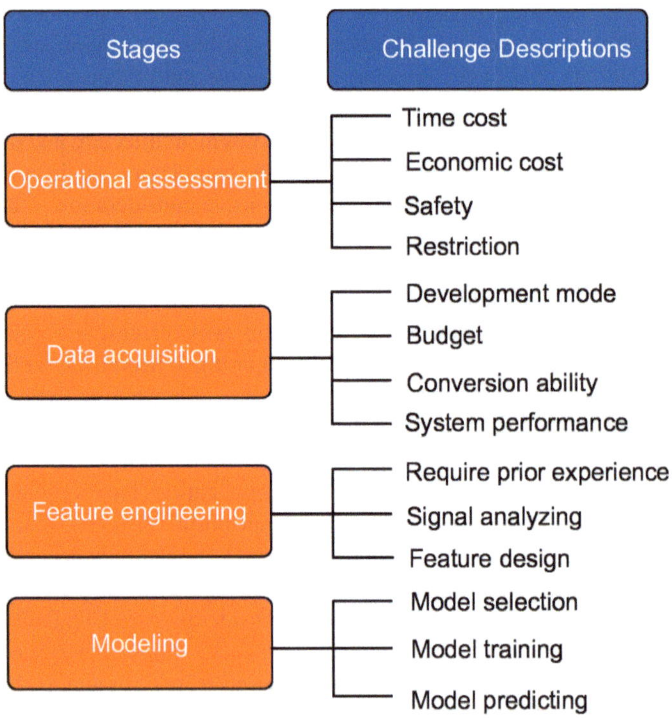

Fig. 7.6 Challenges in the construction of the predictive maintenance system. The challenges of predictive maintenance are summarized from the aspects of operational assessment, data acquisition, feature engineering and modeling construction

existing equipment to gather data. Conversely, for engineering applications, a customized data acquisition system is necessary to meet precise requirements. It has been demonstrated that data analysis at different stages of the machine's life cycle not only enhances the efficiency of data processing but also promotes transparency in monitoring the health status of industrial equipment.

Furthermore, effectively managing the acquired data, including the decision of whether to engage in feature engineering, has emerged as a significant challenge when selecting algorithms. The performance variations among algorithms are especially apparent in specific applications, particularly when dealing with industrial sensor data. Model development should also not be underestimated, as the selection, training, and optimization of models all warrant increased attention. In practical applications, supervised learning has demonstrated impressive outcomes across various fields; however, it relies on accurately annotated data, which poses a substantial challenge for predictive maintenance efforts. With the increasing quantification of industrial big data, unsupervised learning naturally emerges as a highly promising avenue for research and development.

7.3 Applications of Fault Detection and Predictive Maintenance

We delve into the critical application background of fault detection and predictive maintenance in modern industry and technology, and examine their practical application in different sectors, highlighting how these methods contribute to the advancement of modern industry and technology. Readers will gain a clearer understanding of the critical role of fault detection and predictive maintenance in real-world scenarios.

7.3.1 Application Background of Fault Detection and Predictive Maintenance

With the steady development of computing technology for sensing, driving and control systems [63], complex systems such as smart phones, autonomous vehicles, smart grids, automated buildings, and smart cities [64] have been developed [65]. These systems share common characteristics, such as a close integration of hardware and software functions and constraints, often incorporating a mix of diverse components. They typically include sophisticated sensing and drive hardware with embedded computing capabilities. CPS prioritize the harmonious interaction between software and hardware design. This synergy can be enhanced by bolstering awareness of dynamic environmental changes and by enhancing data processing capabilities to handle large volumes of data effectively.

Internet of Things and big data platform, as the application background of CPS, provide important support and technical basis for the development and operation of CPS [66]. By integrating IoT technologies and big data analytics, CPS is better able to achieve a deep understanding of the real world, real-time response and intelligent control. IoT is an important part of CPS, which involves connecting the physical world to the network for real-time monitoring, control and optimization. IoT integrates the physical and digital worlds through sensors, embedded systems, and networked devices [67]. This fusion is the core of CPS, through which intelligent monitoring and control of physical systems can be achieved. Big data platforms involve techniques and tools for processing and analyzing large-scale data sets. In CPS, big data platforms can be used to process the vast amounts of data generated by CPS, including data from sensors, devices, and other sources. By analyzing this data, insights can be gained into the complex relationships between physical and digital components in the CPS.

As per the general definition, CPS are integrated systems constructed through the amalgamation of computer, communication, and control technologies [68]. CPS should possess the ability to promptly detect and regulate any engineering and information interactive systems, while also delivering the pertinent and essential information services. In the context of industrialization and informatization, the value of CPS lies in its capacity to manage a variety of physical devices and unify them within a single network system. Furthermore, physical equipment can access functionalities such as instant messaging, data computation, precise control, self-management, and remote coordination through the integration of information and communication systems [69]. Hence, a CPS can be conceptualized as a network system that also encompasses control system capabilities. Nevertheless, CPS goes beyond the typical functions of control systems, featuring additional attributes like sharing, openness, and information exchange.

CPS represent the fundamental concept underpinning the notion of smart manufacturing. In contrast to traditional factory automation and monitoring/control systems, a typical smart factory comprises three essential components: (1) a physical system, (2) a cyber system equipped with simulation and analytical capabilities, and (3) their seamless integration. CPS places its primary focus on harmonizing these two systems, the physical and cyber, and distinguishing itself by emphasizing simulation and predictive functions. In a smart factory context, numerous manufacturing scenarios and "what-if" cases are simulated and tested using real-world data extracted from the physical system. This approach enables advanced capabilities in terms of modeling, prediction, and analysis for optimizing manufacturing processes and performance [66]. In a typical IoT setup, a sensor-equipped device detects sensor signals and transmits them to a central control server. The server then processes and analyzes this data, formulates an efficient control plan using its analytics capabilities, and subsequently dispatches the corresponding control commands to the machine tools or other IoT devices, enabling coordinated and responsive control actions [70]. Although this scenario works well within a confined environment, it encounters limitations when applied to processes spread across multiple sites. The utilization of

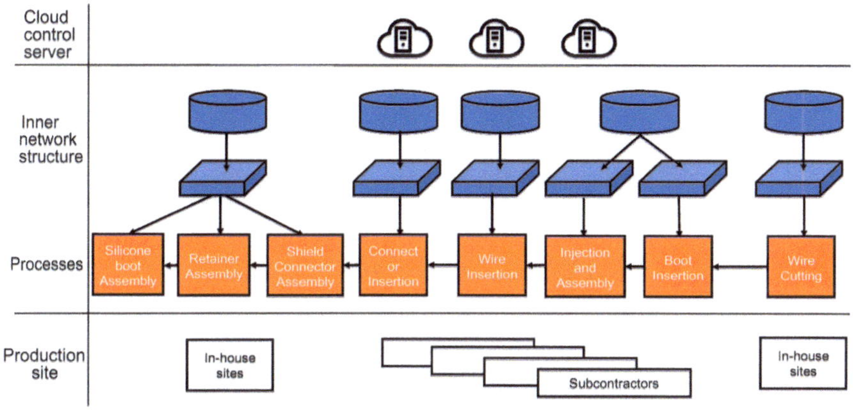

Fig. 7.7 The architecture of the cloud system. The architecture of a cloud system typically comprises a network of servers and storage components that work together to deliver various services over the internet. It includes both front-end components that interface with users and back-end components responsible for data processing, storage, and management

a cloud computing environment proves to be an effective framework for addressing and overcoming these challenges [71]. Figure 7.7 shows the architecture of the suggested cloud control system.

Existing manufacturing knowledge have identified possible defect causes and their related sources/machines shown in Fig. 7.8. To establish and demonstrate the direct relevance and mitigate defects, IoT beacons are installed in the production machines. The primary responsibilities of these IoT beacons include measuring various manufacturing quality factors during the production processes and transmitting this data to the cloud server for analysis and monitoring [72]. The server plays a crucial role in the system, with its primary functions being the analysis of the collected signals and the identification of relationships between these signals and defects. This identified relationship is then leveraged to make more precise decisions in the realm of fault detection and classification, enhancing the overall effectiveness of the system. CPS integrate technologies such as computers, networks, and sensors, enabling devices to monitor, collect, and exchange data in real time. Cloud servers, as powerful platforms for data processing and storage, efficiently analyze this vast amount of data. This integration provides deeper insights and analysis for fault detection and predictive maintenance. It allows us to better predict potential faults and perform timely maintenance, maximizing system stability and security [73].

The IoT cloud system receives signals from various sensors that operate on computer modules powered by Intel chips. The IoT devices in use are connected to the cloud control server through associated hubs and gateways. IoT beacons have the capability to transmit multiple signals and receive multiple commands from the cloud control servers, facilitating a robust and flexible communication system [74]. The developed IoT device serves a dual purpose, effectively performing two key functions: (1) detecting torques of a compression machine: One of the IoT beacons

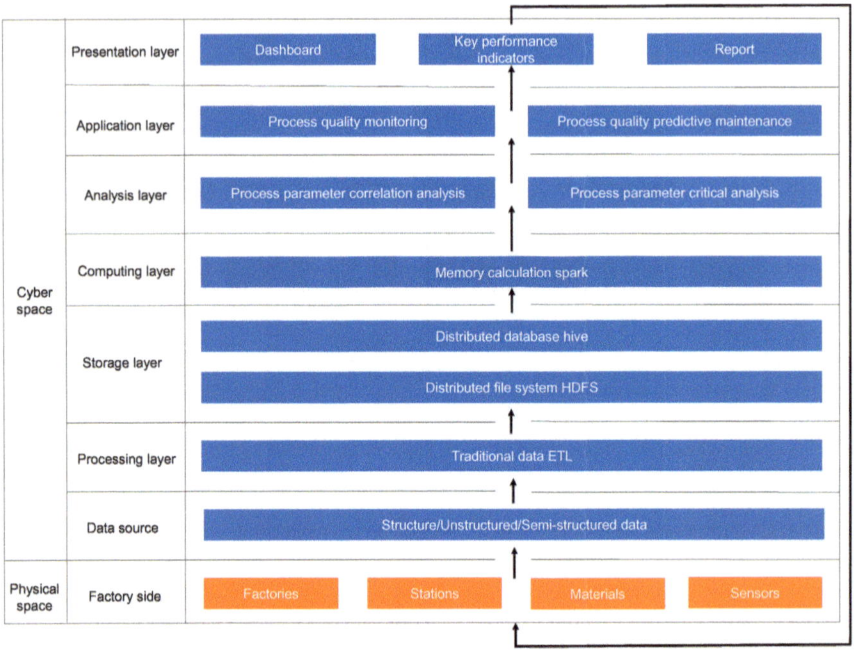

Fig. 7.8 Predictive maintenance big data platform architecture. The physical space encompasses the factory's physical dimensions and layout, while the cyber space encompasses the data source and six distinct layers: processing, storage, computing, analysis, application, and presentation. These layers collectively form the digital infrastructure that supports various functions and operations within the factory

is responsible for sensing the force signals generated by a compression machine. It employs a magnetic-based torque sensor to measure the magnetic field and its orientation within the compression motor. Any changes in the compression torque of the motor affect the associated magnetic field, which is detected by the sensor. Subsequently, the sensor records the degree of this change and transmits this data to the cloud server for analysis. (2) detecting gap between components and related parts: Another IoT beacon employs micro-level proximity sensors to detect gaps or the proximity of products in a machine. These proximity sensors can discern the distance or approach of components or products. The beacon captures these designated signals and relays them to the cloud control server for further processing and monitoring.

The cloud control server assumes various roles and responsibilities, which can be categorized into several key missions: (1) Information database for processing data: The server serves as a repository for processing data, housing a comprehensive database of information related to the production processes. (2) Monitoring processing status: It continuously monitors and tracks the status of various processing activities, ensuring everything is running smoothly. (3) Identifying relationships between defects and causes: The server is equipped with analytics capabilities to uncover hidden connections between defects and their potential root causes, aiding in root cause analysis. (4) Issuing alarms for abnormal conditions: When abnormal states or conditions are detected, the server triggers alarms and notifications to alert relevant parties to take appropriate action. (5) Simulating processes and risk mitigation: The server has the capability to simulate processes under various scenarios, allowing for risk assessment and mitigation strategies to be developed proactively. The cloud control server serves as a core component of the comprehensive CPS within the newly innovated processes. It comprises three main parts: the data gathering unit, the storage unit, and the analytical unit, all working in tandem to facilitate efficient control, monitoring, and decision-making.

Monitoring the online status of IoT systems poses a significant challenge, primarily due to the complexity of managing big data and analyzing sensor data [75]. By harnessing the power of the Internet of Things, users can monitor their machines' real-time status and make necessary adjustments, ultimately enhancing the efficiency of smart factories that rely on various motors. However, it's crucial to be mindful of two primary security threats: malware and fake data injection attacks. In the era of big data, a growing trend is to leverage database resources to gain a competitive edge in various domains [76]. People are increasingly recognizing the hidden insights within data, with the aim of extracting greater value. Oftentimes, they can deduce possible causes or reasons from the outcomes of data analysis, contributing to a deeper understanding of complex systems and processes [77]. Therefore, the utilization of big data analytics and cloud computing facilitates the investigation of potential correlations between anomalies in manufacturing processes and the key influencing factors behind product anomalies. This process involves extracting valuable insights and knowledge, ultimately equipping engineers with the means to effectively monitor and resolve issues when they encounter similar process anomalies in the future. Furthermore, this approach offers opportunities to enhance yields by promptly detecting, preventing, and predicting process anomalies that may have already occurred.

The market offers a plethora of mature tools for cloud computing, including a diverse range of management and data analysis solutions like Microsoft Azure, Google Cloud Platform, and many others [78]. Nevertheless, when a system needs to maintain a flexible architecture on the platform, it becomes necessary to construct the system through a customized interface. The devices and interfaces rely on WiFi as a communication bridge, and the collected data is structured within a database and processed using SQL after being embedded with equipment-independent codes. Typically, the image module of each subsystem is constructed using the simulation platform, and a set of management algorithms is developed to enhance the configuration of the management unit.

The digital factory is an emerging concept [79]. The concept is to utilize information technology for the digital integration of production. Traditional production planning control systems, often designed in conjunction with computer-aided design, are typically centered around a relatively static central control system [80]. To enhance the openness and flexibility of the production process, one approach is to establish a connection between manufacturing equipment and the Internet, thereby increasing the self-control capabilities of production equipment through the use of CPS [81]. Hence, it is imperative to construct a virtual platform capable of connecting and communicating with real-world objects. Furthermore, the creation of a self-contained, testable virtual system, facilitated by a CPS-based automated management platform, streamlines the efficient integration, simulation, and testing of IoT devices.

Digital factories offer real-time information transmission from the production line through to the product's completion and delivery. This encompasses aspects such as product scheduling, production scheduling, equipment monitoring, inspection maintenance, coordination and management of both direct and indirect production staff, management of manufacturing rules, meticulous material control, and tracking of actual costs [82]. Furthermore, every significant processing event occurring on the production line is meticulously documented. This serves a dual purpose. Firstly, it enables product traceability, providing a detailed electronic record of the production process. Secondly, as the volume of production experiences grows, this data can be analyzed and isolated for further insights and improvements.

Manufacturing big data can typically be categorized into three main types: device data, product data, and command data. Leveraging a cloud platform, the collected manufacturing big data can be subjected to further analysis and harnessed for the development of innovative applications, including active preventive maintenance, line optimization, and energy efficiency improvements. This wealth of big data information can also be employed to unearth hidden insights related to defective products and faulty equipment, as well as to predict key influencing factors. The big data platform for fault detection and predictive maintenance is established with the vision of a smart factory in mind.

Fault detection and predictive maintenance harness the real-time operational conditions of equipment, materials, and systems to enhance and optimize manufacturing operations [83]. A comprehensive predictive maintenance management program integrates a range of cost-effective tools, including vibration monitoring, process parameter monitoring, temperature logging, tribology, and visual inspection, to gather data on the actual operation of critical plant systems. This data is then used to schedule all maintenance activities. An essential requirement for effective fault detection and predictive maintenance is the acquisition of a substantial volume of data from every aspect of the manufacturing process. However, this need for extensive data collection can also pose a significant challenge when implementing fault detection and predictive maintenance in manufacturing. The greater the volume of data collected, the more accurate the predictions for maintenance intervals of machines, materials, tools, and other critical components in the manufacturing process. As the amount of manufacturing data grows, so does the demand for data processing capabilities, making the establishment of robust big data platforms a critical necessity.

Remote monitoring of sensor data in the cloud architecture. To build a smart factory, three aspects need to be considered: (1) data acquisition, (2) data persistence, and (3) data processing. To effectively support fault detection and predictive maintenance systems, big data platforms must be capable of remote distribution across the globe and offer high performance, scalability, and fault tolerance. In the data collection process, traditional data preprocessing methods such as extract, transform, and load (ETL) are employed. ETL involves extracting representative data from the data source, transforming it, and then loading it into a designated data transmission process. The destination typically serves as a data repository for business intelligence analysis. Apache Hive, a data warehousing tool built on the Hadoop architecture, is utilized for data extraction and transformation. For data persistence, the Hadoop Distributed File System (HDFS) is chosen due to its fault tolerance and horizontal scalability. Even in the event of failures, HDFS continues to efficiently handle the storage of large volumes of data from manufacturing plants in a distributed manner. In terms of data processing, the system leverages Apache Spark, a memory-based computational framework. Spark temporarily stores intermediate data in memory during operation, significantly boosting execution speed. According to official documentation, Spark can process data up to 100 times faster than traditional methods. Furthermore, Spark is an extensible system that can scale horizontally in a fault-tolerant manner, making it an ideal choice for processing and analyzing large datasets collected from manufacturing plants. The core design of the system revolves around three key elements. First, data preprocessing encompasses ETL steps. The Hive distributed database collaborates with HDFS for data storage and retrieval. Subsequently, Spark is employed to construct a robust data processing mechanism. Finally, the system is realized through the application of big data exploration technology, focusing on process parameter correlation analysis and critical process parameter analysis. The architecture of predictive maintenance is shown in Fig. 7.8.

7.3.2 *Case Studies of Fault Detection and Predictive Maintenance*

We present a collection of real-world examples showcasing the effectiveness of fault detection and predictive maintenance strategies in various industries. The case studies highlight how organizations have leveraged data analytics and advanced technologies to detect faults early, prevent system failures, and optimize maintenance practices. Based on case studies, fault detection and predictive maintenance offer valuable insights into the practical implementation of data-driven fault detection methods, demonstrating their potential in enhancing system reliability and maintenance efficiency within complex industrial systems.

The architecture of an efficient network manufacturing system, along with its detailed implementation, is thoroughly described. This system involves the use of a CPS with cyber manufacturing capabilities, implemented through embedded system technologies and relevant analytics. As a practical test bed, the manufacturing and assembly processes are specifically examined for high-intensity discharge (HID) cable modules, a key component used in vehicles. These processes encompass a combination of manual tasks and automated processes. Furthermore, the production flows include internal manufacturing operations within a factory as well as intermediate outsourcing flows with several subcontractors. Given that various vehicles employ different headlight cable modules, these processes deal with a wide array of component modules. This complexity highlights that the overall processes constitute a multi-product manufacturing system. Based on these exemplary processes, an effective CPS architecture is proposed, outlining how cyber manufacturing is achieved using the implemented systems.

The manufacturing processes related to High-Intensity Discharge HID cables, as well as the challenges associated with fault detection, have been the subjects of thorough study and investigation [67]. The processes involved in manufacturing HID car headlight modules, which are integral components of vehicle lighting systems, are crucial. These modules are packaged and provided to OSRAM©, and the final assembled light systems are subsequently installed in a wide range of vehicle types. Figure 7.9 explains the detailed processes for HID cable modules. The manufacturing of HID wires involves the supply of HID wire material to the factory, which is then cut to fixed specifications. Subsequently, various processes, including wire trimming, insertion, tapping, and more, are carried out to connect a connector to the wire. These processes are adapted to accommodate the different requirements of various types of vehicles, as each may demand different headlight lamps, HID

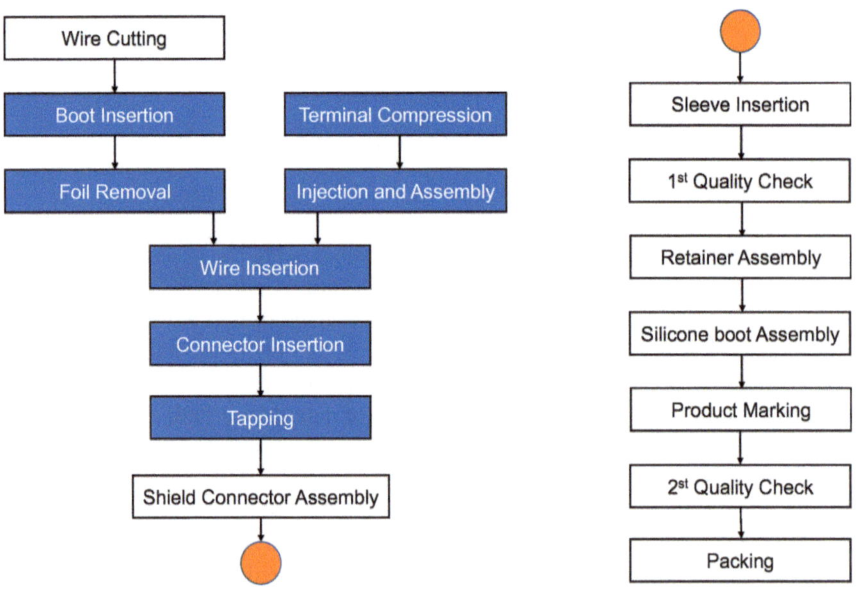

Fig. 7.9 HID cable manufacturing and assembly processes. The overall process includes several outbound workflows. The outbound workflow is performed by several cable-related subcontractors. Subcontracted work is mainly labor-intensive work, such as wire stripping, tapping and so on

wires, and connectors. During the inspection stage, any detected defective modules are discarded and not recycled. The HID wire modules that pass inspection proceed to subsequent stages, where a retainer and silicon boots are assembled. These modules are marked with production information such as Lot Number, manufacturing date, and other details for product identification. The final inspection stage involves testing for issues like insufficient soldering, poor marking, and other quality-related concerns. Subsequently, the accepted HID wire module is assembled with the headlight lamp module. Installing sensors in the subcontractors' factories can be a challenging task [67]. Despite the installation of sensors, the tracking of sensor signals and the ability to conduct continuous quality management activities are limited.

To address these limitations, a database system and analytic functions have been implemented to enhance the tracking of sensor signals and facilitate continuous quality management activities [84]. Data from Intel© IoT Analytics is stored in the cloud-based database system. Subsequently, this data is subjected to analysis and employed for simulations and predictions. The simulation and production modules have been developed using the monitoring and control modules. These modules primarily serve the purpose of assessing the performance of processes, taking into account current information and potential future scenarios. For instance, the detection of abnormal statuses may result in additional delays due to necessary maintenance activities [85]. In scenarios where the delivery of HID cable modules to a subsequent manufacturing site is delayed, the extent of tardiness is quantified using a simulation model. A new manufacturing plan is then devised to minimize these delays. Numerous control strategies can be employed to reduce tardiness or optimize other system performance metrics. Simulation is considered a crucial prerequisite stage for implementing these control strategies and is regarded as a fundamental function in CPS. A stochastic Petri Net-based model has been leveraged to represent a cloud-based manufacturing model and facilitate its simulation. This model aids in assessing and improving the overall manufacturing processes [86]. Petri Net-based CPS approaches offer significant benefits in terms of monitoring and simulation. However, they do come with certain limitations. These limitations include: (1) Complicated representations: Petri Net-based models can become complex when representing numerous statuses or states within the system. (2) Discontinuity with other control strategies: The integration of Petri Net-based models with other control strategies can be challenging and may result in discontinuities. Furthermore, since the processes handle various types of cable modules, a multi-products production model must be used. When applying the Petri Net methodology to model such complex processes, a colored stochastic Petri Net model is required, which can increase the computational load for analysis.

The study presents a framework for a novel and efficient CPS along with its architecture tailored for manufacturing processes involving multiple products and a mix of internal tasks and outsourcing flows. This framework is designed to enhance the performance and control of manufacturing operations in complex, multi-product environments [87]. The challenges stemming from internal and outsourcing processes have been effectively addressed through the utilization of a cloud computing environment. Furthermore, issues related to multi-product manufacturing are resolved by employing several simulation and control functions within the CPS. To facilitate the use of the cloud computing environment, IoT devices have been developed. These implemented IoT beacons play a critical role in detecting the status of machines and processes, transmitting this data to the cloud control server. The data transmitted is then utilized to more clearly identify cause-effect relationships for fault detection and classification, aided by an embedded learning model employing deep learning techniques. These analytical processes significantly contribute to the enhancement of process quality. Within this context, main defect types and their possible causes have been assumed based on accumulated manufacturing knowledge and existing analyses, further improving the overall understanding of the processes.

Figure 7.10 provides the relationship between major defect types and the related possible causes with the processes. Additionally, the cyber and cloud environment enables simulations that take into account various manufacturing scenarios. Given that the provided processes involve numerous connections and handle multiple products, a multi-products queuing network model has been adopted as a fundamental simulation model. Through this modeling and its subsequent simulation, the performance of the processes can be predicted under various what-if scenarios. These results then inform decision-making in subsequent stages. The research has validated the effectiveness of cyber manufacturing and its architecture through the development of IoT devices, transmission protocols and architectures, the implementation of the cloud control server, and the analytical functions. The proposed framework and implementations are applicable to a wide range of industrial fields and processes.

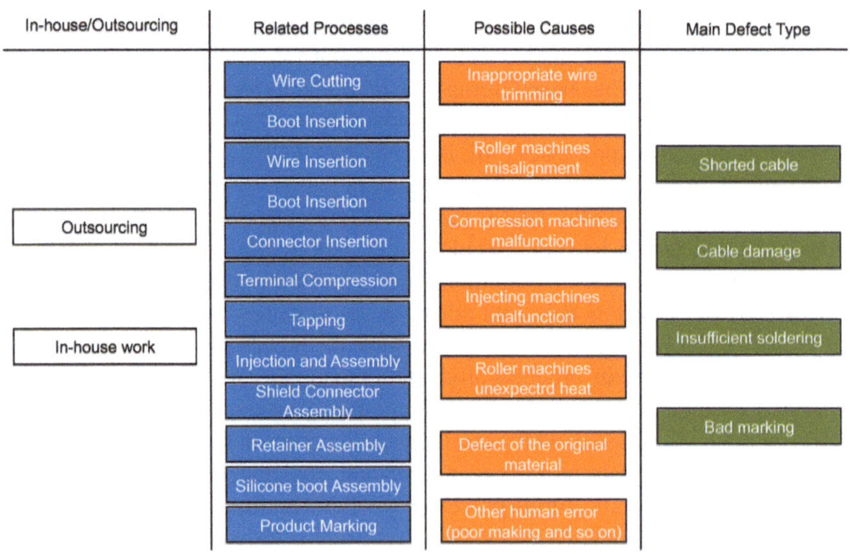

Fig. 7.10 Defect types and the related causes/processes. Understanding the different defect types, their root causes and related processes can help identifying potential problems, improving production or development processes, and ultimately improve product quality and efficiency

In future studies, there is a need for more accurate and rapid analytics at the cloud layer. The granular level of monitoring, an increased number of sensing points, and shorter sensing periods can result in a substantial volume of data and exponential computational loads. This situation could render a developed cyber-physical system ineffective. To prevent such a scenario, more efficient techniques for handling big data and related architectures are being considered.

Vehicular CPS have the primary goal of improving road safety and overall transportation efficiency through advanced monitoring, control, and management mechanisms. These systems leverage technology to create a safer and more efficient transportation environment [88]. Vehicular CPS offer several advantageous characteristics, including real-time capabilities, low computational demands on onboard processors, and the ability to adapt to a wide range of working conditions and parameter variations. These properties make vehicular CPS well-suited for enhancing transportation safety and efficiency.

In recent years, both academics and engineers have dedicated their efforts to developing more dependable architectures for vehicular CPS. This reflects the growing importance of enhancing the reliability and performance of CPS in the realm of transportation [89]. A multilevel framework has been put forward to diagnose faults in hardware, software, and the interaction between hardware and software within vehicular cyber physical systems. Monitoring occurs at each level to pinpoint the origin of faults. The dendrogram architecture is employed for effective fault isolation. However, the data and information are segregated by levels, and the interactions can be intricate. Mechanism model-based frameworks have also been developed to analyze end-to-end response times across different layers at the component level, enhancing the overall understanding of system behavior [90]. In vehicular CPS, the devices are interconnected through a local area network (LAN) with a ring-shaped topology. A data-driven architecture for fault detection and control has been introduced. This framework is implemented at the system level within the vehicles and supports various data-driven monitoring, control, and fault detection strategies. It enhances the capabilities of these systems to operate more efficiently and reliably.

Monitoring and fault detection approaches based only on the measurement data are highly demanded [91]. The increasing complexity of modern vehicles has introduced significant challenges related to the safety and reliability of their control systems. Managing these complexities while ensuring the safety and reliability of vehicular systems is a critical concern [92]. In the context of vehicular CPS, it's crucial to address small and local malfunctions that could potentially propagate and escalate into severe systematic failures. To tackle this challenge, a total principle component regression (TPCR) model and a TPCR-based data-driven fault detection approach were introduced. Unlike the standard principal component regression, the TPCR model divides the output space into two subspaces, allowing for the differentiation of whether the detected faults impact key performance indicators. However, the TPCR-based approach is non-recursive and suitable only for static processes, rendering it unsuitable for automotive applications. Taking into account practical constraints and the trends in the automotive industry, a comprehensive architecture for fault detection and control in vehicular CPS has been proposed. This architecture consists of distinct sub-modules, including the automobile's physical layer, the cyber layer, the control system, and the fault detection system, providing a holistic framework for enhancing the reliability and safety of vehicular CPS.

The ensemble learning scheme involves several stages. First, real-time measurements are collected. Second, test statistics are generated from a pool of local models, and these statistics are compared to corresponding thresholds. The decision logic is influenced by environment recognition techniques. To account for new working conditions and maintain adaptability, a training block is included. Third, the test statistics from the local models undergo further processing. Each local model represents a portion of the input space due to its limited number of measurements and, consequently, only contains partial distribution information. From a pattern recognition perspective, each local model acts as a local classifier or detector, which can be considered weak detectors because they are built using relatively small datasets. In the fourth stage, an ensemble learning strategy is employed to combine these weak detectors into a strong fault detector, and the results from the strong fault detector are then evaluated. This approach helps enhancing the reliability and accuracy of fault detection in complex systems.

This study delved into the domain of fault detection and control in vehicular CPS, a pivotal area within the context of Industry 4.0 [93]. The study focused on creating a modular, adaptable, and extendable data-driven architecture for addressing abnormalities in vehicular systems. This architecture utilized ensemble learning techniques and recursive approaches to enable reliable online tracking of faults. A new fault detection approach called R-TPCR was developed and implemented. The study compared R-TPCR with other non-recursive schemes for suspension system fault detection using the Carsim simulation platform. Results demonstrated that R-TPCR outperformed non-recursive schemes in terms of fault detection rates and sensitivity to faults. Additionally, the recursive implementation improved online efficiency and reduced memory usage. When compared to advanced subspace aided recursive methods, R-TPCR showed comparable fault detection rates but with more stability and lower false alarm rates. The simulation results indicated that R-TPCR was robust to stochastic noises and performed optimally under step faults.

The Cloud Layer of the architecture is built on Microsoft Azure HDInsight, utilizing Microsoft R Server, enterprise R-analysis software, the Apache Hadoop cloud ecosystem framework, and the Apache Spark ecosystem framework. This combination of technologies and platforms allows for advanced data processing, analysis, and management in the cloud environment [94]. The study focuses on analyzing data from gearboxes, specifically data collected during the manufacturing process of both good and bad products. The dataset comprises a total of 388,744 data points, and it is structured to align with the gearbox manufacturing process. To enhance the authenticity of the data, the study assumes a 90% yield rate based on manufacturing statistics. The collected information includes measurements related to current consumption, positive and negative rotational speeds, and product weight. Additionally, simulation data is generated randomly using a Gaussian distribution. This dataset serves as the basis for the case study analysis.

Setting the support level and confidence level too low in the association rule analysis leads to the generation of an excessive number of association rules, resulting in prolonged search times for meaningful rules. Conversely, when these levels are set excessively high, the rules discovered tend to be skewed toward common-sense relationships. Through association rule analysis, the rules pertaining to the occurrence of each anomaly code are identified, as well as the relevant parameter rules that prove valuable when the exception code is triggered. As a whole, the most significant parameter rules can be clearly deduced from the association rules. The findings indicate a strong correlation between employee class (staff group) and the day of the week (weekend vs. weekdays). This suggests that differences in employee efficiency are observed on various working days, with a higher likelihood of anomalies occurring during the week. Furthermore, the results reveal that the part batch number and supplier log significantly contribute to product abnormalities. It is observed that products from the same batch are more prone to experiencing anomalies, a phenomenon influenced by the factory (F) and the product model (Model).

In the analysis, variable importance for the stochastic model is determined by observing the mean decrease in Gini. The greater the drop in the Gini index, the more important the variable is considered in the model [95]. The same principle applies to node purity. When examining the top five parameters of high importance, it becomes evident that the importance indices of the two parameters, staff and product Log, significantly surpass those of other parameters. The remaining three parameters, namely staff Group, Date, and supplier Log, also exhibit relatively high importance indices.

The data is collected from the automated guided vehicle (AGV) as it moves within its environment. This data includes information about the AGV's displacement, its current position, the angle of rotation, and its battery level [96]. The AGV records information about the items stored on its shelves, which is used for warehousing management, logistics management, and documenting the processing results of each item. All this information is organized and transmitted to the main console of the AGV through a Wi-Fi network. From there, it is routed to a remote monitor console using another Wi-Fi channel. The CPS functions as the monitoring console within this architecture. It receives all the AGV data and uses it to create a production traceability record. In the final stage, the production history is analyzed, and prealarm monitoring is performed based on the recorded data. Additionally, the numerical data collected is sent to the IoT platform via a Wi-Fi network in the CPS. As depicted in Fig. 7.11, The proposed IoT platform consists of three parts: the evaluation module, the early warning module, and the mirror module.

The research lies in the establishment of automated factory monitoring and warning systems [97]. A large volume of data collected by the IoT platform is utilized to create a database, which is then employed to establish a set of real-time warning modules. These modules are used to monitor logistics management on the AGV shelves as well as warehousing management. The long-term historical data can also be leveraged for analyzing the relationship between inbound and outbound logistics on the shelves and their interactions with workers. Based on this information, the system can continuously monitor and issue warnings when observed values reach

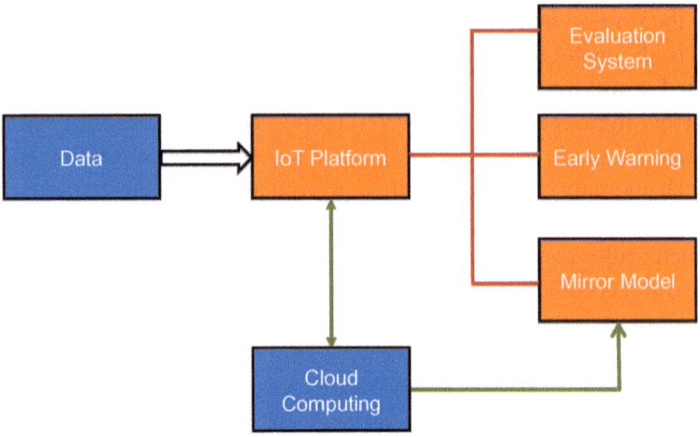

Fig. 7.11 IoT platform of the system. The IoT platform consists of three parts: the evaluation module, the early warning module, and the mirror module

predefined thresholds. These warnings are intended to notify the administrator so that appropriate actions can be taken, achieving the functionality of early warning. Two simulated early warning systems are implemented in this work: a battery simulation warning system and a manufacturing execution system.

Lithium batteries offer several advantages compared to other rechargeable batteries. These advantages include: (1) High energy density: Lithium batteries have a high energy density, which means they can store a significant amount of energy in a relatively compact and lightweight package. This makes them ideal for applications where space and weight are critical factors. (2) High operating voltage: Lithium batteries typically have a higher operating voltage compared to other batteries. A single lithium battery usually operates at around 3.8 V, which is higher than the 1.2 V of many other battery types. This higher voltage makes it easier to achieve the required voltage levels for various applications. (3) Wide operating temperature range: Lithium batteries can function effectively in a broad range of temperatures, making them suitable for both hot and cold environments. (4) Long service life: Lithium batteries tend to have a longer service life compared to many other battery types. They can withstand a large number of charge-discharge cycles before their capacity significantly degrades. (5) Better discharge capacity: Lithium batteries can maintain their charge for extended periods without significant self-discharge. This means they remain nearly fully charged, even after months of storage. A simple mathematical model has been developed to describe the discharge behavior of lithium batteries. This model accurately predicts the remaining charge during discharge, enhancing the predictability and reliability of these batteries in various applications.

The competitiveness of enterprises is sure to be enhanced through the import of information and communications technology to establish a smart factory, strengthen quality management, establish a production traceability system, and create links with upstream and downstream manufacturers in the production and marketing supply chain [98]. Smart factories integrate various automation technologies [99]. A manufacturing execution system comprises numerous sub-systems, such as in-process management, statistical process control, equipment management, anomaly alarm, and more. It incorporates features like post-operation barcoding, materialized electronic management, and transparent production traceability. This represents a shift from the error-prone traditional handwritten methods to precise fool-proof digital operations. Electronic data aids in product traceability throughout the production process, while data analysis for process improvement can reduce the time required for data collection. This, in turn, enables managers to allocate more time to devise optimal courses of action.

The manufacturing execution system operates from the initiation of production orders to the final product completion. It delivers real-time, accurate information concerning products produced in the production line, encompassing details like product numbers, shelf numbers, battery status, staff assignments, and production line records. This factory-generated data is provided promptly, often in report or other forms, and the results can change rapidly in response to current conditions. The primary objective is to reduce non-value-added activities, such as machine malfunctions, while maximizing productive output. Furthermore, all significant processing events within the production line are documented. This information serves a dual purpose: it can be used for product traceability, and it enables electronic tracking of the production process. As the volume of production records grows, this data becomes an invaluable resource. In the event of a product issue in the market, rather than taking several days for staff to retrieve historical information, electronic data is instantly accessible, allowing businesses to respond swiftly to abnormal events. Additionally, it can provide real-time information about on-site emergencies during production activities and can alert managers via multiple communication channels.

The study takes warehousing, detecting, and packaging as the three main workstations in the production line [100]. The warehousing function primarily involves documenting the materials' inflow and outflow while establishing warning thresholds for each material. When the material quantity reaches the alert threshold, the system issues a warning signal and takes early warning measures to facilitate status monitoring and control. Detection involves recording the tasks or actions executed on each product at the detection station, with all detection activities documented in the production traceability. As the production traceability accumulates, the system can analyze and detect any abnormalities during the production process. Additionally, the study addresses staff movement and process tracking. At the end of the production line, during the packaging process, a similar approach is applied as in warehousing. It includes recording the total weight of the packaging materials and defining a warning threshold. If the total weight falls outside the predetermined warning range, the system issues a warning signal and initiates early warning actions to support monitoring and alerting.

In the battery simulation warning system, we conducted tests on a 3S1P lithium battery pack while subjecting it to specific conditions. We aimed to discharge the battery from a voltage of 12.5 V down to 9 V under the load of a 4 W bulb and a 5 W DC motor. During this discharge, the output voltage of the lithium battery decreases gradually with time, and the remaining capacity is in a direct proportional relationship with the voltage. By considering the battery discharge voltage on the y-axis and the percentage of battery discharge on the x-axis (often referred to as capacity), the discharge behavior can be depicted as a continuous declining curve. Lithium batteries exhibit a characteristic discharge curve, with the initial and middle sections of the curve showing a gradual decline. However, in the later stages of discharge, the curve drops rapidly, typically occurring when the battery voltage reaches about 9.5–9.0 V. This sharp decrease in capacity is why controlling the remaining capacity of a lithium battery can be challenging. To address this, we establish a warning threshold based on the declining voltage curve. When the value provided by the system monitoring reaches this warning threshold, the system will generate a warning signal and notify the system administrator to take pre-emptive actions. This mechanism helps ensuring the safe and effective use of the lithium battery.

We conducted a comprehensive investigation into successful cases of CPS fault detection and predictive maintenance. These cases encompass a wide range of applications, including data aggregation, control, monitoring, and early warning systems, all of which contribute to the overall efficiency of management systems. The objective of these cases is to leverage manufacturing big data, cloud computing, and predictive maintenance solutions to benefit factories globally. In these cases, advanced techniques such as correlation analysis and critical analysis of process parameters are realized through extensive data mining technology. These methodologies allow for the detection of faults and the implementation of predictive maintenance measures. By harnessing the power of data, these solutions help enhancing the reliability and performance of various systems and processes.

7.3.3 Challenges in Cases

Despite the successes highlighted in the case studies, several challenges remain in the domain of fault detection and predictive maintenance. First,acquiring high-quality and sufficient data for training accurate fault detection models can be challenging, particularly in legacy systems or environments with limited data collection infrastructure. Second, scaling up data-driven fault detection systems to handle large-scale, real-time data streams can pose technical and computational challenges. Ensuring timely processing and decision-making in dynamic operational environments remains an ongoing concern. Then, integrating data-driven fault detection solutions with existing maintenance practices, legacy systems, and heterogeneous data sources may require careful planning and coordination. Finally, while data-driven models often demonstrate high accuracy, their interpretability and explainability can be limited, making it challenging for maintenance personnel to understand and trust the generated insights and recommendations.

Addressing these challenges requires ongoing research and innovation in data acquisition, preprocessing, algorithm development, and system integration. Additionally, collaboration between domain experts, data scientists, and maintenance professionals is crucial to ensure the effective implementation and adoption of data-driven fault detection and predictive maintenance approaches. Despite these challenges, the case studies presented in this chapter demonstrate the immense potential of data-driven techniques in transforming fault detection and predictive maintenance practices, paving the way for improved system reliability, reduced downtime, and enhanced operational efficiency.

We present a compilation of real-world case studies showcasing the effectiveness of fault detection and predictive maintenance strategies across various industries. These case studies demonstrate how organizations have leveraged data analytics, ML, and advanced technologies to improve system reliability, reduce maintenance costs, and enhance operational efficiency. And the chapter highlights the limitations of traditional reactive maintenance approaches and emphasizes the value of proactive maintenance strategies. By utilizing data-driven techniques, organizations can detect faults early, predict potential failures, and optimize maintenance schedules, thereby minimizing downtime and improving overall system performance.

The case studies we present cover a variety of areas. It involves data acquisition, pre-processing, feature engineering, and the application of ML algorithms in fault detection and predictive maintenance tasks. The integration of domain knowledge with data-driven techniques further enhances the accuracy and interpretability of fault detection systems.The benefits achieved by organizations include increased equipment reliability, reduced maintenance costs, improved safety, and optimized operational uptime. These case studies illustrate the transformative potential of data-driven approaches in improving maintenance practices and driving operational excellence.

7.4 Summary

The chapter focuses on a data-driven approach to tackle challenges in fault detection and predictive maintenance within CPS. The core emphasis lies in harnessing the power of data to enable efficient monitoring and early detection of faults, as well as predicting maintenance needs before potential failures occur.

By leveraging data-driven methodologies, researchers and practitioners can effectively analyze large datasets generated by CPS components, such as sensors, actuators, and control systems. The chapter explores various data-driven techniques. In practical implementation, the proposed set of frameworks and methods proved to be suitable for hardware platforms, thus proving their applicability and effectiveness in the real world.

Overall, the chapter presents a comprehensive exploration of data-driven approaches in fault detection and predictive maintenance for CPS. The incorporation of these techniques not only enables better understanding and monitoring of CPS processes but also aids in mitigating potential faults and ensuring smoother operations, thereby enhancing the reliability and resilience of these complex systems.

Competing Interests

1. The research is supported by the National Key Research and Development Project under Grant 2023YFB3307300, the National Natural Science Foundation of China under Grants 62373014, 62422301, 62125301, and 92467205.
2. The authors have no conflicts of interest to declare that are relevant to the content of this chapter.

References

1. S. Bernardi, J. Merseguer, and D. C. Petriu, "A dependability profile within marte," *Software & Systems Modeling*, vol. 10, pp. 313–336, 2011.
2. G. J. Holzmann, "The model checker spin," *IEEE Transactions on software engineering*, vol. 23, no. 5, pp. 279–295, 1997.
3. N. Navet and L. Fejoz, "Cpal: High-level abstractions for safe embedded systems," in *Proceedings of the International Workshop on Domain-Specific Modeling*, pp. 35–41, 2016.
4. A. Benveniste, P. Caspi, S. A. Edwards, N. Halbwachs, P. Le Guernic, and R. De Simone, "The synchronous languages 12 years later," *Proceedings of the IEEE*, vol. 91, no. 1, pp. 64–83, 2003.
5. A. El-Hokayem, Y. Falcone, and M. Jaber, "Modularizing behavioral and architectural crosscutting concerns in formal component-based systems–application to the behavior interaction priority framework," *Journal of logical and algebraic methods in programming*, vol. 99, pp. 143–177, 2018.
6. I. C. Bertolotti, T. Hu, and N. Navet, "Model-based design languages: a case study," in *2017 IEEE 13th International Workshop on Factory Communication Systems (WFCS)*, pp. 1–6, IEEE, 2017.
7. L. Fejoz, B. Regnier, P. Miramont, and N. Navet, "Simulation-based fault injection as a verification oracle for the engineering of time-triggered ethernet networks," in *ERTS 2018*, 2018.
8. Y. Xu, I. Koren, and C. M. Krishna, "Adaft: A framework for adaptive fault tolerance for cyber-physical systems," *ACM Transactions on Embedded Computing Systems (TECS)*, vol. 16, no. 3, pp. 1–25, 2017.
9. V. Sfyrla, G. Tsiligiannis, I. Safaka, M. Bozga, and J. Sifakis, "Compositional translation of simulink models into synchronous bip," in *International Symposium on Industrial Embedded System (SIES)*, pp. 217–220, IEEE, 2010.
10. M. Kölbl, S. Leue, and H. Singh, "From sysml to model checkers via model transformation," in *Model Checking Software: 25th International Symposium, SPIN 2018, Malaga, Spain, June 20-22, 2018, Proceedings 25*, pp. 255–274, Springer, 2018.
11. M. Voelter, D. Ratiu, B. Schaetz, and B. Kolb, "mbeddr: an extensible c-based programming language and ide for embedded systems," in *Proceedings of the 3rd annual conference on Systems, programming, and applications: software for humanity*, pp. 121–140, 2012.
12. G. Bloom, G. Cena, I. C. Bertolotti, T. Hu, N. Navet, and A. Valenzano, "Event notification in can-based sensor networks," *IEEE Transactions on Industrial Informatics*, vol. 15, no. 10, pp. 5613–5625, 2019.
13. K. Ding, A. Morozov, and K. Janschek, "More: Model-based redundancy for simulink," in *Computer Safety, Reliability, and Security: 37th International Conference, SAFECOMP 2018, Västerås, Sweden, September 19-21, 2018, Proceedings 37*, pp. 250–264, Springer, 2018.
14. J. Brunel, D. Doligez, R. R. Hansen, J. L. Lawall, and G. Muller, "A foundation for flow-based program matching: using temporal logic and model checking," in *Proceedings of the 36th annual ACM SIGPLAN-SIGACT symposium on Principles of programming languages*, pp. 114–126, 2009.

15. E. A. Lee, "The past, present and future of cyber-physical systems: A focus on models," *Sensors*, vol. 15, no. 3, pp. 4837–4869, 2015.
16. D. Ameller, X. Franch, C. Gómez, S. Martínez-Fernández, J. Araújo, S. Biffl, J. Cabot, V. Cortellessa, D. M. Fernández, A. Moreira, *et al.*, "Dealing with non-functional requirements in model-driven development: A survey," *IEEE Transactions on Software Engineering*, vol. 47, no. 4, pp. 818–835, 2019.
17. A. Ataíde, J. P. Barros, I. S. Brito, and L. Gomes, "Towards automatic code generation for distributed cyber-physical systems: A first prototype for arduino boards," in *2017 22nd IEEE International Conference on Emerging Technologies and Factory Automation (ETFA)*, pp. 1–4, IEEE, 2017.
18. X. Koutsoukos, G. Karsai, A. Laszka, H. Neema, B. Potteiger, P. Volgyesi, Y. Vorobeychik, and J. Sztipanovits, "Sure: A modeling and simulation integration platform for evaluation of secure and resilient cyber–physical systems," *Proceedings of the IEEE*, vol. 106, no. 1, pp. 93–112, 2017.
19. J. Sztipanovits, T. Bapty, X. Koutsoukos, Z. Lattmann, S. Neema, and E. Jackson, "Model and tool integration platforms for cyber–physical system design," *Proceedings of the IEEE*, vol. 106, no. 9, pp. 1501–1526, 2018.
20. H. Zhang, D. Ge, J. Liu, and Y. Zhang, "Multifunctional cyber-physical system testbed based on a source-grid combined scheduling control simulation system," *IET Generation, Transmission & Distribution*, vol. 11, no. 12, pp. 3144–3151, 2017.
21. A. Cervin, D. Henriksson, B. Lincoln, J. Eker, and K.-E. Arzen, "How does control timing affect performance? analysis and simulation of timing using jitterbug and truetime," *IEEE control systems magazine*, vol. 23, no. 3, pp. 16–30, 2003.
22. S. M. Sundharam, N. Navet, S. Altmeyer, and L. Havet, "A model-driven co-design framework for fusing control and scheduling viewpoints," *Sensors*, vol. 18, no. 2, p. 628, 2018.
23. F. Pasqualetti, F. Dörfler, and F. Bullo, "Attack detection and identification in cyber-physical systems," *IEEE transactions on automatic control*, vol. 58, no. 11, pp. 2715–2729, 2013.
24. V. Reppa, M. M. Polycarpou, and C. G. Panayiotou, "Distributed sensor fault diagnosis for a network of interconnected cyberphysical systems," *IEEE Transactions on Control of Network Systems*, vol. 2, no. 1, pp. 11–23, 2014.
25. C. Liu, P. Zheng, and X. Xu, "Digitalisation and servitisation of machine tools in the era of industry 4.0: a review," *International journal of production research*, pp. 1–33, 2021.
26. T. Mens and P. Van Gorp, "A taxonomy of model transformation," *Electronic notes in theoretical computer science*, vol. 152, pp. 125–142, 2006.
27. L. Ming and J. Zhao, "Review on chemical process fault detection and diagnosis," in *2017 6th international symposium on advanced control of industrial processes (AdCONIP)*, pp. 457–462, IEEE, 2017.
28. I. Hwang, S. Kim, Y. Kim, and C. E. Seah, "A survey of fault detection, isolation, and reconfiguration methods," *IEEE transactions on control systems technology*, vol. 18, no. 3, pp. 636–653, 2009.
29. X. Du, "Fault detection using bispectral features and one-class classifiers," *Journal of Process Control*, vol. 83, pp. 1–10, 2019.
30. H. J. Chang, D. S. Song, P. J. Kim, and J. Y. Choi, "Spatiotemporal pattern modeling for fault detection and classification in semiconductor manufacturing," *IEEE transactions on semiconductor manufacturing*, vol. 25, no. 1, pp. 72–82, 2011.
31. K. Severson, P. Chaiwatanodom, and R. D. Braatz, "Perspectives on process monitoring of industrial systems," *Annual Reviews in Control*, vol. 42, pp. 190–200, 2016.
32. J. Shang, M. Chen, H. Ji, and D. Zhou, "Recursive transformed component statistical analysis for incipient fault detection," *Automatica*, vol. 80, pp. 313–327, 2017.
33. P. Zhou, S.-W. Lu, and T. Chai, "Data-driven soft-sensor modeling for product quality estimation using case-based reasoning and fuzzy-similarity rough sets," *IEEE Transactions on Automation Science and Engineering*, vol. 11, no. 4, pp. 992–1003, 2014.
34. W. Li and S. J. Qin, "Consistent dynamic pca based on errors-in-variables subspace identification," *Journal of Process Control*, vol. 11, no. 6, pp. 661–678, 2001.

35. Q. Liu, S. J. Qin, and T. Chai, "Decentralized fault diagnosis of continuous annealing processes based on multilevel pca," *IEEE Transactions on Automation Science and Engineering*, vol. 10, no. 3, pp. 687–698, 2013.
36. G. Li, S. J. Qin, and D. Zhou, "A new method of dynamic latent-variable modeling for process monitoring," *IEEE Transactions on Industrial Electronics*, vol. 61, no. 11, pp. 6438–6445, 2014.
37. T. Kanungo, D. M. Mount, N. S. Netanyahu, C. D. Piatko, R. Silverman, and A. Y. Wu, "An efficient k-means clustering algorithm: Analysis and implementation," *IEEE transactions on pattern analysis and machine intelligence*, vol. 24, no. 7, pp. 881–892, 2002.
38. G. Li, S. J. Qin, and T. Yuan, "Nonstationarity and cointegration tests for fault detection of dynamic processes," *IFAC Proceedings Volumes*, vol. 47, no. 3, pp. 10616–10621, 2014.
39. A. Triki-Lahiani, A. B.-B. Abdelghani, and I. Slama-Belkhodja, "Fault detection and monitoring systems for photovoltaic installations: A review," *Renewable and Sustainable Energy Reviews*, vol. 82, pp. 2680–2692, 2018.
40. R. Al-Mashhadani, G. Alkawsi, Y. Baashar, A. A. Alkahtani, F. H. Nordin, W. Hashim, and T. S. Kiong, "Deep learning methods for solar fault detection and classification: a review," *solar cells*, vol. 11, p. 12, 2021.
41. A. Mellit and S. Kalogirou, "Artificial intelligence and internet of things to improve efficacy of diagnosis and remote sensing of solar photovoltaic systems: Challenges, recommendations and future directions," *Renewable and Sustainable Energy Reviews*, vol. 143, p. 110889, 2021.
42. J. Vicente-Gabriel, A.-B. Gil-González, A. Luis-Reboredo, P. Chamoso, and J. M. Corchado, "Lstm networks for overcoming the challenges associated with photovoltaic module maintenance in smart cities," *Electronics*, vol. 10, no. 1, p. 78, 2021.
43. F. P. Gasparin, A. J. Bühler, G. A. Rampinelli, and A. Krenzinger, "Statistical analysis of i–v curve parameters from photovoltaic modules," *Solar energy*, vol. 131, pp. 30–38, 2016.
44. M. Dhimish, V. Holmes, B. Mehrdadi, and M. Dales, "Comparing mamdani sugeno fuzzy logic and rbf ann network for pv fault detection," *Renewable energy*, vol. 117, pp. 257–274, 2018.
45. M. M. Rahman, J. Selvaraj, N. Rahim, and M. Hasanuzzaman, "Global modern monitoring systems for pv based power generation: A review," *Renewable and Sustainable Energy Reviews*, vol. 82, pp. 4142–4158, 2018.
46. M. Fadhila, K. D. Eddine, M. Mounir, and E. Adil, "Smart sensors materials based diagnosis of induction machine taking rotor faults into account," *Sensor Letters*, vol. 16, no. 3, pp. 182–187, 2018.
47. M. Hussain, D. K. Soother, I. H. Kalwar, T. D. Memon, Z. A. Memon, K. Nisar, and B. S. Chowdhry, "Stator winding fault detection and classification in three-phase induction motor.," *Intelligent Automation & Soft Computing*, vol. 29, no. 3, 2021.
48. S. B. Lee, A. Naeini, S. Jayaram, G. C. Stone, and M. Šašić, "Surge test-based identification of stator insulation component with partial discharge activity for low voltage ac motors," *IEEE Transactions on Industry Applications*, vol. 56, no. 3, pp. 2541–2549, 2020.
49. M. D. Prieto, G. Cirrincione, A. G. Espinosa, J. A. Ortega, and H. Henao, "Bearing fault detection by a novel condition-monitoring scheme based on statistical-time features and neural networks," *IEEE Transactions on Industrial Electronics*, vol. 60, no. 8, pp. 3398–3407, 2012.
50. Q. Cao, F. Giustozzi, C. Zanni-Merk, F. de Bertrand de Beuvron, and C. Reich, "Smart condition monitoring for industry 4.0 manufacturing processes: An ontology-based approach," *Cybernetics and Systems*, vol. 50, no. 2, pp. 82–96, 2019.
51. L. Xiao, Y. Li, X. Huang, and X. Du, "Cloud-based malware detection game for mobile devices with offloading," *IEEE Transactions on Mobile Computing*, vol. 16, no. 10, pp. 2742–2750, 2017.
52. J. Xie, Z. Li, Z. Zhou, and S. Liu, "A novel bearing fault classification method based on xgboost: The fusion of deep learning-based features and empirical features," *IEEE Transactions on Instrumentation and Measurement*, vol. 70, pp. 1–9, 2020.
53. M. Bauer, J. W. Cox, M. H. Caveness, J. J. Downs, and N. F. Thornhill, "Finding the direction of disturbance propagation in a chemical process using transfer entropy," *IEEE transactions on control systems technology*, vol. 15, no. 1, pp. 12–21, 2006.

54. S. Yoon and J. F. MacGregor, "Fault diagnosis with multivariate statistical models part i: using steady state fault signatures," *Journal of process control*, vol. 11, no. 4, pp. 387–400, 2001.
55. J. Lee, E. Lapira, B. Bagheri, and H. an Kao, "Recent advances and trends in predictive manufacturing systems in big data environment," *Manufacturing Letters*, vol. 1, no. 1, pp. 38–41, 2013.
56. H. Zhang, X. Tian, X. Deng, and Y. Cao, "Multiphase batch process with transitions monitoring based on global preserving statistics slow feature analysis," *Neurocomputing*, vol. 293, pp. 64–86, 2018.
57. M. Shcherbakov and C. Sai, "A hybrid deep learning framework for intelligent predictive maintenance of cyber-physical systems," *ACM Transactions on Cyber-Physical Systems (TCPS)*, vol. 6, no. 2, pp. 1–22, 2022.
58. F.-N. Yang and H.-Y. Lin, "Development of a predictive maintenance platform for cyber-physical systems," in *2019 IEEE International Conference on Industrial Cyber Physical Systems (ICPS)*, pp. 331–335, IEEE, 2019.
59. J. Lee, H. D. Ardakani, S. Yang, and B. Bagheri, "Industrial big data analytics and cyber-physical systems for future maintenance & service innovation," *Procedia cirp*, vol. 38, pp. 3–7, 2015.
60. M. P. Lambán, P. Morella, J. Royo, and J. C. Sánchez, "Using industry 4.0 to face the challenges of predictive maintenance: A key performance indicators development in a cyber physical system," *Computers & Industrial Engineering*, vol. 171, p. 108400, 2022.
61. R. F. Babiceanu and R. Seker, "Big data and virtualization for manufacturing cyber-physical systems: A survey of the current status and future outlook," *Computers in industry*, vol. 81, pp. 128–137, 2016.
62. X. Bampoula, G. Siaterlis, N. Nikolakis, and K. Alexopoulos, "A deep learning model for predictive maintenance in cyber-physical production systems using lstm autoencoders," *Sensors*, vol. 21, no. 3, 2021.
63. Y. Jiang and S. Yin, "Recent advances in key-performance-indicator oriented prognosis and diagnosis with a matlab toolbox: Db-kit," *IEEE Transactions on Industrial Informatics*, vol. 15, no. 5, pp. 2849–2858, 2018.
64. Y. Qin, X. Wang, and J. Zou, "The optimized deep belief networks with improved logistic sigmoid units and their application in fault diagnosis for planetary gearboxes of wind turbines," *IEEE Transactions on Industrial Electronics*, vol. 66, no. 5, pp. 3814–3824, 2018.
65. X. Liu, P. Song, C. Yang, C. Hao, and W. Peng, "Prognostics and health management of bearings based on logarithmic linear recursive least-squares and recursive maximum likelihood estimation," *IEEE Transactions on Industrial Electronics*, vol. 65, no. 2, pp. 1549–1558, 2017.
66. R. Zhao, D. Wang, R. Yan, K. Mao, F. Shen, and J. Wang, "Machine health monitoring using local feature-based gated recurrent unit networks," *IEEE Transactions on Industrial Electronics*, vol. 65, no. 2, pp. 1539–1548, 2017.
67. Y. Wang, Z. Wei, and J. Yang, "Feature trend extraction and adaptive density peaks search for intelligent fault diagnosis of machines," *IEEE Transactions on Industrial Informatics*, vol. 15, no. 1, pp. 105–115, 2018.
68. V. Gopakumar, S. Tiwari, and I. Rahman, "A deep learning based data driven soft sensor for bioprocesses," *Biochemical Engineering Journal*, vol. 136, pp. 28–39, 2018.
69. W. Ahmad, S. A. Khan, M. M. Islam, and J.-M. Kim, "A reliable technique for remaining useful life estimation of rolling element bearings using dynamic regression models," *Reliability Engineering & System Safety*, vol. 184, pp. 67–76, 2019.
70. B. Farooq, J. Bao, J. Li, T. Liu, and S. Yin, "Data-driven predictive maintenance approach for spinning cyber-physical production system," *Journal of Shanghai Jiaotong University (Science)*, vol. 25, pp. 453–462, 2020.
71. R. Zhao, R. Yan, Z. Chen, K. Mao, P. Wang, and R. X. Gao, "Deep learning and its applications to machine health monitoring," *Mechanical Systems and Signal Processing*, vol. 115, pp. 213–237, 2019.

72. Y. Jiang and S. Yin, "Recent advances in key-performance-indicator oriented prognosis and diagnosis with a matlab toolbox: Db-kit," *IEEE Transactions on Industrial Informatics*, vol. 15, no. 5, pp. 2849–2858, 2018.
73. F. Shen, J. Zheng, L. Ye, and X. Ma, "Lstm soft sensor development of batch processes with multivariate trajectory-based ensemble just-in-time learning," *IEEE Access*, vol. 8, pp. 73855–73864, 2020.
74. L. Ye, H. Guan, X. Yuan, and X. Ma, "Run-to-run optimization of batch processes with self-optimizing control strategy," *The Canadian Journal of Chemical Engineering*, vol. 95, no. 4, pp. 724–736, 2017.
75. R. Zhao, R. Yan, Z. Chen, K. Mao, P. Wang, and R. X. Gao, "Deep learning and its applications to machine health monitoring," *Mechanical Systems and Signal Processing*, vol. 115, pp. 213–237, 2019.
76. Y. Liu, C. Yang, Z. Gao, and Y. Yao, "Ensemble deep kernel learning with application to quality prediction in industrial polymerization processes," *Chemometrics and Intelligent Laboratory Systems*, vol. 174, pp. 15–21, 2018.
77. L. Wang, M. Törngren, and M. Onori, "Current status and advancement of cyber-physical systems in manufacturing," *Journal of Manufacturing Systems*, vol. 37, pp. 517–527, 2015.
78. F. Shen, J. Zheng, L. Ye, and D. Gu, "Quality-relevant monitoring of batch processes based on stochastic programming with multiple output modes," *Processes*, vol. 8, no. 2, 2020.
79. M. Onel, C. A. Kieslich, Y. A. Guzman, C. A. Floudas, and E. N. Pistikopoulos, "Big data approach to batch process monitoring: Simultaneous fault detection and diagnosis using nonlinear support vector machine-based feature selection," *Computers & Chemical Engineering*, vol. 115, pp. 46–63, 2018.
80. A. Caggiano, T. Segreto, and R. Teti, "Cloud manufacturing framework for smart monitoring of machining," *Procedia Cirp*, vol. 55, pp. 248–253, 2016.
81. M. V. Shcherbakov, A. V. Glotov, and S. V. Cheremisinov, "Proactive and predictive maintenance of cyber-physical systems," in *Cyber-Physical Systems: Advances in Design & Modelling*, pp. 263–278, Springer, 2019.
82. M. Pech, J. Vrchota, and J. Bednář, "Predictive maintenance and intelligent sensors in smart factory: Review," *Sensors*, vol. 21, no. 4, 2021.
83. B. Vogel-Heuser and D. Hess, "Guest editorial industry 4.0–prerequisites and visions," *IEEE Transactions on automation Science and Engineering*, vol. 13, no. 2, pp. 411–413, 2016.
84. Z. M. Cinar, A. A. Nuhu, Q. Zeeshan, and O. Korhan, "Digital twins for industry 4.0: A review," in *Industrial Engineering in the Digital Disruption Era* (F. Calisir and O. Korhan, eds.), (Cham), pp. 193–203, Springer International Publishing, 2020.
85. Q. Jiang, X. Yan, and B. Huang, "Review and perspectives of data-driven distributed monitoring for industrial plant-wide processes," *Industrial & Engineering Chemistry Research*, vol. 58, no. 29, pp. 12899–12912, 2019.
86. T. Nasir, M. Asmaela, Q. Zeeshana, and D. Solyalib, "Applications of machine learning to friction stir welding process optimization," *Jurnal Kejuruteraan*, vol. 32, no. 1, pp. 171–186, 2020.
87. A. N. Pedersen, M. Borup, A. Brink-Kjær, L. E. Christiansen, and P. S. Mikkelsen, "Living and prototyping digital twins for urban water systems: Towards multi-purpose value creation using models and sensors," *Water*, vol. 13, no. 5, 2021.
88. R. Rai, M. K. Tiwari, D. Ivanov, and A. Dolgui, "Machine learning in manufacturing and industry 4.0 applications," *International Journal of Production Research*, vol. 59, no. 16, pp. 4773–4778, 2021.
89. W. Zhang, D. Yang, and H. Wang, "Data-driven methods for predictive maintenance of industrial equipment: A survey," *IEEE Systems Journal*, vol. 13, no. 3, pp. 2213–2227, 2019.
90. J. Wan, S. Tang, D. Li, S. Wang, C. Liu, H. Abbas, and A. V. Vasilakos, "A manufacturing big data solution for active preventive maintenance," *IEEE Transactions on Industrial Informatics*, vol. 13, no. 4, pp. 2039–2047, 2017.
91. Y. Yoo, S. H. Park, and J.-G. Baek, "A clustering-based equipment condition model of chemical vapor deposition process," *International Journal of Precision Engineering and Manufacturing*, vol. 20, no. 10, pp. 1677–1689, 2019.

92. M. V. Shcherbakov, A. V. Glotov, and S. V. Cheremisinov, *Proactive and Predictive Maintenance of Cyber-Physical Systems*, pp. 263–278. Cham: Springer International Publishing, 2020.
93. Z. M. Cinar, Q. Zeeshan, D. Solyali, and O. Korhan, "Simulation of factory 4.0: A review," in *Industrial Engineering in the Digital Disruption Era* (F. Calisir and O. Korhan, eds.), (Cham), pp. 204–216, Springer International Publishing, 2020.
94. S. Bose, B. Natarajan, C. M. Scoglio, N. N. Schulz, D. M. Gruenbacher, and S. Das, "Shipboard power systems reconfiguration—a cyber-physical framework for response time analysis," *IEEE Transactions on Industrial Informatics*, vol. 10, no. 1, pp. 439–449, 2013.
95. M. A. Djeziri, R. Merzouki, and B. O. Bouamama, "Robust monitoring of an electric vehicle with structured and unstructured uncertainties," *IEEE transactions on vehicular technology*, vol. 58, no. 9, pp. 4710–4719, 2009.
96. R. Du, C. Chen, B. Yang, N. Lu, X. Guan, and X. Shen, "Effective urban traffic monitoring by vehicular sensor networks," *IEEE transactions on Vehicular Technology*, vol. 64, no. 1, pp. 273–286, 2014.
97. S. Kelkar and R. Kamal, "Adaptive fault diagnosis algorithm for controller area network," *IEEE transactions on Industrial Electronics*, vol. 61, no. 10, pp. 5527–5537, 2014.
98. H. Wang, Z. Man, H. Kong, Y. Zhao, M. Yu, Z. Cao, J. Zheng, and M. T. Do, "Design and implementation of adaptive terminal sliding-mode control on a steer-by-wire equipped road vehicle," *IEEE Transactions on Industrial Electronics*, vol. 63, no. 9, pp. 5774–5785, 2016.
99. H. Li, X. Jing, H.-K. Lam, and P. Shi, "Fuzzy sampled-data control for uncertain vehicle suspension systems," *IEEE transactions on cybernetics*, vol. 44, no. 7, pp. 1111–1126, 2013.
100. C. Wu, Z. Hu, J. Liu, and L. Wu, "Secure estimation for cyber-physical systems via sliding mode," *IEEE transactions on cybernetics*, vol. 48, no. 12, pp. 3420–3431, 2018.

Chapter 8
Cybersecurity in Data-Driven Cyber Physical System

Abstract In this chapter, we give an overview of cybersecurity in data-driven CPS. We first review the fundamentals of cybersecurity. The properties of cybersecurity in CPS are analyzed. Then, we summarize the attacks in different layers of CPS. Based on this, we analyze the requirements of cybersecurity in data-driven CPS and the importance of meeting these requirements. After that, cybersecurity challenges in data-driven CPS are proposed from a data-driven perspective.

CPS are network of interconnected computing devices, physical components, and infrastructure that work together to control and monitor physical processes such as power grids, transportation systems, and manufacturing plants. The physical and computational parts comprise the main body of the CPS. Predefined actions can be executed correctly and manage the entire system's state by the feedback loop inside the system. Cyber attacks can target individuals, organizations, or even entire nations, and they exploit vulnerabilities in computer systems or human behavior to achieve their objectives.

CPS are vulnerable to cyber attacks as they couple the digital world with the physical world. Several factors contribute to such vulnerability. The first factor is the increased connectivity from different subsystems. CPS rely on the integration of physical and digital systems to automate and control various operations. This integration requires a high level of connectivity between different devices and systems, which increases the potential attack surface for cybercriminals. Then, the components of CPS exhibit a high level of complexity, which also makes CPS vulnerable to uncertainty from different dimensions as attacks might be triggered from the diverse possibility of layers in CPS. At last, during the design phase of CPS, more emphasis was placed on functionality rather than security, leading to numerous security vulnerabilities that often remain hidden within the system.

Furthermore, due to the data-driven nature of CPS, there is a higher demand for security and stability in data transfer, making network security an indispensable component of data-driven CPS. These systems interact with the physical world and cybersecurity becomes crucial in ensuring their effective performance. In the real world, the distribution of CPS geographically results in varying degrees of vulnerability across different parts. The characteristics of interactions between CPS and

F. Li et al., *Data-Driven Cyber Physical Systems*,
https://doi.org/10.1007/978-981-96-8709-1_8

the physical world imply that external interference and attacks have varying degrees of system impact. Such attacks from different sources can sometimes cause serious damage to CPS themselves or to the physical entities affected by CPS.

Traditionally, data had a minimal role in cybersecurity arrangements, as it was managed in an ad hoc manner and manually handled by security analysts. However, with the rise of increasingly complex and frequent cybersecurity attacks, these conventional approaches have proven inadequate in mitigating threats and reducing risks to businesses and organizations. In addition, the data flow in CPS provides enormous potential for countering cyber attacks in the era of big data. As shown in Fig. 8.1, the data flow in CPS involves collecting, communicating, processing, and using data to interact with the physical world in new and innovative ways. It is a big challenge to deal with such a large volume of data and extract efficient and trustworthy features to protect the whole system throughout the entire CPS. By utilizing big data, machine learning, and artificial intelligence, organizations can conduct comprehensive analyses of both current and historical data to establish a baseline of normal activity. With this information, cybersecurity arrangements can be optimized to fit into the CPS.

This chapter will introduce the types of attacks that occur in the different layers of CPS. Then, we discuss the requirements of cybersecurity in CPS from a different point of view. After that, the importance of cybersecurity and the challenges that are desired to cope with are provided for obtaining a safer system. We analyze the techniques that detect the attack in a data-driven way.

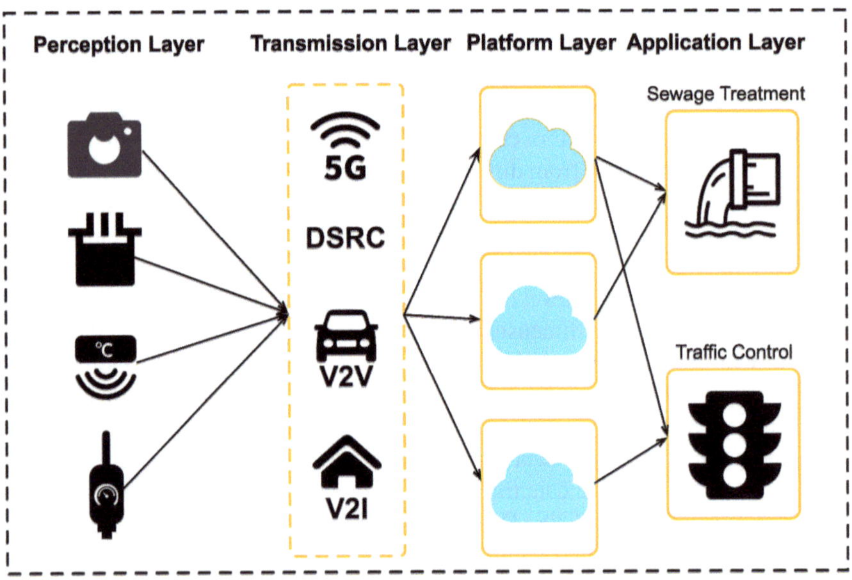

Fig. 8.1 The data in CPS flows successively into the perception layer, transmission layer, platform layer, and application layer

8.1 Cyber Attacks in Data-Driven CPS

A cyber attack refers to a deliberate and unauthorized attempt to exploit computer systems, networks, or digital devices, with the intention of causing damage, stealing information, disrupting operations, or gaining unauthorized access [1]. Attacks pose a significant threat due to their potential to cause severe damage to the physical environment. It is of great significance to recognize that vulnerabilities exist at every layer of the CPS, rendering them susceptible to both passive and active attacks. Moreover, CPS encounter a wider range of attacks compared to traditional information systems, as the scope extends beyond the confines of the CPS itself. Notably, attacks may originate from the network, with the Internet serving as a transmission layer [2]. This expanded attack surface necessitates a comprehensive understanding of potential risks. Specifically, the perception layer of CPS is particularly vulnerable to attacks targeting individual nodes, such as sensors and actuators. These attacks may compromise the integrity of data collected from the physical environment, leading to distorted or inaccurate information being processed within the system. Additionally, the transmission layer faces characteristic challenges, including data leakage, data integrity compromise, and potential security issues during the transmission of data between various components of the CPS. The above vulnerabilities not only jeopardize the confidentiality and integrity of the transmitted data but also pose a risk to the overall system performance and reliability.

Furthermore, attacks at the application layer can have serious consequences, with unauthorized access being a significant concern. Unauthorized access can potentially lead to user privacy loss, as malicious actors may expose or manipulate sensitive information. Such privacy breaches not only impact individuals but can also have broader implications, eroding trust in the system and undermining its overall functionality. Building a robust security framework that addresses the vulnerabilities across all layers of the CPS is imperative. While specific attacks may target individual layers, it is essential to recognize that some attacks can exploit weaknesses across multiple layers simultaneously. Consequently, a holistic and comprehensive defensive strategy must be implemented to ensure the protection and resilience of CPS against potential threats [3].

Moreover, the interconnectivity and interdependence of different components within CPS further amplify the potential impact of attacks. A successful breach in one layer can cascade through the entire system, compromising the functionality of CPS and introducing vulnerabilities to other layers. For instance, an attack on the perception layer, such as manipulating sensor data, can arouse erroneous inputs being processed at higher layers, resulting in faulty decisions and actions. Similarly, a breach at the application layer could provide unauthorized access that enables an attacker to manipulate control commands, compromising the integrity of the entire system.

To effectively address these evolving threats, a proactive and adaptive approach to CPS security is required. Such an approach entails continuous monitoring and analysis of system behavior to identify anomalous patterns or deviations from normal operation. By leveraging advanced technologies like machine learning and artificial intelligence, CPS can establish baseline models of normal behavior and promptly detect and respond to suspicious activities or potential attacks. In addition, robust encryption and authentication mechanisms should be implemented to ensure secure communication and data integrity within CPS. The approaches of cybersecurity include secure protocols for data transmission, strong authentication mechanisms to prevent unauthorized access, and encryption techniques to protect sensitive information from interception or tampering.

As technology advances and the cybersecurity landscape is everchanging, the attack methods continue to evolve and improve. We can shift our focus toward the specific categories of attacks that present a significant threat to CPS. In CPS, four potential modes of attack are available to an adversary: physical-only attack, cyber-only attack, physical-enabled cyber-attack, and cyber-enabled physical attack [4]. These attacks can occur at various layers, each with its unique vulnerabilities and potential consequences. The various attacks in CPS, along with their respective targets, are succinctly encapsulated in Fig. 8.2. Understanding the breadth and depth of these attacks is crucial for developing effective defence strategies. There are the perception layer, transmission layer, platform layer, and application layer in the abstract structure of CPS. We consider attacks at different layers separately in the following section.

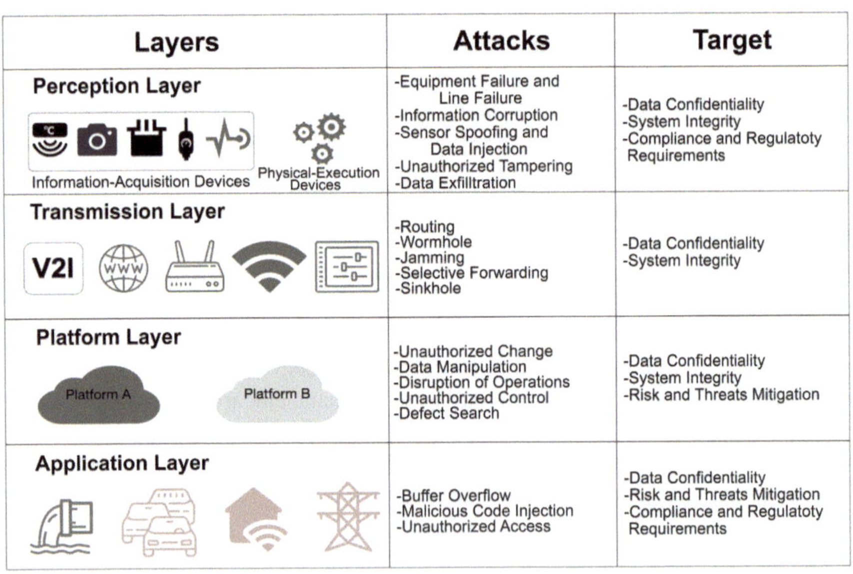

Fig. 8.2 Different types of attacks in each layer of CPS violate different targets of cyber security

8.1.1 Attacks at the Perception Layer

The perception layer, also called the recognition layer [5] or sensor layer [6], is the first layer of CPS that collects multi-source information from the physical world. Multiple terminal equipment is comprised of the perception network of the perception layer such as actuators, cameras, sensors, global position systems (GPSs), laser scanners, intelligent devices, radio frequency identification (RFID) tags with 2-D bar code labels, and readers [3, 7–9]. Such terminal equipment can be divided into two types: information-acquisition devices and physical-execution devices. Information-acquisition devices like sensors collect information such as temperature, pressure, motion, or sound, while physical-execution devices like actuators enable the system to interact with the physical world by controlling devices or initiating actions based on the sensor data. There are six potential weaknesses that might originate the cyber attacks in the CPS.

(1) **Lack of authentication and encryption**: When the authentication mechanisms and encryption protocols for sensors and actuators in the perception layer lack robustness, attackers can easily impersonate legitimate devices or intercept and manipulate transmitted data. This exposes the system to significant risks of unauthorized access and data tampering.
(2) **Inadequate data validation**: Inadequate data validation techniques within the perception layer make it vulnerable to processing inaccurate or malicious data. Without proper validation, the system may accept manipulated sensor readings, leading to erroneous decisions and actions based on flawed information.
(3) **Weak physical security measurement**: The absence of adequate physical security measures for sensor nodes and actuators renders them susceptible to physical tampering or unauthorized access. If attackers can easily gain physical access to these devices, they can manipulate them, alter their functionality, or introduce malicious components, thereby compromising the integrity of the perception layer.
(4) **Limited sensor integrity verification**: When the perception layer lacks mechanisms to verify the integrity and authenticity of sensor data, detecting and mitigating attacks that manipulate or spoof data becomes challenging. The absence of sensor integrity verification allows attackers to inject false or misleading information into the system, resulting in incorrect perceptions of the physical environment.
(5) **Dependency on insecure communication networks**: Relying on insecure communication networks, especially when transmitting data over the internet or other untrusted networks, exposes the perception layer to interception, eavesdropping, and man-in-the-middle attacks. Attackers can exploit vulnerabilities in the network infrastructure to gain unauthorized access to the CPS or tamper with transmitted data.

Such potential defects are the main reasons that the cyber attack appears in the perception layer putting a threat to the safety of the CPS. Furthermore, the capabilities of terminal devices in CPS are critical for system functionality. However, it is noteworthy that these devices are predominantly situated in external and outdoor environments, rendering them susceptible to physical attacks, including tampering with components or device replacement. Consequently, terminal devices are highly vulnerable to a diverse range of attack vectors.

Common attacks encountered at the perception layer encompass equipment failure, line failure, witch, electromagnetic interference, perceptual data corruption [2], differential power analysis [10], information disclosure, information tracking, tampering, sensing information leakage, physical destruction and energy-exhaustion attacks. Cybersecurity attacks at the perception layer present significant risks to the integrity and functionality of the system. Various types of attacks can compromise the security of the information-acquisition devices and the physical-execution devices in CPS. These attacks include:

Equipment failure and line failure: Malicious entities can exploit vulnerabilities in the hardware or communication channels, causing equipment failures or line failures. These incidents can disrupt the collection and transmission of data, compromising the accuracy and reliability of the system. The classical attack of such form is the *node capture* and false node.

Information corruption: Attackers may manipulate or alter the sensor data, introducing false or misleading information into the perception layer. This can lead to distorted perceptions of the physical environment, potentially resulting in incorrect decisions and actions. The related attack of information corruption is the Node Outage.

Sensor spoofing and data injection: Adversaries can spoof sensor identities or inject malicious data into the system, deceiving the CPS into making incorrect assessments or initiating inappropriate actions. By impersonating legitimate sensors or introducing false data, attackers can undermine the trustworthiness of the perception layer.

Unauthorized access and tTampering: Attackers may gain unauthorized access to sensor devices or actuators, enabling them to manipulate their behavior or compromise their functionality. Unauthorized access can lead to data manipulation, system disruption, or unauthorized control over physical processes. For disrupting the system, the path-based DOS is a typical attack that floods into a large number of packets to break down the devices.

Data exfiltration: Cybercriminals may exploit vulnerabilities in the communication channels or sensor devices to extract sensitive information from the perception layer. By intercepting or capturing data, attackers can compromise the confidentiality and privacy of the system.

The attacks encountered at the perception layer in CPS possess distinct characteristics that lead to the risks of interference. Attackers employ techniques such as manipulation and spoofing to deceive the CPS by introducing false or misleading sensor data, which can lead to erroneous decisions or actions. Exploiting vulnerabilities in equipment, communication channels, or software, these attacks disrupt data collection, transmission, or processing, compromising the overall functionality of the system. Adversaries corrupt or inject malicious data into the perception layer, distorting the data of the physical environment and jeopardizing the accuracy and reliability of system operations. Unauthorized access and tampering with sensor devices or actuators enable attackers to compromise their behavior and functionality, resulting in system disruptions or unauthorized control. Additionally, cybercriminals exploit vulnerabilities to extract sensitive information from the perception layer, leading to breaches of confidentiality and privacy, thereby compromising the system's security. Furthermore, physical destruction or manipulation of sensor or actuator devices, as well as energy-based attacks, can cause disruptions in data collection, exhaust energy resources, and introduce interference, impacting the operational effectiveness of the system. Understanding these distinct characteristics is crucial in implementing effective security measures and countermeasures to mitigate risks and ensure the integrity, confidentiality, and reliability of CPS at the perception layer. Robust authentication, encryption, anomaly detection, and intrusion prevention mechanisms can help mitigate these risks and ensure the resilience and security of the perception layer in CPS.

8.1.2 Attacks at the Transmission Layer

In the transmission layer of CPS, cyber attacks manifest in the form of data leakage during information transmission. This vulnerability is primarily due to the openness of the transmission media, particularly prevalent in wireless communication. Malicious actors capitalize on this openness to intercept transmitted messages through the radio interface. They exploit the captured data by modifying and retransmitting it or exchanging information across heterogeneous networks, effectively impersonating legitimate users.

As multiple characteristics in CPS, the reasons that contribute to the vulnerability of the transmission layer are complex. The transmission layer in CPS often relies on various communication channels, including wireless communication. On the one hand, the openness of these transmission media can make them susceptible to interception, eavesdropping, and unauthorized access by malicious actors. Moreover, attacks on the transmission layer can result in data leakage during information transmission. The openness of the transmission media, especially in wireless communication, increases the risk of attackers capturing transmitted messages, modifying them, or exchanging information between heterogeneous networks. This can lead to data integrity issues and unauthorized access to sensitive information. On the other hand, the transmission layer of CPS suffers from remote access vulnerabilities. The transmission layer involves remote access mechanisms among numerous network nodes. If not properly secured, these mechanisms can introduce vulnerabilities that attackers can exploit. Remote access vulnerabilities are the reason for unauthorized access, data manipulation, or traffic congestion, increasing the chances of successful attacks.

Because CPS relies on network infrastructure for communication between its various components, such as sensors, actuators, and controllers, vulnerabilities in the network infrastructure, such as insecure protocols, weak authentication mechanisms, or inadequate encryption can provide advantageous opportunities for attackers to undertake covert infiltration. Various types of attacks are commonly encountered at this layer, including the response and Sybil attacks, traffic analysis, tampering, exhaustion, collision, black hole attacks, flooding, trap doors, sink node attacks, direction misleading sinkhole attacks, wormhole attacks, wrong path selection, tunneling [2, 11], and illegal access [12, 13]. The forms of attacks at the transmission layer can be divided as follows.

Routing: This attack creates routing loops within the network, potentially leading to disrupted data transmission, increased transmission delays, or the elongation of the data path [10, 11].

Wormhole: The wormhole attack creates artificial "holes" in the network by falsely advertising paths through which all data packets are routed [14].

Jamming: In a jamming attack, the attacker intentionally disrupts wireless communication between sensor nodes and the remote base station. This interference can introduce noise or transmit signals on the same frequency, potentially causing a Denial of Service (DoS) situation due to deliberate network interference [11, 15, 16].

Selective forwarding: This attack involves compromised nodes selectively dropping and discarding data packets while forwarding only specific ones. In some instances, a compromised node may cease forwarding packets to their intended destination or only forward chosen messages, making it appear legitimate [11].

Sinkhole: In a sinkhole attack, an attacker falsely advertises itself as the best routing path for network traffic. This can be used as a precursor to other attacks, such as selective forwarding and spoofing [11].

8.1.3 Attacks at the Platform Layer

A CPS platform provides automatic discovery of connected devices and systems and enhances risk management through various protective features. It provides centralized management, monitoring, and control over complex networks. However, a compromised CPS platform also represents a grave threat. With access to such a platform, an attacker could manipulate both cyber and physical assets. They could mask unauthorized changes to devices and data, bypass controls, pivot to access operational systems, and ultimately disrupt critical infrastructure. To launch an attack at the platform layer, the adversary first gains visibility into the target network by monitoring communications and scanning devices. They identify assets, connections, and potential vulnerabilities in the CPS platform. Next, the attacker can exploit a vulnerability to gain initial access, whether through a software flaw, stolen credential, or misconfiguration, etc. The platform, as an integration of multiple CPS, is always a main target of the intruders. The CPS platform represents a high-value target. By compromising such a platform, and the attacker gains access to the components that control and monitor critical operational systems and infrastructure. This provides opportunities for disruption, damage, and unauthorized changes that could severely impact an organization's physical processes and assets. The sensitivity of these systems makes a CPS platform an attractive target.

There are three reasons that contribute to the vulnerability in the platform layer. First, platforms incorporate and expose more attack surfaces. First, A CPS platform adds additional software, network connections, and other potential vulnerabilities that did not previously exist when systems were managed individually. These new codebases, connections, and access points create more opportunities for adversaries to find and exploit weaknesses to gain entry. Rigorous testing and patching are required to address vulnerabilities in a timely manner. Second, the platform of CPS relies on sensitive credentials and data to operate. A CPS platform gathers and relies on critical information like system logins, passwords, network configurations, and details on connected physical assets to function. This sensitive information could allow attackers to access operational environments and make unauthorized changes if accessed. Strong access controls and encryption are essential to protecting data and credentials. Third, platforms increase the potential for software and configuration issues. Any new system introduces the possibility of software bugs, flaws, or misconfigurations that could create security risks. With a CPS platform, any such issues could impact multiple connected assets and systems. Extensive testing in staging environments is needed to identify any problems before platforms are deployed. However, vulnerabilities are often discovered only after deployment, requiring prompt updates and patching. In summary, the platform of CPS aims to leverage the integration level of different components of CPS and boost the system's efficiency.

At the platform layer of CPS, several cyber attacks can pose significant risks. These attacks target the integrated CPS platform itself and can have severe consequences for the entire system. Here are some notable cyber attacks at the platform layer of CPS.

Compromised Platform Access: Attackers may gain unauthorized access [17, 18] to the CPS platform by exploiting software vulnerabilities, stolen credentials, misconfigurations, or other means. Once inside, they can manipulate both cyber and physical assets, bypass security controls, and make unauthorized changes.

Data manipulation: By compromising the CPS platform, attackers can manipulate and tamper with critical data. This includes altering sensor readings, modifying control parameters, or falsifying operational information. Such data manipulation can have serious implications for decision-making processes and the overall functioning of the CPS.

Disruption of operations: Attackers targeting the CPS platform can disrupt operations by causing system failures, introducing malware or malicious code, or triggering cascading effects across interconnected devices and systems. This disruption can impact critical infrastructure sectors such as power grids, transportation networks, or manufacturing facilities.

Unauthorized control: By compromising the CPS platform, attackers can gain control over the centralized management, monitoring, and control functions. This allows them to manipulate cyber and physical assets, override safety mechanisms, or interfere with normal operations. Unauthorized control can lead to catastrophic events and safety hazards.

Exploitation of platform vulnerabilities: The CPS platform introduces new attack surfaces, including additional software, network connections, and integrated components. Attackers can exploit vulnerabilities in these elements to gain entry, escalate privileges, or move laterally within the platform to access sensitive systems and infrastructure.

To mitigate these risks, it is crucial to prioritize security measures at the platform layer of CPS. This includes regular vulnerability assessments, robust access controls, encryption of sensitive data, continuous monitoring, and prompt patching and updates. Implementing strong security practices can help safeguard the CPS platform and protect against potential cyber threats and their cascading effects on physical systems.

8.1.4 Attacks at the Application Layer

At the application layer of CPS, various attacks can have detrimental effects on user data, privacy, and device security. The application layer deals with user interactions, information processing, and the execution of applications within the CPS environment. Attackers targeting the application layer aim to exploit vulnerabilities, compromise user privacy, gain unauthorized access, or inject malicious code.

One prominent attack at the application layer is the buffer overflow. This attack takes advantage of vulnerabilities in software that can lead to buffer overflow situations. By exploiting these vulnerabilities, attackers can inject malicious code or overwrite critical data in the application's memory space. Buffer overflow attacks can have severe consequences, including system crashes, unauthorized access to sensitive information, and even remote code execution. Such attacks pose significant risks to the integrity and availability of user data, compromising the overall functionality of the CPS application.

Another type of attack at the application layer is the deployment of malicious code. Attackers launch various forms of malicious code, such as viruses and worms, to target user applications within the CPS environment. The presence of malicious code can significantly impact system performance, causing network slowdowns, application crashes, or data corruption. These attacks not only disrupt normal operations but can also lead to data loss, privacy breaches, and compromised user devices. As malicious code spreads through the CPS network, it poses a continuous threat to the overall stability and security of the system. The consequences of attacks at the application layer are far-reaching. User privacy leakage is a critical concern when sensitive user information is gathered and processed within the CPS application. If attackers successfully breach the application's security, they can gain unauthorized access to personal data, user habits, or even health conditions. The exposure of such private information can result in severe privacy breaches, identity theft, and potential harm to individuals. Safeguarding user privacy within the application layer is paramount to maintaining trust and ensuring the secure functioning of the CPS system.

Unauthorized Access is another prevalent attack at the application layer. Attackers aim to bypass authentication mechanisms or exploit vulnerabilities to gain unauthorized entry into the CPS application. Once inside, attackers can manipulate user data, compromise system resources, or even control connected devices. Unauthorized access attacks not only violate user privacy but also put critical infrastructure and physical assets at risk. Implementing robust access control measures, strong authentication mechanisms, and regular security audits are essential to mitigate the risks associated with unauthorized access.

In addition to these attacks, other common threats at the application layer include the forgery of Database and Control Commands. Attackers may attempt to manipulate or forge commands to gain control over the CPS application or disrupt its normal operations. By tampering with database records or control commands, attackers can cause significant damage, manipulate system behavior, or render the application unreliable. Ensuring the integrity and authenticity of data and control commands is crucial to preventing such attacks and maintaining the secure functioning of the CPS application.

Comprehensive security measures must be implemented to protect against attacks at the application layer. These include regular vulnerability assessments, secure coding practices, input validation, encryption of sensitive data, and intrusion detection systems. Furthermore, user awareness and education about potential risks and best security practices are vital to fostering a secure CPS environment.

8.2 Requirements of Cybersecurity

8.2.1 Objective of Cybersecurity

The objective of cybersecurity in CPS is to safeguard the interconnected components and networks from potential threats, vulnerabilities, and attacks. It aims to protect the integrity, confidentiality, and availability of data, ensure the reliable operation of CPS applications, and mitigate the risks associated with unauthorized access and malicious activities. Cybersecurity in CPS serves several key objectives.

Protecting system integrity: Maintaining the integrity of CPS systems is the primary objective of cybersecurity. This involves preventing unauthorized modifications, tampering, or manipulation of critical components, software, and data. By ensuring the authenticity and trustworthiness of system resources, cybersecurity measures help to preserve the accuracy and reliability of CPS operations.

Ensuring data confidentiality: Ensuring data confidentiality is the fundamental of data management. It involves implementing encryption mechanisms, access controls, and secure data storage practices to prevent unauthorized disclosure, interception, or tampering. By maintaining data confidentiality, cybersecurity measures safeguard user privacy, sensitive information, and proprietary data.

Ensuring system availability: Ensuring the availability of CPS systems and preventing disruptions that can impact critical operations are the requirements for system stability. By implementing robust defenses against denial-of-service (DoS) attacks, network congestion, or system failures, cybersecurity helps to maintain the continuous and reliable functioning of CPS applications.

Mitigating risks and threats: Mitigating risks and threats that may compromise the security of CPS systems is the premise of maintaining cybersecurity. The processes involve proactive monitoring, threat intelligence, and vulnerability assessments to identify potential weaknesses and address them before they can be exploited by attackers. By promptly responding to emerging threats, cybersecurity measures minimize the likelihood and impact of successful attacks.

Ensuring compliance and regulatory requirements: Ensuring compliance with relevant laws, regulations, and industry standards is the objective related to commerce laws. The measures include data protection regulations, privacy requirements, and specific security guidelines for critical infrastructure. By adhering to these standards, cybersecurity measures help organizations demonstrate due diligence and maintain the trust of stakeholders.

Subsequently, this section will outline the various requirements for cybersecurity pertaining to different components. As shown in Fig. 8.3, the comprehensive requirements of cybersecurity are comprised from four parts: hardware security, software

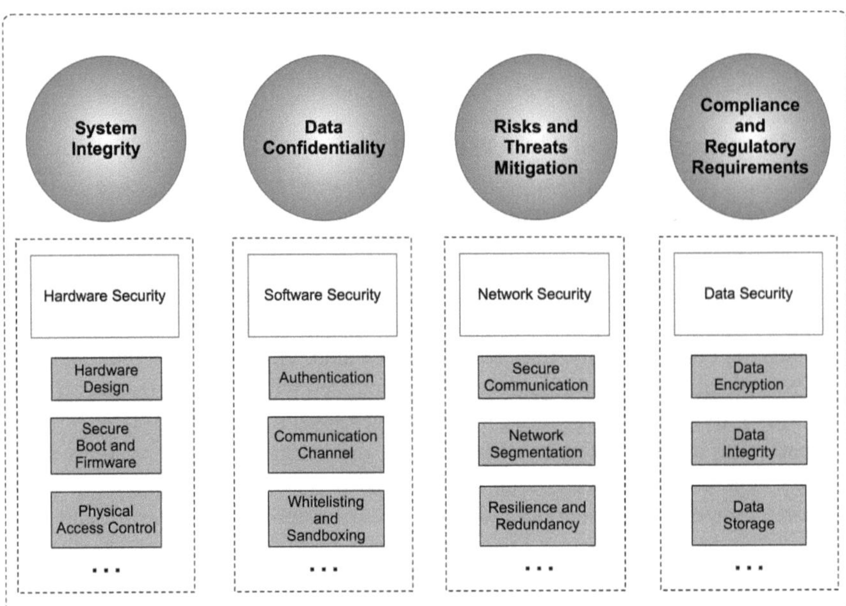

Fig. 8.3 Overall framework of DDCPS ensures the cybersecurity requirements to sustain the safety and smoothness of CPS

security, network security and data security. Each requirement is linked to the corresponding objective that CPS are made for. Along with these requirements, the overall framework of DDCPS was established and functions for the goals of CPS.

8.2.2 Hardware Security

Hardware security is a critical aspect of cybersecurity in CPS as it focuses on protecting the underlying physical components and devices from potential threats and vulnerabilities. Ensuring the security of hardware components is essential to maintain the integrity and reliability of CPS operations. To adequately address hardware security needs, the following critical elements must be considered:

Secure hardware design: The design and manufacturing processes of CPS hardware components should incorporate security considerations from the outset. This involves implementing secure hardware architectures, utilizing trusted computing modules, and incorporating tamper-resistant mechanisms to prevent unauthorized access, tampering, or reverse engineering of the hardware.

Secure boot and firmware: To protect against unauthorized modifications and ensure the integrity of the CPS system, secure boot mechanisms should be implemented. Secure boot verifies the authenticity and integrity of the system firmware and software during the boot-up process, preventing the execution of malicious or tampered code. Additionally, regular firmware updates and secure firmware distribution mechanisms should be employed to address any identified vulnerabilities or exploits.

Physical access controls: Physical security measures play a crucial role in protecting CPS hardware components. Restricted access to critical infrastructure, secure storage of hardware devices, and tamper-evident mechanisms help preventing unauthorized physical access and tampering. Monitoring and logging physical access events contribute to traceability and aid in identifying potential security breaches.

Hardware-based encryption: Hardware security modules (HSMs) and cryptographic accelerators provide hardware-based encryption capabilities, protecting sensitive data and communication channels within CPS systems. Hardware-based encryption helps secure data transmission, storage, and processing, ensuring the confidentiality and integrity of critical information.

Hardware testing and verification: Robust testing and verification processes are essential to identify and address potential vulnerabilities in CPS hardware. This includes techniques such as fault injection testing, side-channel analysis, and penetration testing to evaluate the resistance of hardware components against various attack vectors. Regular security assessments and hardware updates based on testing results help maintaining a strong security posture.

By addressing hardware security requirements in CPS, organizations can establish a foundation of trust and resilience in their systems. To bolster cybersecurity, CPS can use secure hardware designs, trusted supply chains, secure boot mechanisms, physical access controls, hardware-based encryption, and thorough testing to mitigate hardware risks.

8.2.3 *Software Security*

Ensuring the security of all software, including the operating system, applications, and firmware, is a crucial step for CPS to take, such as preventing attackers from exploiting vulnerabilities for attacks, preventing tampering, and protecting sensitive information from being stolen. Software security as a crucial component of cybersecurity focuses on protecting the software components and applications that enable the system's functionality. Ensuring software security in CPS is essential to prevent unauthorized access, mitigate vulnerabilities, and safeguard critical operations. The following key aspects are important in addressing software security requirements:

Access control and authentication: Implementing strong access controls and authentication mechanisms is crucial to prevent unauthorized access to CPS software components. User authentication, role-based access control (RBAC), and secure user management practices should be employed to ensure that only authorized individuals can interact with the system. Multi-factor authentication and strong password policies add an additional layer of security.

Secure communication channels: Secure communication channels are essential to protect data transmission within CPS systems. Implementing encryption protocols (such as SSL/TLS) and secure communication standards helps ensuring the confidentiality and integrity of data exchanged between different software components. Secure network configurations, firewalls, and intrusion detection systems further enhance the security of communication channels.

Application whitelisting and sandboxing: Employing application whitelisting techniques allows organizations to define a trusted list of authorized software applications that can run within the CPS environment. This helps preventing the execution of unauthorized or malicious software. Sandboxing techniques isolate applications from critical system resources, limiting their potential impact if compromised.

Secure software updates: Secure software update mechanisms are crucial to maintain the security of CPS software components. Organizations should establish secure channels for distributing software updates and validating the authenticity and integrity of updates before installation. Secure update processes help mitigating the risk of deploying compromised or malicious software.

By addressing software security requirements in CPS, organizations can enhance the overall cybersecurity of their systems. Implementing secure software development practices, managing vulnerabilities effectively, enforcing robust access controls and authentication mechanisms, employing secure communication protocols, securing applications through whitelisting and sandboxing, and implementing secure software update processes significantly reduce the risk of unauthorized access, data breaches, and system compromises in CPS.

8.2.4 *Network Security*

Network security plays a pivotal role in ensuring the cybersecurity of CPS. As CPS networks interconnect various components and devices, they become potential targets for malicious activities. Safeguarding these networks requires the implementation of effective network security measures. Here are some key aspects of network security in CPS.

Secure communication: Establishing secure communication protocols is vital to protect data transmitted within CPS networks. Encryption techniques like transport layer Security (**TLS**) or secure shell (**SSH**) can be employed to ensure the confidentiality and integrity of data during transmission. By encrypting the communication channels, unauthorized access and tampering can be prevented.

Access control: Strong access control mechanisms are essential to restrict unauthorized access to CPS networks. Robust authentication methods, such as username/password combinations, two-factor authentication, or digital certificates, help verify the identity of users and devices before granting network access. Implementing access control policies ensures that only authorized entities can connect to the network.

Network segmentation: Network segmentation involves dividing the CPS network into distinct segments or zones, each with its own security policies. This helps isolate critical components and restricts lateral movement within the network. By segregating resources based on functionality or security requirements, the impact of potential security breaches can be contained, limiting the scope of attacks.

Network resilience and redundancy: Designing CPS networks with resilience in mind is crucial for maintaining operational continuity. Redundancy measures, such as implementing redundant network links, backup systems, or alternate communication paths, help ensure that critical operations can continue despite network disruptions. Building network resilience enhances the overall security posture of CPS networks.

By implementing comprehensive network security measures, the administrators of CPS can mitigate risks, protect sensitive data, and ensure their systems' reliable and secure operation. Safeguarding the network infrastructure is essential for preventing unauthorized access, maintaining data integrity, and safeguarding critical infrastructure from cyber threats.

8.2.5 *Data Security*

Ensuring data security is a vital component of cybersecurity in CPS. Data security involves safeguarding sensitive and critical information from unauthorized access, disclosure, modification, or destruction. With the vast amount of data collected, processed, and transmitted within CPS, robust data security measures are essential. Here are key considerations for ensuring data security in CPS:

Data encryption: Encryption plays a key role in securing data in transit and at rest. Encryption algorithms transform sensitive data into ciphertext, which can only be decrypted with the appropriate encryption key. Strong encryption techniques, such as advanced encryption standard (AES), ensure the confidentiality and integrity of data, even if it falls into the wrong hands.

Data integrity: Maintaining data integrity is essential to ensure that data remains unaltered and trustworthy. Hash functions and digital signatures can be used to verify the integrity of data. Hash functions generate unique hash values for data, and any modifications to the data will result in different hash values. Digital signatures provide a way to verify the authenticity and integrity of data by using cryptographic algorithms.

Secure data storage: Proper storage of data is critical to its security. Implementing secure storage solutions, such as encrypted databases or secure file systems, helps protecting data at rest. Encryption of stored data ensures that even if physical storage devices are compromised, the data remains protected from unauthorized access.

Data backup and recovery: Regular data backups are essential to mitigate the risk of data loss due to various incidents, such as hardware failures, natural disasters, or cyber attacks. Implementing backup strategies, both onsite and offsite, helps ensuring data availability and facilitating recovery in case of data breaches or system failures.

Data lifecycle management: Effective data lifecycle management involves managing data from its creation to deletion in a secure manner. This includes proper data classification, defining data retention policies, and securely disposing of data that is no longer needed. By categorizing data based on its sensitivity and applying appropriate security controls, organizations can protect critical data throughout its lifecycle.

Secure data transmission: Protecting data during transmission is crucial to prevent unauthorized interception or tampering. Secure communication protocols, such as TLS or VPNs, ensure encrypted and authenticated data transmission between CPS components. Implementing secure communication channels prevents eavesdropping, data manipulation, and unauthorized access during data transfer.

Auditing and monitoring: Implementing robust auditing and monitoring mechanisms helps detecting and responding to data security incidents promptly. By monitoring data access, modifications, and user activities, organizations can identify any anomalous behavior or security breaches. Logging and analyzing data-related events provide valuable insights for forensic investigations and security incident response.

Data security aims to protect sensitive data, maintain data integrity, and comply with privacy regulations. Safeguarding data at all stages, from collection to storage and transmission, is essential for maintaining the confidentiality, integrity, and availability of data in CPS environments.

8.3 Importance of Cybersecurity in Data-Driven CPS

Security originates from the Latin roots that signify "without care" and encompasses the attributes of a system that grants individuals a sense of freedom from concerns. It encompasses a collection of measures designed to guarantee that a system can achieve its intended objectives while minimizing any unintended adverse effects. As new features are introduced to a system, security measures are the mechanisms that protect the functionality of data-driven CPS. The importance of cybersecurity in data-driven CPS cannot be overstated. As data plays a central role in decision-making, control, and optimization in these systems, ensuring the security of data transmission in cyberspace is of great significance. There are five aspects of why cybersecurity is important in data-driven CPS: data integrity and accuracy, privacy and confidentiality, system resilience and availability, and regulatory requirements. First, we introduce the importance of data integrity. Cybersecurity measures such as data authentication, validation, and anomaly detection ensure the integrity and accuracy of data in data-driven CPS. By protecting against the cyber attacks mentioned before, cybersecurity safeguards the reliability of the data used for decision-making and control actions. Second, cybersecurity in data-driven CPS is also related to the privacy and confidentiality of the system. Cybersecurity in data-driven CPS ensures the protection of sensitive information, including personal data and proprietary information. Through encryption, access controls, and secure data storage, cybersecurity safeguards the privacy and confidentiality of the data, preventing unauthorized access and data breaches. Third, we elaborate on the significance of cybersecurity for system resilience and availability. At last, the cyber security in data-driven CPS is also made to meet the regulatory and compliance requirements.

8.3.1 *Data Integrity and Accuracy*

Data integrity refers to the accuracy, consistency, and reliability of data throughout its lifecycle. In other words, the data integrity in CPS means that there are no unauthorized modifications of data [19]. As the large scale of devices in CPS, the "thermostat model" can be used to describe the data integrity. As shown in Fig. 8.4, the reality in CPS is that deviations from the ideal perfect state of data integrity are inevitable. The illustration shows the actual state of data within the system, which typically hovers around the desired ideal state. While the ideal perfect state may only exist in theoretical or hypothetical scenarios, efforts are made to maintain the data integrity as close to this ideal state as possible in data-driven CPS. In data-driven CPS, where data play a crucial role in decision-making and control processes, ensuring the integrity of data is of paramount importance.

Data integrity is important in CPS due to its direct impact on the overall system reliability, performance, and safety. The quality of the data is important as the following reasons:

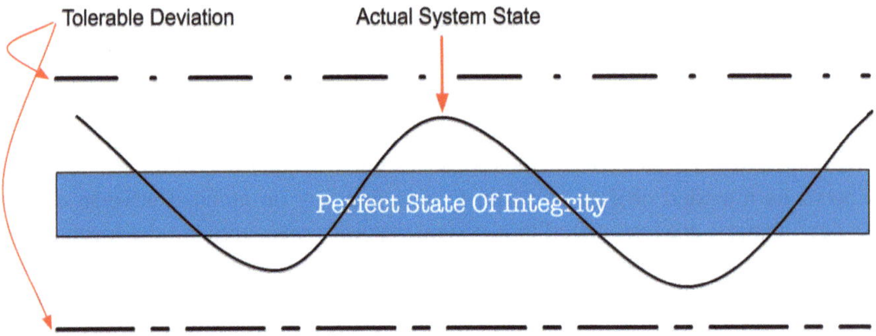

Fig. 8.4 The "thermostat model" of data integrity in DDCPS

1. **Reliable decision-making**: CPS rely on accurate and trustworthy data for real-time decision-making. Inaccurate or compromised data can lead to incorrect system behavior, resulting in suboptimal or even dangerous outcomes. Many industrial application [20–23] in CPS have shown that data integrity ensures that the information used for decision-making is correct, reliable, and consistent, thus enhancing the system's overall performance and effectiveness.
2. **System safety**: In safety-critical CPS applications such as autonomous vehicles [24, 25], industrial control systems [26, 27], data integrity is paramount. Any alteration, corruption, or unauthorized access to data can have severe consequences, jeopardizing the safety of human lives, infrastructure, or sensitive operations. Ensuring data integrity minimizes the risk of system failures or accidents caused by compromised or erroneous data.
3. **Trust in System Operations**: CPS often involves the integration of multiple components, networks, and stakeholders. Data integrity builds trust among these entities by assuring them that the data being exchanged and utilized is accurate and reliable. Trust is crucial for collaboration, interconnectivity, and coordination within CPS, enabling efficient and secure operation across diverse domains.

Essentially, the success and reliability of DDCPS hinge significantly on the integrity of the data they utilize. Data integrity in DDCPS enables accurate decision-making, enhances system safety, and leverages trust in system operations. By prioritizing data integrity, CPS can achieve optimal performance, security, and safety in various domains, ranging from transportation and energy to healthcare and manufacturing.

8.3.2 Privacy and Confidentiality

Privacy, as defined by the National Institute of Standards and Technology (NIST), is "Assurance that the confidentiality of, and access to, certain information about an entity is protected" [28]. In some application scenarios, the privacy and confidentiality of the data in CPS are of great significance as one of the assessments of

CPS safety. On one hand, the privacy of each distributed data is of great importance. In healthcare CPS, sensitive data such as patient health records, medical images, and diagnostic results require privacy protection to prevent unauthorized access and ensure confidentiality [29]. Similarly, intelligent transportation systems [30] rely on the privacy and confidentiality of data related to vehicle location, driving behavior, and traffic monitoring to ensure security and optimize traffic flow. Energy CPS [31] involve protecting sensitive information on energy usage patterns, consumer data, and transaction records to ensure a stable energy supply and prevent fraudulent activities. Industrial control systems handle sensitive data on production equipment, performance monitoring, and product specifications, necessitating data privacy and confidentiality to protect trade secrets and maintain production process security.

On the other hand, in DDCPS, there may be sensitive or proprietary information that needs to be protected from unauthorized access or disclosure as the confidentiality property of CPS. This includes trade secrets, intellectual property, industrial processes, etc. Safeguarding the confidentiality of such data is essential to prevent economic losses, intellectual property theft, or competitive disadvantages. For detecting the attacks that aim to sneak the data, there have been many works [32–34] to recognize the attacks. As shown in Fig. 8.5, attackers can take advantage of the vulnerability of the CPS to breach the wall of data transmission.

In brief, privacy and confidentiality are of paramount importance in industrial applications of data-driven CPS. Industrial CPS deal with sensitive information, such as proprietary algorithms, customer data, and operational strategies. Protecting privacy and ensuring confidentiality prevents unauthorized access, data leakage, and potential financial losses or reputational damage. Compliance with data protection

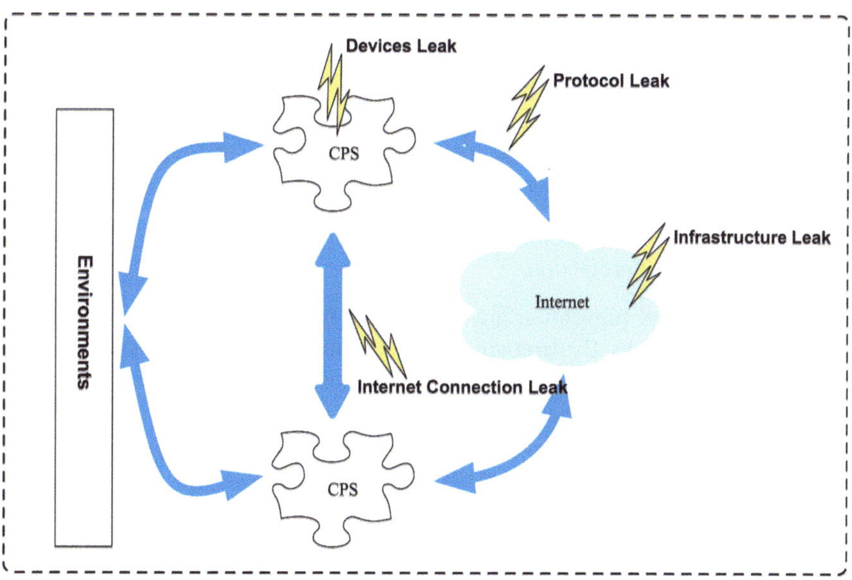

Fig. 8.5 Privacy leak points can arouse in different components of DDCPS

regulations, safeguarding intellectual property, and protecting trade secrets are vital aspects of privacy and confidentiality in this context. By prioritizing data security, organizations build trust among customers, partners, and stakeholders, contributing to a secure environment for data-driven operations and maintaining a competitive edge in the market.

8.3.3 System Resilience and Availability

System resilience refers to the ability of a system to adapt, recover, and continue functioning effectively in the face of disturbances, failures, or unexpected events [35]. It is the system's capacity to withstand and respond to various environments such as cyber attacks and network failures while maintaining its essential functions and performance. Moreover, system resilience in CPS refers to the ability of these interconnected systems to maintain their critical functions and operational capabilities in the face of disruptions or attacks. CPS often operate in complex and dynamic environments, where failures or disturbances can have significant consequences, including safety hazards, financial losses, or disruptions to critical services.

In the context of data-driven CPS, system resilience becomes crucial due to the reliance on data for decision-making and control processes. Ensuring the integrity, availability, and confidentiality of data is essential for maintaining system resilience. Cybersecurity plays a vital role in protecting CPS from cyber threats and unauthorized access that can compromise data integrity and disrupt system operations.

By implementing robust security measures, such as access controls [36, 37], encryption [38, 39], intrusion detection systems [13, 40], and real-time monitoring [41, 42], CPS can enhance their resilience against cyber attacks and data breaches. Proactive measures like regular vulnerability assessments, incident response planning, and disaster recovery strategies also contribute to system resilience by enabling rapid detection, response, and recovery from security incidents. The approaches that ensure the system resilience and availability in data-driven CPS can be divided into two types (Fig. 8.6):

1. Threat detection and prevention
2. Data integrity and protection.

Threat detection and prevention are to keep the robustness of the CPS which is resistant to cyber attacks. By deploying advanced intrusion detection systems [40, 43, 44], firewalls [45, 46], and security monitoring tools, organizations can proactively identify and mitigate potential cyber threats. This allows for the early detection of unauthorized access attempts, malware infections, or other security breaches that could compromise system resilience and availability.

Additionally, data integrity and protection are essential in preserving system resilience and availability. Ensuring the confidentiality and privacy of sensitive data prevents unauthorized access and potential tampering, maintaining the trustworthiness of the system. Implementing encryption techniques and access controls can

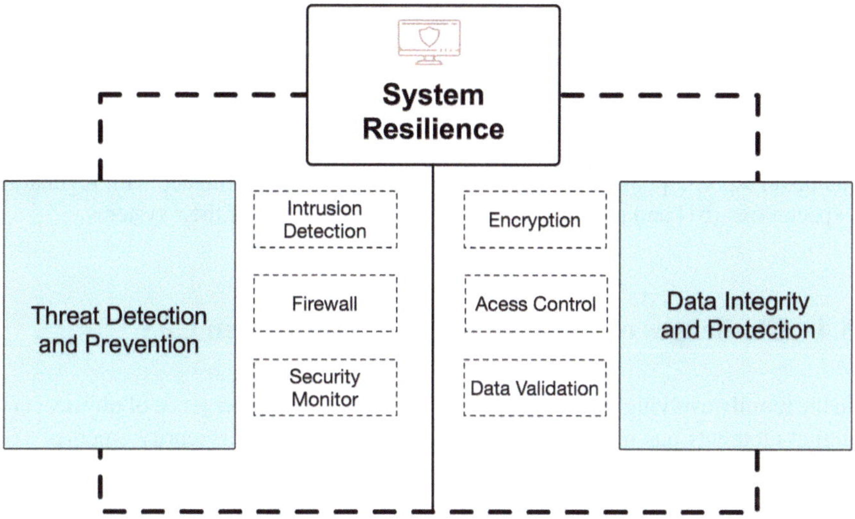

Fig. 8.6 System resilience of DDCPS depends on two kinds of operation. One is threat detection and prevention which defenses the external infringement and internal interference

safeguard data from unauthorized disclosure or modification, reducing the risk of data corruption and ensuring the availability of accurate and reliable information for decision-making.

Cybersecurity measures that focus on threat detection and data integrity are essential for maintaining system resilience and availability in data-driven CPS. By safeguarding the system against cyber threats, protecting data integrity, and having effective response plans in place, organizations can ensure the continuous operation and availability of their data-driven CPS, minimizing the impact of disruptions and maintaining critical services.

8.3.4 Regulatory Requirements

Regulatory bodies impose specific requirements and standards regarding data privacy, security, and protection. Cybersecurity measures help organizations ensure compliance with these regulations, such as the General Data Protection Regulation (GDPR) [47] or industry-specific standards like the Health Insurance Portability and Accountability Act (HIPAA) [48]. By implementing robust cybersecurity practices, organizations can meet regulatory obligations and avoid penalties or legal consequences.

Similar to the privacy mentioned before, regulatory requirements also emphasize the protection of personal information and require organizations to maintain strict data privacy measures. Cybersecurity plays a crucial role in safeguarding this data from unauthorized access, ensuring compliance with privacy regulations, and maintaining the safety of the CPS. Second, Regulatory requirements often focus on risk management and mitigation strategies. Cybersecurity also helps organizations identify, assess, and mitigate risks associated with data-driven CPS [49, 50]. By implementing security controls, conducting regular risk assessments, and developing incident response plans, organizations can demonstrate compliance with regulatory expectations [51] and protect the integrity and availability of their systems.

8.4 Challenges of Cybersecurity in Data-Driven CPS

In the rapidly evolving landscape of DDCPS, where the convergence of physical and digital elements has revolutionized various industries, cybersecurity emerges as a critical concern. As organizations increasingly rely on data-driven CPS to optimize operations, enhance efficiency, and drive innovation [52], they face a multitude of challenges that threaten the security and integrity of these systems. In this chapter, we delve into the multifaceted challenges that organizations encounter in safeguarding the cybersecurity of data-driven CPS. By understanding these challenges, we can better navigate the complexities and proactively address the vulnerabilities that arise in this dynamic environment.

8.4.1 Data-Driven Techniques for Attack Detection and Mitigation

Traditional intrusion detection and prevention systems (IDPS) are essential for detecting and preventing cyber threats in data-driven CPS [53]. These systems typically rely on signature-based monitoring [54], which involves comparing network traffic and system events against a database of known attack signatures. While this approach is effective in defending against known attacks and detecting threats aimed at disrupting system functionality, it has certain limitations. One significant limitation is the inability of traditional IDPSs to identify attacks that do not have known signatures. As cyber threats evolve and attackers become more sophisticated, they constantly develop new attack techniques that may not match any existing signatures in the database. Consequently, these systems fail to detect novel and emerging threats, leaving the data-driven CPS vulnerable to advanced persistent threats (APTs) [55] and low and slow attack vectors.

Moreover, traditional IDPSs are primarily designed as point-based solutions [56], focusing on the defense of individual targets within the system. This approach may overlook the holistic view of the CPS environment, as it fails to consider the interconnectedness and interdependencies of various components [57]. As a result, attackers may exploit vulnerabilities in one area of the system to gain unauthorized access or disrupt critical operations elsewhere.

To address these limitations, there is a need for data-driven techniques of intrusion detection and mitigation in CPS. Data-driven techniques should go beyond signature-based monitoring and incorporate techniques such as anomaly detection, machine learning, and behavior analysis. By leveraging these advanced techniques, cybersecurity systems can adapt to evolving threats, detecting unknown attack patterns, and providing a more comprehensive defense against APTs and low and slow-attack vectors. Additionally, adopting a holistic approach that considers the entire CPS ecosystem enables a more robust and effective defense strategy, ensuring the resilience and security of CPS applications.

However, implementing a data-driven approach to attack detection comes with challenges from the following areas:

Data quality and scalability: Data-driven techniques heavily rely on the availability of high-quality and diverse data for accurate analysis [58]. However, in CPS, data collection can be challenging due to resource constraints, network limitations, and the distributed nature of the system. Ensuring data quality, including accuracy, completeness, consistency, and timeliness, becomes important to obtain reliable insights and make informed decisions [58]. Large-scale CPS may face data quality issues due to various reasons such as sensor malfunctions [59], calibration issues, poor sensor quality, environmental effects, external noise [60], network or communication errors, and real-time scheduling problems [61, 62]. These challenges are greater in large-scale CPS, such as environmental and noise monitoring systems, which rely on various sensors and devices connected by extended networks and operate under noisy and dynamic conditions [63]. Additionally, as CPS generate massive volumes of data from numerous sensors and devices, scalability becomes a challenge in processing and analyzing this data in real time.

Complex data integration: Data integration is an important challenge in data-driven techniques for attack detection and mitigation because it is essential for providing a comprehensive and accurate view of the security landscape [64]. In order to effectively detect and mitigate attacks, organizations must be able to analyze data from multiple sources, including network traffic, system logs, and threat intelligence feeds. This data must be integrated in a way that ensures its accuracy, consistency, and completeness so that security teams can identify patterns and correlations that may indicate an attack. However, this can be a challenging task due to the large volume and complexity of data generated by various systems and devices, as well as the need to ensure that the data is protected from unauthorized access or tampering. Additionally, the dynamic nature of cyber threats means that data integration must be able to adapt quickly to changing conditions in order to provide timely and actionable information. All of these factors make data integration a critical challenge in data-driven techniques for attack detection and mitigation.

Anomaly detection and false positives: Detecting attacks in CPS requires identifying anomalous behavior that deviates from normal system operation. However, distinguishing genuine attacks from legitimate system behavior can be challenging, leading to false positives or false negatives. The dynamic nature of CPS, where legitimate system operations may exhibit variations due to changing conditions or legitimate user actions, adds to the complexity of accurately identifying anomalies. Balancing the detection of genuine attacks while minimizing false positives is a critical challenge in data-driven attack detection techniques.

Real-time responsiveness: Timeliness is crucial in responding to attacks and mitigating their impact in CPS. Traditional data-driven techniques may encounter challenges in providing real-time responsiveness due to the volume, velocity, and variety of data in CPS. Analyzing large-scale and high-velocity data streams in real time to detect and respond to attacks requires efficient algorithms, scalable processing frameworks, and optimized data storage and retrieval mechanisms.

Privacy and security of data: While data-driven techniques leverage data for attack detection, ensuring the privacy and security of sensitive information becomes a significant challenge. CPS often involve personal or confidential data, and its collection, storage, and processing need to comply with privacy regulations and safeguard against unauthorized access or data breaches. Balancing the need for data-driven analysis with privacy and security requirements poses a continuous challenge in CPS cybersecurity.

8.4.2 Data Trustworthiness and Policy-Based Sharing

Data trustworthiness refers to the reliability, authenticity, and integrity of data in a CPS environment. It involves ensuring the accuracy, completeness, and consistency of data, as well as verifying its origin and history through data provenance. Policy-based sharing involves the formulation and enforcement of policies that govern the access, usage, and sharing of data in a CPS environment. These policies define rules and guidelines for data access, data sharing permissions, and data usage restrictions.

Data trustworthiness and policy-based sharing play crucial roles in ensuring the security and effectiveness of data-driven CPS systems. They enable the reliable and secure exchange of data between different components and stakeholders within the system. Data Trustworthiness helps prevent malicious data injection, data tampering, and data poisoning attacks. Ensuring the authenticity and integrity of data helps maintain the reliability and accuracy of decision-making processes in CPS. Policy-based Sharing, on the other hand, allows for controlled access to data, ensuring that sensitive information is only accessible to authorized entities. It helps protect data privacy, confidentiality, and availability in the system. To achieve data trustworthiness and policy-based sharing, there are some key points as follows:

Data provenance: Capturing and maintaining accurate and comprehensive data provenance is challenging in CPS environments with distributed data sources and complex data transformations. Ensuring the accuracy and reliability of data lineage information is crucial but requires overcoming technical and operational hurdles.

Policy complexity: Developing effective policies and analyzing their impact can be challenging. Policy conflicts, granularity, and enforcement mechanisms need to be addressed to ensure consistent and enforceable policies across different CPS components.

Secure information sharing: Enabling secure and controlled information sharing requires robust encryption, authentication, and access control mechanisms. The challenge lies in balancing data sharing needs with data protection requirements to prevent unauthorized disclosure or tampering.

Reproducible experiments: Conducting repeatable experiments and validating research findings in dynamic CPS environments can be challenging. Replicating experimental conditions, data availability, and operational constraints pose difficulties in establishing reliable benchmarks and validating research outcomes.

8.4.3 Risk-Based Security Metrics

A risk-based approach to security metrics involves assessing and prioritizing risks based on their potential impact on the CPS. It aims to identify vulnerabilities, evaluate their likelihood of exploitation, and determine the potential consequences to prioritize security efforts effectively. The comprehensive game theoretical risk assessment framework applies game theory principles to assess risks in data-driven CPS. The framework considers various actors, strategies, and interactions to model and analyze the security landscape comprehensively. To access such risk-based security metrics, the data-driven CPS require to address the following points:

Complexity of system interdependencies: Data-driven CPS consist of interconnected components and networks with complex dependencies. Assessing risks across these interdependencies requires understanding the intricate relationships and potential cascading effects of vulnerabilities and attacks.

Uncertainty in adversarial behavior: Modeling the behavior and intentions of adversaries in a game-theoretical framework introduces uncertainties. Adversaries may adapt their strategies, collaborate, or employ sophisticated techniques, making it challenging to accurately predict their actions and assess associated risks.

Scalability and computational complexity: Analyzing risks and performing comprehensive game theoretical assessments at scale can be computationally intensive. As CPS grow in size and complexity, the framework needs to handle large-scale simulations and computations efficiently.

Availability of accurate and timely data: Gathering reliable and up-to-date data on system characteristics, vulnerabilities, and threat intelligence is crucial for accurate risk assessment. However, acquiring relevant data and ensuring its accuracy, completeness, and timeliness can be challenging in dynamic and distributed CPS environments.

Adoption and Integration: Integrating the risk assessment framework into existing data-driven CPS may face challenges in terms of compatibility, acceptance by stakeholders, and practical implementation. Ensuring that the framework aligns with existing security practices and standards is essential for its successful adoption.

8.5 Data-Driven Techniques of Cybersecurity in CPS

Data-driven cybersecurity technologies have many advantages over traditional technologies in cyber-physical systems. Data-driven techniques can achieve real-time threat detection [65, 66] and rapid response [67, 68]. Data-driven cybersecurity technology can monitor and analyze data traffic, logs, and events in systems in real-time to quickly detect potential threats and attacks. With rapid response and automated security measures, threats can be responded to quickly, the impact of attacks on the system can be reduced, and the availability and integrity of the system can be maintained. Secondly, data-driven techniques are capable of powerful threat detection. Data-driven cybersecurity techniques leverage methods such as machine learning [69], deep learning [70], and data analytics [69] to uncover hidden attack patterns and anomalous behavior from vast amounts of data. These technologies can identify new and unknown threats that are difficult to detect with traditional security tools, improving the accuracy and efficiency of threat detection. Thirdly, data-driven cybersecurity technologies can improve defenses through continuous learning and adaptation. Machine learning models can be continuously trained and optimized based on new data and threat intelligence to continuously improve their ability to identify and defend against emerging threats. This adaptability allows the system to better adapt to the evolving threat landscape. Fourthly, data-driven cybersecurity technologies can process and analyze large-scale data sets to extract valuable information and insights from them [71, 72]. Through data analysis and visualization techniques, security teams can better understand the security posture of systems, identify potential vulnerabilities and threats, and take appropriate action to address them. Finally, data-driven cybersecurity technologies can automate many security tasks and processes, reducing the risk and error of human operations [65, 73]. Automated threat detection, incident response, and security decisions improve response speed and accuracy and reduce the burden on security teams to focus more on critical security issues.

8.5.1 Data-Driven Attack Detection and Migitation

Data-driven attack detection and mitigation is an approach in cybersecurity that leverages data analytics, machine learning, and other computational techniques to identify and respond to cyber attacks in real time. This approach involves analyzing large volumes of data collected from various sources within CPS to detect patterns, anomalies, and indicators of malicious activity. One key aspect of data-driven attack detection and mitigation is the use of advanced algorithms and models to process and analyze the collected data. These algorithms can learn from historical data and detect patterns that may indicate ongoing or potential attacks. Machine learning techniques, such as supervised and unsupervised learning, anomaly detection, and behavioral

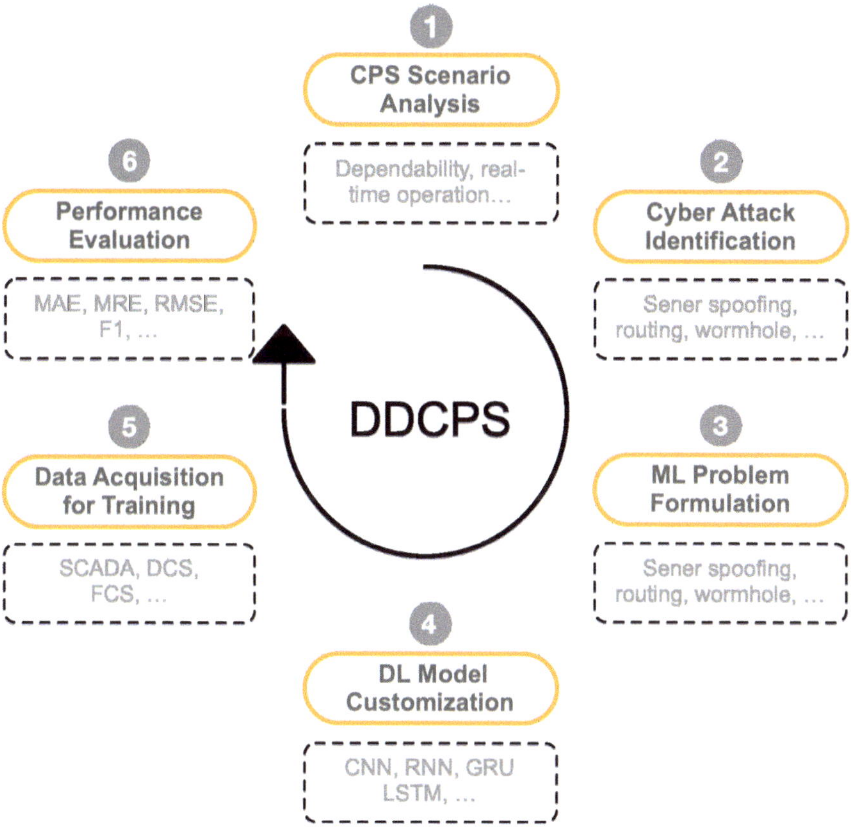

Fig. 8.7 The basic steps of DDCPS analysis the potential attacks in the system through DL methods

analysis, are important in identifying deviations from normal system behavior and detecting previously unknown attacks.

8.5.1.1 Basic Steps of the Data-Driven Approach in Attack Detection and Mitigation

The data-driven approach in attack detection and mitigation involves several basic steps, each contributing to the overall process of data-driven CPS against cyber threats [74]. These steps provide a systematic framework for addressing cybersecurity challenges in CPS scenarios, as shown in Fig. 8.7.

- **Step I: CPS scenario analysis**
  The first step is to conduct a comprehensive analysis of the CPS scenario. This analysis takes into account various factors such as dependability, real-time operation,

fault tolerance, and cybersecurity. Dependability encompasses service availability and reliability to minimize system downtime. Real-time operation is crucial for maintaining system functionality amidst rapidly changing inputs and environmental conditions. Fault tolerance ensures critical system components have sufficient backups to prevent the shutdown. With the increasing connection of CPS to cyberspace, cybersecurity requirements are becoming more prominent to enhance system control and overall quality of service. The analysis identifies key characteristics of CPS intrusion detection, including physical process monitoring, closed control loops, attack sophistication, and legacy technology [13]. By understanding these characteristics, a suitable cybersecurity problem can be formulated.

- **Step II: Cyber attack identification**
  Once the CPS scenario is defined, the next step involves identifying the cyber attacks associated with the CPS characteristics. Physical process monitoring enables the detection of anomalies in the system by constantly monitoring the physical properties governed by the laws of physics. Closed control loops indicate that CPS events are more regular and predictable, facilitating the detection of attacks like replay attacks. The consideration of attack sophistication is essential due to the potential impact and rewards of successful cyber attacks, including theft of sensitive information and intelligence. Legacy technology, represented by existing mechanical and hydraulic controls, poses challenges in interacting with software-defined control. By identifying and understanding these cyber attacks, the groundwork is laid for further analysis and mitigation.
- **Step III: ML problem formulation**
  The third step involves translating the cybersecurity problem into the domain of machine learning (ML). ML is a process by which a computer program learns from experience to improve its performance on specific tasks. In this step, The tasks are defined, which can be classification, clustering, regression, or other ML tasks. Classification tasks aim to allocate outputs to predefined attack categories while clustering tasks involve grouping outputs into clusters representing normal or attack traffic. Regression tasks focus on predicting numerical values.
- **Step IV: DL Model customization**
  In this step, DL models are customized based on the chosen architecture and parameter optimization. Different DL models, such as autoencoders for data representation and CNNs for classification, are selected based on the specific needs of the research problem. The configuration of the DL model considers the available data and trade-offs between layer size and depth. By understanding DL algorithms and CPS cybersecurity data, insights are gained into customizing the models for improved performance.
- **Step V: Data acquisition for training**
  Data acquisition is a critical step for training DL models. The quality and quantity of data directly impact the effectiveness of solving the research problem. Simulation methods and existing datasets collected by other researchers serve as sources of data. These datasets, such as the SWaT dataset, SCADA IDS dataset, and UNSW-NB15 dataset, contain various cyberattacks commonly observed in CPS.

The inclusion of diverse attack scenarios enriches the training data and enhances the model's ability to detect and mitigate cyber threats.

- **Step VI: Performance evaluation**
  The final step involves evaluating the performance of the DL models to determine if they meet the objectives. Performance metrics are used to measure the model's performance based on the tasks at hand. For prediction or regression tasks, metrics such as mean absolute error (MAE), mean relative error (MRE), root mean squared error (RMSE), and mean absolute percentage error (MAPE) assess the model's accuracy. Through rigorous evaluation, the effectiveness and efficiency of the DL models in detecting and mitigating cyber attacks in CPS are assessed.

8.5.1.2 Machine Learning Model for Attack Detection

Machine learning models have emerged as powerful tools for detecting and mitigating cyber attacks in the complex and interconnected realm of CPS. These models leverage their ability to analyze vast amounts of data, recognize patterns, and identify anomalies to enhance the security posture of CPS environments. By harnessing the potential of machine learning, CPS can benefit from scalable, adaptable, and real-time attack detection capabilities.

One of the key advantages of machine learning models is their scalability [75, 76]. CPS environments generate substantial volumes of data, ranging from network traffic to sensor readings and system logs. Machine learning models excel in handling this data deluge, enabling comprehensive analysis and detection of subtle attack patterns that may be concealed within the data [77, 78]. Their ability to process and analyze large datasets in real-time allows for efficient monitoring and detection of attacks across dynamic CPS. Machine learning models also possess remarkable adaptability [79], a critical feature in the ever-evolving landscape of cyber attacks. Traditional rule-based systems often struggle to keep pace with the sophistication and rapid evolution of attack techniques. However, machine learning models can learn from historical data [80], continuously update their knowledge, and detect emerging attack patterns. This adaptability ensures that CPS remain resilient against evolving threats, providing timely and effective detection capabilities.

Pattern recognition is another fundamental advantage of machine learning models for CPS attack detection [81]. By analyzing historical data, these models learn the normal behavior of CPS and identify deviations that may indicate potential attacks. This ability to recognize both known and unknown attack patterns enables proactive identification of malicious activities, even if specific attack signatures have not been explicitly programmed into the models. By capturing the complex relationships and interdependencies within CPS, machine learning models offer contextual understanding [82], allowing them to detect abnormal behaviors that span across multiple subsystems. Machine learning models have a strong capability for anomaly detection [83], which is particularly relevant for CPS attack detection. By comparing real-time data with learned patterns, these models can identify instances that significantly deviate from the expected behavior. This anomaly detection approach raises red flags and triggers further investigation, potentially leading to the early detection and mitigation of attacks. Moreover, by reducing false positives, machine learning models optimize the use of resources and alleviate the burden on security personnel, ensuring that attention is focused on genuine threats.

Finally, real-time detection is a critical requirement in CPS, where the timely identification and response to cyber attacks can prevent potential physical damage or disruption to critical infrastructure. Machine learning models operate in real-time, enabling the continuous monitoring and rapid detection of attacks. Their ability to process data streams in real-time ensures that potential threats are detected promptly, allowing for swift and effective countermeasures to be implemented. The following is a description of some popular machine learning techniques.

Extreme learning machine

Extreme learning machine (ELM) is a machine learning algorithm that has gained attention for its efficient and fast learning capabilities. ELM is a single-layer feedforward neural network algorithm that differs from traditional neural networks in its training approach. The structure of the ELM consists of multiple layers that work together to perform efficient and fast learning. The structure of ELM includes an input layer, a hidden layer, an activation function, and an output layer. The input layer receives the feature vectors or input data, with each neuron representing a specific attribute or feature. The hidden layer transforms the input data into a higher-dimensional feature space through a set of randomly initialized neurons connected to all neurons in the input layer. An activation function is applied to each hidden neuron to introduce non-linearity and calculate the neuron's output based on the weighted sum of inputs. Finally, the output layer produces the final predictions or classifications, with one or multiple neurons representing different class labels. The connections between the hidden and output layers utilize a random weight assignment approach, with the weights determined analytically using a linear regression method. The fixed weight matrix allows for quick convergence and efficient real-time analysis, making ELM suitable for applications such as attack detection in data-driven CPS. The structure of ELM is as shown in Fig. 8.8.

Unlike traditional neural networks that iteratively adjust the weights during training, ELM randomly assigns weights to the hidden layer neurons. This random weight initialization allow for significantly faster training times compared to iterative optimization algorithms like backpropagation.

Some research has been done on the potential of ELM in attack detection. Bacha et al. [84] proposed an anomaly-based intrusion detection system (IDS) for IoT environments using ELM. This work utilized kernel principal component analysis for dimensionality reduction and employed kernel ELM for binary and multiclass classification. Kushwah et al. [85] investigated the use of ELM, including a basic ELM method based on random features and a kernel-based ELM method, for the

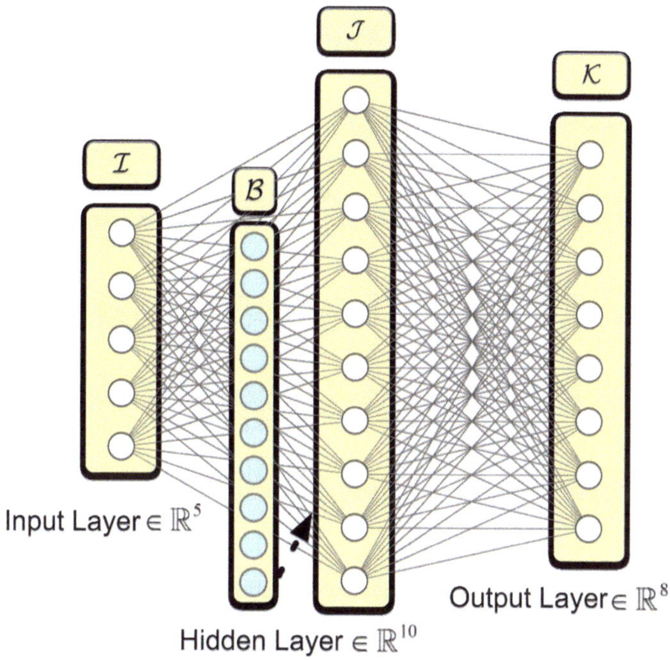

Fig. 8.8 A typical ELM is composed of input layer, hidden layer, and output layer, and the bias vector $\mathcal{B}$ sampled from Gaussian distribution is added to each component of the hidden layer

classification and detection of intrusions. More researches such as [86–91] provide more applications of ELM in data-driven CPS.

The use of ELM offers advantages in terms of efficiency, accuracy, and adaptability. The studies showed improved detection performance, faster training and testing times, and high accuracy rates compared to other methods such as SVM. ELM-based approaches, combined with techniques like kernel methods and dimensionality reduction, exhibit promise for effective attack detection and classification.

Random forest

The random forest algorithm combines multiple decision trees to make predictions and is a popular machine learning technique. It operates by constructing an ensemble of decision trees, where each tree is trained on a random subset of the data and features. The predictions of the individual trees are then aggregated to obtain the final prediction. Random forest is known for handling complex data, capturing nonlinear relationships, and mitigating overfitting. A typical random forest application in data-driven CPS can be shown in Fig. 8.9.

In the context of attack detection, random forest has been widely applied due to its effectiveness in identifying malicious activities within network traffic or system logs. It leverages the ensemble of decision trees to analyze various features and their interactions, enabling the detection of anomalies or patterns associated with cyber

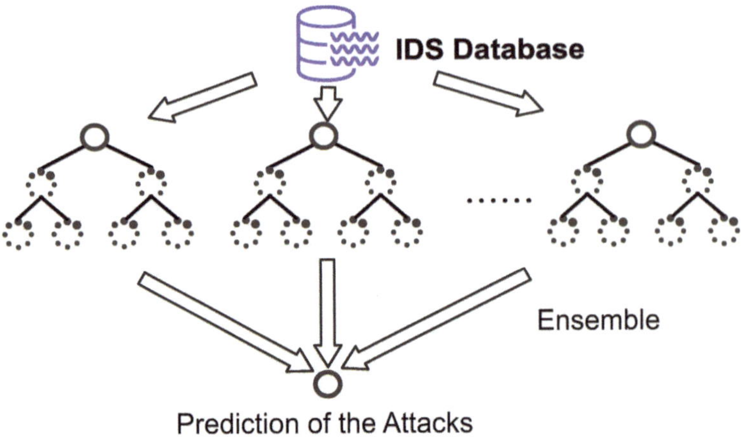

Fig. 8.9 Random forest model was used in ensemble style to predict the attack class from IDS database

attacks. Random forest has demonstrated its capability in accurately identifying and detecting various types of attacks, making it a valuable tool in bolstering the security of CPS and IoT networks. Primartha et al. [92] presented comparative analyses of IDS performance, specifically using the random forest classifier. Both studies employ public intrusion datasets such as NSL-KDD, UNSW-NB15, and GPRS. The experimental results consistently demonstrated the superiority of random forest over other ensemble models and single classifiers in terms of accuracy and false alarm rate. In addition, the random forest algorithm was also applied in the synthetic minority over-sampling technique for better detecting the attack of CPS [93, 94]. More applications [95–98] focus on specific attack types like DDOS in CPS.

The application of random forest in attack detection in CPS offers several advantages. It can effectively handle high-dimensional data and capture complex relationships and interactions between features. Random forest is robust against noise and outliers, reducing the risk of false positives and false negatives. Additionally, the ensemble nature of random forest allows for interpretability by providing insights into the importance of features, aiding in the understanding of the underlying attack characteristics within CPS.

Support vector machine

SVM is a widely used method rooted in statistical learning theory. It offers the advantage of achieving accurate classification outcomes even when training data is limited. By mapping the original dataset, which may contain nonlinearly separable samples, to a higher-dimensional or infinite-dimensional feature space, SVM aims to transform the data into a linearly separable form. Subsequently, it identifies the optimal classification boundary within this transformed feature space. The utilization of a kernel function in SVM effectively addresses the challenge of dimensionality, which arises from high-dimensional mappings, and enhances the model's capability to handle high-dimensional, small-sample datasets.

There are several advantages of SVM for coping with attack detection in CPS.SVM can effectively handle high-dimensional feature spaces, making it well-suited for analyzing complex data in CPS. As attacks in CPS often involve multiple variables and characteristics, SVM's ability to handle high dimensionality allows it to capture intricate relationships and patterns [99, 100]. Moreover, SVM provides interpretability by identifying support vectors, which are the data points closest to the decision boundary. These support vectors play a crucial role in defining the decision surface and provide insights into the distinguishing characteristics of attacks.

8.5.1.3 Deep Learning Model for Attack Detection

Deep neural networks (DNNs) have emerged as a powerful and promising approach for attack detection in DDCPS. With the increasing complexity and diversity of cyber threats, DNNs offer the ability to automatically learn intricate patterns and representations from large-scale data, making them well-suited for tackling the challenges posed by attack detection in CPS. The structure of neural networks is based on the extension of the perceptron machine, while DNN can be understood as a neural network with many hidden layers. The layers of a neural network within a DNN can be divided into three categories, the input layer, the hidden layer, and the output layer. Generally speaking, the first layer is the input layer, the last layer is the output layer, and all the layers in between are hidden layers.

In data-driven CPS, where vast amounts of heterogeneous data are generated from interconnected devices and systems, DNNs can efficiently process and analyze this data. The multi-layered architecture of DNNs allows them to learn hierarchical features, capturing both low-level and high-level abstractions, which are essential for distinguishing normal behavior from malicious activities. There are various fields in which DNNs have been applied in cyber attack detection. In the context of in-vehicle network security, a DNN-based model was introduced [101] to understand the communication patterns among electronic control units. The electronic control units, which collectively control and monitor a vehicle's status during maneuvers, communicate primarily through the controller area network protocol. However, this protocol does not support authentication or integrity checks, making it challenging to ensure cybersecurity. The situation is further complicated by the potential injection of fake packets into the open communication channel, posing severe cybersecurity risks. Much work about DNNs has been done in the mitigation of such risks. On the other hand, the security of electric power grids is another area where DNNs have shown promise. Some work [102–104] uses DNNs to learn the patterns of network traffic within the grid. The cybersecurity of an electric power grid is largely contingent on accurate state estimation, which is fundamental to the critical control processes of the grid. These examples illustrate the potential of DNNs in enhancing cybersecurity across different domains.

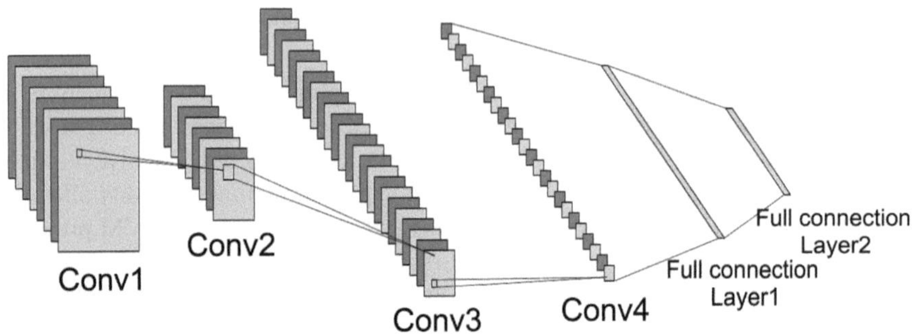

Fig. 8.10 CNN used to convolutional filter to downsample the input into a smaller scale and capture higher dimension features

DNNs' capacity to handle high-dimensional and sequential data is particularly advantageous in CPS attack detection [105, 106]. They can effectively process time-series data from sensors and network traffic, identifying temporal patterns associated with attacks and providing real-time detection capabilities. Moreover, more variants [107–110] in DNN have great potential for dealing with cyber attack detection in specific scenarios. The applications of the variants are expounded below.

Convolutional neural networks (CNNs)

Convolutional neural networks (CNNs) have emerged as a promising machine learning technique for cyber attack detection in data-driven cyber-physical systems [111]. Inspired by visual cortexes, convolutional neural networks employ multiple layers of convolutions and pooling to automatically extract spatial features and patterns from raw input data. During training, a CNN repeatedly passes input through the convolutional, pooling and fully connected layers, updating weights to minimize the loss, as shown in Fig. 8.10. CNN is well-suited for analyzing the high-dimensional time-series data generated by distributed sensors and controllers in CPS environments [112].

CNNs employ multiple convolutional and pooling layers to automatically learn patterns and extract features from raw time series and network data generated across the distributed components of a cyber physical system. A CNN-based anomaly detector can be trained on normal system data to establish a performance baseline. At runtime, the CNN model analyzes real-time streams of measurements, signals, and packets, flagging any significant deviations from the learned baseline as a potential cyber attack. Carefully designed CNN architectures provide adaptive attack detection without relying on predefined signatures [113, 114]. Ongoing research has demonstrated successful applications of CNN-based cyber attack detection in industrial control systems, power grids, connected vehicles and other cyber-physical system domains. CNN models have proven capable of identifying a wide range of sophisticated attack types by finding anomalies hidden within massive volumes of diverse system data. The unique capabilities of convolutional neural networks offer the exciting possibility of enhancing security and resilience for critical infrastructure facing rising cyber threats. Some CNN-based models were proposed for attack detection in CPS. Kravchik et al. [115] utilized a 1D CNN detected attacks in an industrial water treatment plant dataset. The CNN model achieved high precision, recall, and AUC scores, demonstrating effective attack detection. A deeper CNN called Inception-ResNet [116] was proposed to detect false message injection attacks in a vehicle's controller area network. The CNN analyzed in-vehicle network packets to identify malicious attack packets. It had low false negative and error rates in detecting denial of service, fuzzy, spoofing and other attack types. The unique capabilities of convolutional neural networks have the potential to significantly enhance cyber attack detection and system resilience for data-driven cyber-physical systems across critical infrastructure sectors [112].

Recurrent Neural networks and long short-term memory Models

Recurrent neural networks (RNNs) and Long short-term memory (LSTM) provide effective solutions to the problem of modeling sequential data in a range of CPSs, such as video, speech, vibration signals, etc.

RNNs are a type of neural network architecture designed to collect time-dependent features in sequential data [117]. RNNs are characterized by their ability to maintain hidden states and use them as memory to process sequences. In the context of CPS, RNNs can effectively handle time-series data from sensors or network traffic, allowing them to capture temporal patterns and dependencies in the data. In addition, LSTM is a specialized variant of RNNs that addresses the issue of vanishing gradients in traditional RNNs, enabling them to learn long-range dependencies in sequences. LSTM includes gating mechanisms that control the flow of information, facilitating the storage and retrieval of relevant information over extended periods of time. The structures of RNN and its variant, LSTM, are shown in Fig. 8.11.

The advantage of RNNs and LSTM in cyber attack detection lies in their ability to handle temporal dynamics and dependencies in CPS data [118]. By considering the sequential nature of the data, RNNs and LSTM can capture patterns and behaviors that may not be evident in isolated data points. This temporal context is particularly crucial in cyber attack detection, where attackers may exhibit sophisticated and evolving strategies over time.

Furthermore, RNNs and LSTM can efficiently process large-scale and complex data in CPS, allowing for real-time attack detection. They are capable of handling high-dimensional data [119, 120], such as network logs or sensor readings while learning and representing essential features for detecting different types of cyber attacks. The memory-retention properties of LSTM enable them to store and update information over extended sequences, making them particularly effective in recognizing patterns associated with stealthy or long-term attacks that span multiple data points.

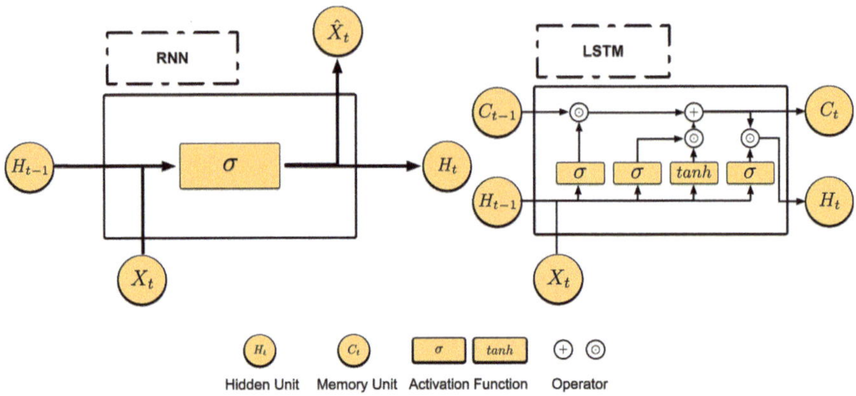

Fig. 8.11 The structures of RNN and LSTM equip the model with the ability to remember sequence data in the time dimension

The applications of RNNs and LSTM models in cyber attack detection span various domains, including water treatment plants, gas pipelines, smart grids, aviation, and industrial networks. In the domain of water treatment plants, an LSTM-based model [121] was proposed to identify anomalous sensor behaviors in water treatment plants. The model, trained on the SWaT dataset, was designed to detect anomalies and identify attacked sensors simultaneously, aiming for a low false positive rate. Despite the lengthy training time, it demonstrated the potential of deep learning models in detecting cyber attacks against CPS.

In the domain of gas pipelines, another LSTM-based model [122] was developed to detect unknown or zero-day attacks in gas pipeline control systems. This model combined packet content analysis and temporal dependencies to detect anomalies. The model, trained on a dataset of network logs from the gas pipeline, demonstrated success in detecting cyber attacks in a small SCADA system, outperforming several other classifiers. In the domain of smart grids, an RNN-based model [123] was proposed to detect various cyber attacks on smart grids. The model, trained on three different datasets, outperformed benchmark classifiers like SVM, random forest, and NB in terms of mean false positive rate and accuracy.

In the domain of aviation, an LSTM-based model [124] was proposed to detect anomalous Automatic Dependent Surveillance-Broadcast (ADS-B) messages in aviation. The model, trained on the Flightradar24 dataset, achieved excellent results in detecting spoofed messages, outperforming several baseline anomaly detection methods. In the domain of industrial networks, an LSTM-based model [125] was proposed to predict multimedia data requests in cyber-physical industrial networks. The model, trained on network traffic logs from a factory-intensive area accurately predicted multimedia contents.

The ability of RNNs and LSTM to learn temporal dependencies and handle large-scale data, coupled with their capability to detect evolving attack patterns, makes them valuable tools in enhancing the security and resilience of data-driven CPS against various cyber threats [126, 127].

8.5.2 *Data-Driven Data Confidence Assessment*

In DDCPS, ensuring data quality is of great significance [128]. Noise, distortion [129], and malicious spoofing [130] in sensor readings, process signals, and even network traffic can lead to erroneous outputs from the analytical systems that rely on these data. Therefore, we need to assess the trustworthiness of various types of data in data-driven cyber-physical systems to provide reliable support for subsequent decision analysis [131].

The physical model-based approach is a traditional way to assess the confidence of data [132]. It is based on the modeling of physical knowledge of processes and measurements to determine whether the data conforms to physical laws. However, this approach requires domain knowledge and faces the limitation of incomplete modeling. Moreover, data-driven approaches assess quality by directly analyzing the characteristics of the data itself through algorithms, and the classification of data-driven approaches is shown in Fig. 8.12. Statistical methods can detect anomalies in the data. Machine learning techniques such as class detection enable dynamic confidence assessment of time series data. Deep learning models, such as autoencoders [133], can learn complex mapping relationships in the data and capture hidden anomalies in the data. In addition, contextual information can be combined to improve the confidence assessment. For example, readings from multiple relevant sensors can be fused, or auxiliary data such as environmental parameters and network conditions can be added to enhance the detection of anomalies and attacks. The results of confidence assessment can be used as weights for subsequent data processing or for labeling outing suspicious data. By evaluating the confidence of various types of process data and cyber data, we can improve the robustness and interpretability of data-based cyber-physical systems. This provides credible, high-quality information for mission-critical Entscheidungen, which relies on large amounts of dynamic data.

Statistical method

In the context of DDCPS, statistical methods are instrumental in data confidence assessment, offering quantitative tools to evaluate the data's reliability and quality. Here are some statistical methods commonly used.

Descriptive statistics: In the context of CPS, measures like mean, median, mode, standard deviation, and variance can provide a summary of sensor readings or system states, helping to identify potential anomalies or outliers that might affect data confidence [134, 135].

Probability distributions: Understanding the probability distribution of sensor readings or system states can provide insights into their reliability. For instance, if the data significantly deviates from expected distributions, the deviation might indicate the anomaly data.

Hypothesis testing: Hypothesis testing can be used to assess whether the data supports certain assumptions about the physical system or process [136, 137]. For

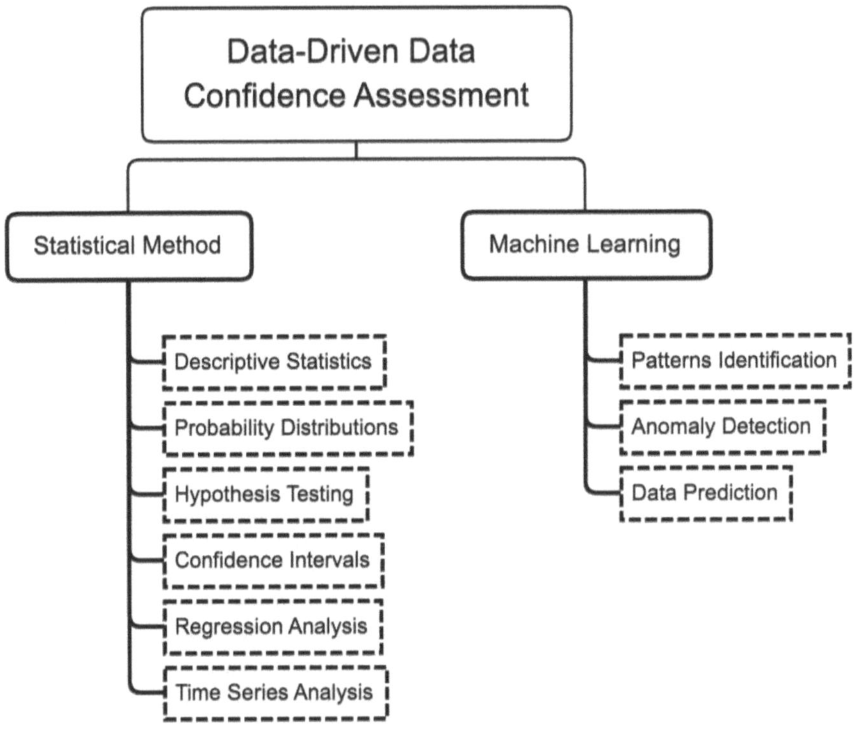

Fig. 8.12 The approach of data-driven data confidence assessment can be divided into two categories: statistical methods and machine learning methods

example, a Bayesian test could be used to determine whether the mean of a sensor reading differs significantly from a hypothesized value [138].

Confidence intervals: Confidence intervals provide a range of values that likely contain the true parameter value with a certain level of confidence [139, 140]. They can be used to assess the reliability of estimates derived from sensor readings or system states.

Regression analysis: Regression analysis can be used to model the relationship between different variables in the CPS, such as the relationship between different sensor readings or between sensor readings and system states [141, 142].

Time series analysis: For data that is collected over time, such as sensor readings or system states, time series analysis can be used to identify trends, cycles, and other patterns [143–145]. This can help assessing the reliability of the data and predicting future values.

Machine learning

Machine learning methods are increasingly being used for data confidence assessment in DDCPS. These methods can help to identify patterns, detect anomalies, and make predictions about the data, thereby providing a measure of its reliability and quality.

Identifying patterns: Unsupervised learning algorithms, such as clustering algorithms (e.g., K-means [146, 147], Hierarchical Clustering [148, 149]) and dimensionality reduction techniques (e.g., Principal Component Analysis [150, 151]), can identify patterns and structures in the data. Deep learning methods, such as CNNs for image data and RNNs for time-series data, can learn complex patterns in the data. Semi-supervised learning methods, which use a combination of labeled and unlabeled data, can also be used to identify patterns in the data [152, 153].

Detecting anomalies: Anomaly detection methods, such as One-Class SVM [154], isolation forest [155–158], and local outlier factor [159], can identify unusual or suspicious data points, which can indicate potential errors or attacks on the CPS. Autoencoders, a type of deep learning model, can also be used for anomaly detection by learning to reconstruct normal data and then measuring the reconstruction error for new data points.

Predictions of data: Supervised learning algorithms, such as SVM [160], decision trees [161], and neural networks [162], can be trained on labeled data to predict outcomes. The accuracy of these predictions can provide a measure of data confidence. Ensemble methods [163], such as random forests and gradient boosting, combine multiple models to make more accurate and robust predictions. Reinforcement learning, where an agent learns to make decisions by interacting with its environment and receiving rewards or penalties, can be used to make predictions about the best actions to take in a given state [164, 165].

In a data-driven CPS, these ML methods can be used to assess data confidence at various stages, from data collection and preprocessing to model training and validation. By providing a measure of data quality and reliability, they can help ensure that the CPS operates effectively and makes accurate predictions.

8.5.3 Risk Assessment Metrics

The utilization of risk-based security metrics is of utmost importance in the evaluation and administration of cybersecurity risks within DDCPS. These metrics provide a quantitative measure of the likelihood and potential impact of security threats, helping to prioritize security efforts and resources. Here are some methods as shown in Fig. 8.13 which are commonly used to obtain risk-based security metrics in CPS:

Attack graphs and attack trees: Attack graphs and attack trees are graphical models that represent the possible paths an attacker can take to compromise a system [166, 167]. They can be used to calculate the probability of a successful attack and the potential impact, providing a measure of security risk.

Bayesian networks: Bayesian networks are probabilistic graphical models that represent the conditional dependencies between different variables [168, 169]. They can be used to model the complex relationships between different security factors and calculate the probability of different security events.

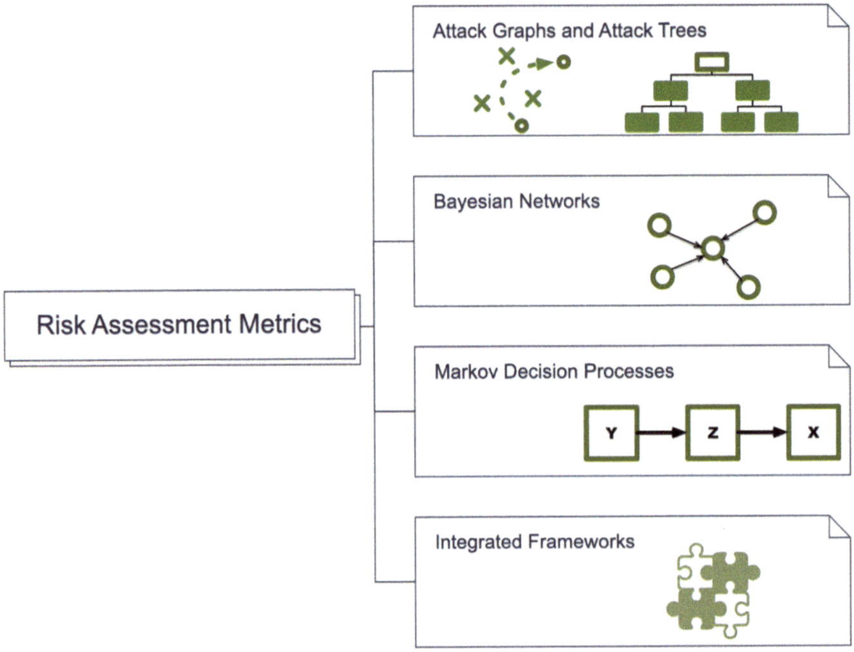

Fig. 8.13 Risk assessment metrics for DDCPS consist of attack graphs and attack trees, Bayesian networks, Markov decision processes, and integrated frameworks

Markov decision processes (MDPs): MDPs are mathematical models used for decision making in situations where outcomes are partly random and partly under the control of a decision maker [29, 170]. They can be used to model the behavior of an attacker and calculate the expected impact of different security controls.

Integrated frameworks: Frameworks such as OCTAVE (Operationally Critical Threat, Asset, and Vulnerability Evaluation) [171], FAIR (Factor Analysis of Information Risk) [172], and NIST's Risk Management Framework [173] provide structured approaches to risk assessment. They can help to identify and prioritize risks, assess the effectiveness of security controls, and calculate risk-based security metrics.

8.6 Summary

In this chapter, we undertake an in-depth analysis of cybersecurity as it pertains to DDCPS. We elucidate the nature of cyber attacks aimed at each individual component within DDCPS, presenting an intricate perspective of the vulnerabilities that exist. The subsequent insights pave the way for us to outline the integral requirements necessary to ensure robust cybersecurity within these systems.

Recognizing the paramount importance of cybersecurity for the effective operation and integrity of DDCPS, we delve into a critical analysis of its role. This enables us to explore the significant challenges that stand on the path to achieving full-scale cybersecurity. As we navigate these complexities, we realize the necessity of data-driven techniques that can proactively address these hurdles. Moving forward, we discuss a variety of innovative data-driven methodologies that are instrumental in fortifying the cybersecurity of DDCPS. These cutting-edge techniques, employed to combat and mitigate the risks of cyber threats, serve as a testament to the advances made in this field.

By providing this comprehensive overview, we aim to instill a deep understanding of the interplay between DDCPS and cybersecurity. This lays a solid foundation upon which future research and practical measures can be developed, steering us closer to the goal of fully secured and resilient DDCPS.

Competing Interests

1. The research is supported by the National Key Research and Development Project under Grant 2023YFB3307300, the National Natural Science Foundation of China under Grants 62373014, 62422301, 62125301, and 92467205.
2. The authors have no conflicts of interest to declare that are relevant to the content of this chapter.

References

1. N. institute of standards and technology, "The definition of cyber attack." (2023,10,20).
2. Y. Peng, T. Lu, J. Liu, Y. Gao, X. Guo, and F. Xie, "Cyber-physical system risk assessment," in *2013 Ninth International Conference on Intelligent Information Hiding and Multimedia Signal Processing*, pp. 442–447, IEEE, 2013.
3. T. Lu, J. Lin, L. Zhao, Y. Li, and Y. Peng, "A security architecture in cyber-physical systems: security theories, analysis, simulation and application fields," *International Journal of Security and Its Applications*, vol. 9, no. 7, pp. 1–16, 2015.
4. A. H. El-Kady, S. Halim, M. M. El-Halwagi, and F. Khan, "Analysis of safety and security challenges and opportunities related to cyber-physical systems," *Process Safety and Environmental Protection*, 2023.
5. J. S. Kumar and D. R. Patel, "A survey on internet of things: Security and privacy issues," *International Journal of Computer Applications*, vol. 90, no. 11, 2014.
6. R. Mahmoud, T. Yousuf, F. Aloul, and I. Zualkernan, "Internet of things (iot) security: Current status, challenges and prospective measures," in *2015 10th international conference for internet technology and secured transactions (ICITST)*, pp. 336–341, IEEE, 2015.
7. Y. Ashibani and Q. H. Mahmoud, "Cyber physical systems security: Analysis, challenges and solutions," *Computers & Security*, vol. 68, pp. 81–97, 2017.
8. M. Wu, T.-J. Lu, F.-Y. Ling, J. Sun, and H.-Y. Du, "Research on the architecture of internet of things," in *2010 3rd international conference on advanced computer theory and engineering (ICACTE)*, vol. 5, pp. V5–484, IEEE, 2010.
9. B. Zhang, X.-X. Ma, and Z.-G. Qin, "Security architecture on the trusting internet of things," *Journal of Electronic Science and Technology*, vol. 9, no. 4, pp. 364–367, 2011.

10. K. Zhao and L. Ge, "A survey on the internet of things security," in *2013 Ninth international conference on computational intelligence and security*, pp. 663–667, IEEE, 2013.
11. S. Raza, *Lightweight security solutions for the internet of things*. PhD thesis, Mälardalen University, Västerås, Sweden, 2013.
12. Q. Gou, L. Yan, Y. Liu, and Y. Li, "Construction and strategies in iot security system," in *2013 IEEE international conference on green computing and communications and IEEE internet of things and IEEE cyber, physical and social computing*, pp. 1129–1132, IEEE, 2013.
13. R. Mitchell and I.-R. Chen, "A survey of intrusion detection techniques for cyber-physical systems," *ACM Computing Surveys (CSUR)*, vol. 46, no. 4, pp. 1–29, 2014.
14. N. Gaddam, G. S. A. Kumar, and A. K. Somani, "Securing physical processes against cyber attacks in cyber-physical systems," in *Proc. Nat. Workshop Res. High-Confidence Transp. Cyber-Phys. Systems, Autom., Aviation Rail*, pp. 1–3, 2008.
15. Y. Li, L. Shi, P. Cheng, J. Chen, and D. E. Quevedo, "Jamming attack on cyber-physical systems: A game-theoretic approach," in *2013 IEEE International Conference on Cyber Technology in Automation, Control and Intelligent Systems*, pp. 252–257, IEEE, 2013.
16. P. Maheshwari, "Security issues of cyber physical system: a review," *Int. J. Comput. Appl*, no. 1, pp. 7–11, 2016.
17. E. K. Wang, Y. Ye, X. Xu, S.-M. Yiu, L. C. K. Hui, and K.-P. Chow, "Security issues and challenges for cyber physical system," in *2010 IEEE/ACM Int'l Conference on Green Computing and Communications & Int'l Conference on Cyber, Physical and Social Computing*, pp. 733–738, IEEE, 2010.
18. A. Al-Omary, A. Othman, H. M. AlSabbagh, and H. Al-Rizzo, "Survey of hardware-based security support for iot/cps systems," *KnE Engineering*, pp. 52–70, 2018.
19. R. S. Sandhu, "On five definitions of data integrity.," in *DBSec*, pp. 257–267, Citeseer, 1993.
20. S. Sridhar and G. Manimaran, "Data integrity attacks and their impacts on scada control system," in *IEEE PES general meeting*, pp. 1–6, IEEE, 2010.
21. G. Schuh, C. Reuter, J.-P. Prote, F. Brambring, and J. Ays, "Increasing data integrity for improving decision making in production planning and control," *CIRP Annals*, vol. 66, no. 1, pp. 425–428, 2017.
22. X. Zhang, X. Yang, J. Lin, G. Xu, and W. Yu, "On data integrity attacks against real-time pricing in energy-based cyber-physical systems," *IEEE Transactions on Parallel and Distributed Systems*, vol. 28, no. 1, pp. 170–187, 2016.
23. J. R. Pérez, "Maintaining data integrity," *Quality Progress*, vol. 50, no. 3, p. 14, 2017.
24. R. Changalvala and H. Malik, "Lidar data integrity verification for autonomous vehicle," *IEEE Access*, vol. 7, pp. 138018–138031, 2019.
25. T. Abera, R. Bahmani, F. Brasser, A. Ibrahim, A.-R. Sadeghi, and M. Schunter, "Diat: Data integrity attestation for resilient collaboration of autonomous systems.," in *NDSS*, 2019.
26. Y. Luo, L. Cheng, Y. Liang, J. Fu, and G. Peng, "Deepnoise: Learning sensor and process noise to detect data integrity attacks in cps," *China Communications*, vol. 18, no. 9, pp. 192–209, 2021.
27. C. M. Ahmed, A. Mathur, and M. Ochoa, "Noisense: Detecting data integrity attacks on sensor measurements using hardware based fingerprints," *arXiv preprint* arXiv:1712.01598, 2017.
28. E. Barker, M. Smid, D. Branstad, and S. Chokhani, "A framework for designing cryptographic key management systems," *NIST Special Publication*, vol. 800, no. 130, pp. 1–112, 2013.
29. H. Abie, "Cognitive cybersecurity for cps-iot enabled healthcare ecosystems," in *2019 13th International Symposium on Medical Information and Communication Technology (ISMICT)*, pp. 1–6, IEEE, 2019.
30. Q. Mei, H. Xiong, Y.-C. Chen, and C.-M. Chen, "Blockchain-enabled privacy-preserving authentication mechanism for transportation cps with cloud-edge computing," *IEEE Transactions on Engineering Management*, 2022.
31. C.-S. Shih, J.-J. Chou, N. Reijers, and T.-W. Kuo, "Designing cps/iot applications for smart buildings and cities," *IET Cyber-Physical Systems: Theory & Applications*, vol. 1, no. 1, pp. 3–12, 2016.

32. P. M. Lima, L. K. Carvalho, and M. V. Moreira, "Ensuring confidentiality of cyber-physical systems using event-based cryptography," *Information Sciences*, vol. 621, pp. 119–135, 2023.
33. E. Huedo, R. S. Montero, and I. M. Llorente, "Evaluating the reliability of computational grids from the end user's point of view," *Journal of Systems Architecture*, vol. 52, no. 12, pp. 727–736, 2006.
34. W. Yang, Z. Zheng, G. Chen, Y. Tang, and X. Wang, "Security analysis of a distributed networked system under eavesdropping attacks," *IEEE Transactions on Circuits and Systems II: Express Briefs*, vol. 67, no. 7, pp. 1254–1258, 2019.
35. S. Hosseini, K. Barker, and J. E. Ramirez-Marquez, "A review of definitions and measures of system resilience," *Reliability Engineering & System Safety*, vol. 145, pp. 47–61, 2016.
36. S. V. Sudarsan, O. Schelén, and U. Bodin, "Survey on delegated and self-contained authorization techniques in cps and iot," *IEEE Access*, vol. 9, pp. 98169–98184, 2021.
37. M. Gupta, S. Bhatt, A. H. Alshehri, and R. Sandhu, "Introduction: Requirements for access control in iot and cps," in *Access Control Models and Architectures For IoT and Cyber Physical Systems*, pp. 1–17, Springer, 2022.
38. R. Priyadarshini, A. Quadir Md, N. Rajendran, V. Neelanarayanan, and H. Sabireen, "An enhanced encryption-based security framework in the cps cloud," *Journal of Cloud Computing*, vol. 11, no. 1, p. 64, 2022.
39. Z. Zhang, X. Che, X. Jiao, W. Yu, and L. Wan, "Quadratic optimization using additive homomorphic encryption in cps," in *2022 13th Asian Control Conference (ASCC)*, pp. 1995–2000, IEEE, 2022.
40. S. Han, M. Xie, H.-H. Chen, and Y. Ling, "Intrusion detection in cyber-physical systems: Techniques and challenges," *IEEE systems journal*, vol. 8, no. 4, pp. 1052–1062, 2014.
41. M. Barrère, C. Hankin, A. Barboni, G. Zizzo, F. Boem, S. Maffeis, and T. Parisini, "Cps-mt: A real-time cyber-physical system monitoring tool for security research," in *2018 IEEE 24th International Conference on Embedded and Real-Time Computing Systems and Applications (RTCSA)*, pp. 240–241, IEEE, 2018.
42. G. Hatzivasilis, I. Papaefstathiou, and C. Manifavas, "Real-time management of railway cps secure administration of iot and cps infrastructure," in *2017 6th Mediterranean conference on embedded computing (MECO)*, pp. 1–4, IEEE, 2017.
43. M. M. Althobaiti, K. P. M. Kumar, D. Gupta, S. Kumar, and R. F. Mansour, "An intelligent cognitive computing based intrusion detection for industrial cyber-physical systems," *Measurement*, vol. 186, p. 110145, 2021.
44. S. E. Quincozes, D. Passos, C. Albuquerque, D. Mossé, and L. S. Ochi, "An extended assessment of metaheuristics-based feature selection for intrusion detection in cps perception layer," *Annals of Telecommunications*, vol. 77, no. 7-8, pp. 457–471, 2022.
45. M. S. Haghighi, F. Farivar, and A. Jolfaei, "A machine learning-based approach to build zero false-positive ipss for industrial iot and cps with a case study on power grids security," *IEEE Transactions on Industry Applications*, 2020.
46. A. Wun and H.-A. Jacobsen, "A policy management framework for content-based publish/subscribe middleware," in *Middleware 2007: ACM/IFIP/USENIX 8th International Middleware Conference, Newport Beach, CA, USA, November 26-30, 2007. Proceedings 8*, pp. 368–388, Springer, 2007.
47. P. Regulation, "General data protection regulation," *Intouch*, vol. 25, pp. 1–5, 2018.
48. P. F. Edemekong, P. Annamaraju, and M. J. Haydel, "Health insurance portability and accountability act," 2018.
49. S. Ali, T. Al Balushi, Z. Nadir, O. K. Hussain, S. Ali, T. Al Balushi, Z. Nadir, and O. K. Hussain, "Risk management for cps security," *Cyber Security for Cyber Physical Systems*, pp. 11–33, 2018.
50. Z. Wang and X. Liu, "Cyber security of railway cyber-physical system (cps)–a risk management methodology," *Communications in Transportation Research*, vol. 2, p. 100078, 2022.
51. G. Grispos, W. B. Glisson, and K.-K. R. Choo, "Medical cyber-physical systems development: A forensics-driven approach," in *2017 IEEE/ACM International Conference on Connected*

Health: Applications, Systems and Engineering Technologies (CHASE), pp. 108–113, IEEE, 2017.
52. T. Czvetkó, A. Kummer, T. Ruppert, and J. Abonyi, "Data-driven business process management-based development of industry 4.0 solutions," *CIRP journal of manufacturing science and technology*, vol. 36, pp. 117–132, 2022.
53. K. Scarfone, P. Mell, et al., "Guide to intrusion detection and prevention systems (idps)," *NIST special publication*, vol. 800, no. 2007, p. 94, 2007.
54. E. Bou-Harb, "A brief survey of security approaches for cyber-physical systems," in *2016 8th IFIP International Conference on New Technologies, Mobility and Security (NTMS)*, pp. 1–5, IEEE, 2016.
55. P. Chen, L. Desmet, and C. Huygens, "A study on advanced persistent threats," in *Communications and Multimedia Security: 15th IFIP TC 6/TC 11 International Conference, CMS 2014, Aveiro, Portugal, September 25-26, 2014. Proceedings 15*, pp. 63–72, Springer, 2014.
56. M. L. Mathews, A. Joshi, T. Finin, et al., "Detecting botnets using a collaborative situational-aware idps," in *Second International Conference on Information Systems Security and Privacy*, 2016.
57. L. Cazorla, C. Alcaraz, and J. Lopez, "Cyber stealth attacks in critical information infrastructures," *IEEE Systems Journal*, vol. 12, no. 2, pp. 1778–1792, 2016.
58. A. A. Alwan, M. A. Ciupala, A. J. Brimicombe, S. A. Ghorashi, A. Baravalle, and P. Falcarin, "Data quality challenges in large-scale cyber-physical systems: A systematic review," *Information Systems*, vol. 105, p. 101951, 2022.
59. A. G. Labouseur and C. C. Matheus, "An introduction to dynamic data quality challenges," *Journal of Data and Information Quality (JDIQ)*, vol. 8, no. 2, pp. 1–3, 2017.
60. N. U. Okafor, Y. Alghorani, and D. T. Delaney, "Improving data quality of low-cost iot sensors in environmental monitoring networks using data fusion and machine learning approach," *ICT Express*, vol. 6, no. 3, pp. 220–228, 2020.
61. K. Sha and S. Zeadally, "Data quality challenges in cyber-physical systems," *Journal of Data and Information Quality (JDIQ)*, vol. 6, no. 2-3, pp. 1–4, 2015.
62. J.-E. Kim, T. Abdelzaher, L. Sha, A. Bar-Noy, R. Hobbs, and W. Dron, "On maximizing quality of information for the internet of things: A real-time scheduling perspective," in *2016 IEEE 22nd International Conference on Embedded and Real-Time Computing Systems and Applications (RTCSA)*, pp. 202–211, IEEE, 2016.
63. V. J. Lawson and L. Ramaswamy, "Tau-five: a multi-tiered architecture for data quality and energy-sustainability in sensor networks," in *2016 International Conference on Distributed Computing in Sensor Systems (DCOSS)*, pp. 169–176, IEEE, 2016.
64. W. Guo, Y. Zhang, and L. Li, "The integration of cps, cpss, and its: A focus on data," *Tsinghua Science and Technology*, vol. 20, no. 4, pp. 327–335, 2015.
65. J. Lee, J. Kim, I. Kim, and K. Han, "Cyber threat detection based on artificial neural networks using event profiles," *Ieee Access*, vol. 7, pp. 165607–165626, 2019.
66. S. Singh and S. Silakari, "A survey of cyber attack detection systems," *International Journal of Computer Science and Network Security*, vol. 9, no. 5, pp. 1–10, 2009.
67. H. Rasheed, A. Hadi, and M. Khader, "Threat hunting using grr rapid response," in *2017 International Conference on New Trends in Computing Sciences (ICTCS)*, pp. 155–160, IEEE, 2017.
68. M. W. Condry and C. B. Nelson, "Using smart edge iot devices for safer, rapid response with industry iot control operations," *Proceedings of the IEEE*, vol. 104, no. 5, pp. 938–946, 2016.
69. A. L. Buczak and E. Guven, "A survey of data mining and machine learning methods for cyber security intrusion detection," *IEEE Communications surveys & tutorials*, vol. 18, no. 2, pp. 1153–1176, 2015.
70. M. A. Ferrag, L. Maglaras, S. Moschoyiannis, and H. Janicke, "Deep learning for cyber security intrusion detection: Approaches, datasets, and comparative study," *Journal of Information Security and Applications*, vol. 50, p. 102419, 2020.
71. W. Hurst, M. Merabti, and P. Fergus, "Big data analysis techniques for cyber-threat detection in critical infrastructures," in *2014 28th International Conference on Advanced Information Networking and Applications Workshops*, pp. 916–921, IEEE, 2014.

72. L. F. Dias and M. Correia, "Big data analytics for intrusion detection: an overview," *Handbook of research on machine and deep learning applications for cyber security*, pp. 292–316, 2020.
73. M. F. Haque and R. Krishnan, "Toward automated cyber defense with secure sharing of structured cyber threat intelligence," *Information Systems Frontiers*, pp. 1–14, 2021.
74. J. Zhang, L. Pan, Q.-L. Han, C. Chen, S. Wen, and Y. Xiang, "Deep learning based attack detection for cyber-physical system cybersecurity: A survey," *IEEE/CAA Journal of Automatica Sinica*, vol. 9, no. 3, pp. 377–391, 2021.
75. S. Alghunaim and H. H. Al-Baity, "On the scalability of machine-learning algorithms for breast cancer prediction in big data context," *Ieee Access*, vol. 7, pp. 91535–91546, 2019.
76. G. Paliouras, *Scalability of machine learning algorithms*. PhD thesis, University of Manchester, 1993.
77. L. Zhou, S. Pan, J. Wang, and A. V. Vasilakos, "Machine learning on big data: Opportunities and challenges," *Neurocomputing*, vol. 237, pp. 350–361, 2017.
78. O. Y. Al-Jarrah, P. D. Yoo, S. Muhaidat, G. K. Karagiannidis, and K. Taha, "Efficient machine learning for big data: A review," *Big Data Research*, vol. 2, no. 3, pp. 87–93, 2015.
79. N. Geetha and P. Bridjesh, "Withdrawn: Overview of machine learning and its adaptability in mechanical engineering," 2020.
80. J. Zheng and A. Dagnino, "An initial study of predictive machine learning analytics on large volumes of historical data for power system applications," in *2014 IEEE international conference on big data (Big Data)*, pp. 952–959, IEEE, 2014.
81. C. M. Bishop and N. M. Nasrabadi, *Pattern recognition and machine learning*, vol. 4. Springer, 2006.
82. M. Cummings and A. Stimpson, "Identifying critical contextual design cues through a machine learning approach," *AI Magazine*, vol. 40, no. 4, pp. 28–39, 2019.
83. S. Omar, A. Ngadi, and H. H. Jebur, "Machine learning techniques for anomaly detection: an overview," *International Journal of Computer Applications*, vol. 79, no. 2, 2013.
84. G. S. Kushwah and V. Ranga, "Voting extreme learning machine based distributed denial of service attack detection in cloud computing," *Journal of Information Security and Applications*, vol. 53, p. 102532, 2020.
85. G. S. Kushwah and V. Ranga, "Optimized extreme learning machine for detecting ddos attacks in cloud computing," *Computers & Security*, vol. 105, p. 102260, 2021.
86. C. Cheng, W. P. Tay, and G.-B. Huang, "Extreme learning machines for intrusion detection," in *The 2012 International joint conference on neural networks (IJCNN)*, pp. 1–8, IEEE, 2012.
87. L. Lv, W. Wang, Z. Zhang, and X. Liu, "A novel intrusion detection system based on an optimal hybrid kernel extreme learning machine," *Knowledge-based systems*, vol. 195, p. 105648, 2020.
88. W. L. Al-Yaseen, Z. A. Othman, and M. Z. A. Nazri, "Multi-level hybrid support vector machine and extreme learning machine based on modified k-means for intrusion detection system," *Expert Systems with Applications*, vol. 67, pp. 296–303, 2017.
89. S. Roshan, Y. Miche, A. Akusok, and A. Lendasse, "Adaptive and online network intrusion detection system using clustering and extreme learning machines," *Journal of the Franklin Institute*, vol. 355, no. 4, pp. 1752–1779, 2018.
90. C.-R. Wang, R.-F. Xu, S.-J. Lee, and C.-H. Lee, "Network intrusion detection using equality constrained-optimization-based extreme learning machines," *Knowledge-Based Systems*, vol. 147, pp. 68–80, 2018.
91. G. S. Kushwah and S. T. Ali, "Distributed denial of service attacks detection in cloud computing using extreme learning machine," *International Journal of Communication Networks and Distributed Systems*, vol. 23, no. 3, pp. 328–351, 2019.
92. R. Primartha and B. A. Tama, "Anomaly detection using random forest: A performance revisited," in *2017 International conference on data and software engineering (ICoDSE)*, pp. 1–6, IEEE, 2017.
93. Y. Chang, W. Li, and Z. Yang, "Network intrusion detection based on random forest and support vector machine," in *2017 IEEE international conference on computational science and engineering (CSE) and IEEE international conference on embedded and ubiquitous computing (EUC)*, vol. 1, pp. 635–638, IEEE, 2017.

94. M. G. Karthik and M. M. Krishnan, "Hybrid random forest and synthetic minority over sampling technique for detecting internet of things attacks," *Journal of Ambient Intelligence and Humanized Computing*, pp. 1–11, 2021.
95. M. Idhammad, K. Afdel, and M. Belouch, "Detection system of http ddos attacks in a cloud environment based on information theoretic entropy and random forest," *Security and Communication Networks*, vol. 2018, 2018.
96. J. Cheng, M. Li, X. Tang, V. S. Sheng, Y. Liu, and W. Guo, "Flow correlation degree optimization driven random forest for detecting ddos attacks in cloud computing," *Security and Communication Networks*, vol. 2018, pp. 1–14, 2018.
97. Y. Chen, J. Hou, Q. Li, and H. Long, "Ddos attack detection based on random forest," in *2020 IEEE International Conference on Progress in Informatics and Computing (PIC)*, pp. 328–334, IEEE, 2020.
98. A. A. Najar and S. Manohar Naik, "Ddos attack detection using mlp and random forest algorithms," *International Journal of Information Technology*, vol. 14, no. 5, pp. 2317–2327, 2022.
99. K. S. Sahoo, B. K. Tripathy, K. Naik, S. Ramasubbareddy, B. Balusamy, M. Khari, and D. Burgos, "An evolutionary svm model for ddos attack detection in software defined networks," *IEEE access*, vol. 8, pp. 132502–132513, 2020.
100. J. Ye, X. Cheng, J. Zhu, L. Feng, and L. Song, "A ddos attack detection method based on svm in software defined network," *Security and Communication Networks*, vol. 2018, 2018.
101. M.-J. Kang and J.-W. Kang, "Intrusion detection system using deep neural network for in-vehicle network security," *PloS one*, vol. 11, no. 6, p. e0155781, 2016.
102. M. Ashrafuzzaman, S. Das, Y. Chakhchoukh, S. Shiva, and F. T. Sheldon, "Detecting stealthy false data injection attacks in the smart grid using ensemble-based machine learning," *Computers & Security*, vol. 97, p. 101994, 2020.
103. M. Abbasi, A. Shahraki, and A. Taherkordi, "Deep learning for network traffic monitoring and analysis (ntma): A survey," *Computer Communications*, vol. 170, pp. 19–41, 2021.
104. Z. M. Fadlullah, F. Tang, B. Mao, N. Kato, O. Akashi, T. Inoue, and K. Mizutani, "State-of-the-art deep learning: Evolving machine intelligence toward tomorrow's intelligent network traffic control systems," *IEEE Communications Surveys & Tutorials*, vol. 19, no. 4, pp. 2432–2455, 2017.
105. D. M. Mason, S. Friedensohn, C. R. Weber, C. Jordi, B. Wagner, S. Meng, P. Gainza, B. E. Correia, and S. T. Reddy, "Deep learning enables therapeutic antibody optimization in mammalian cells by deciphering high-dimensional protein sequence space," *BioRxiv*, p. 617860, 2019.
106. Z. Chen, J. Hu, G. Min, A. Y. Zomaya, and T. El-Ghazawi, "Towards accurate prediction for high-dimensional and highly-variable cloud workloads with deep learning," *IEEE Transactions on Parallel and Distributed Systems*, vol. 31, no. 4, pp. 923–934, 2019.
107. A. Makuvaza, D. S. Jat, and A. M. Gamundani, "Deep neural network (dnn) solution for real-time detection of distributed denial of service (ddos) attacks in software defined networks (sdns)," *SN Computer Science*, vol. 2, pp. 1–10, 2021.
108. M. Krishnan, Y. Lim, S. Perumal, and G. Palanisamy, "Detection and defending the xss attack using novel hybrid stacking ensemble learning-based dnn approach," *Digital Communications and Networks*, 2022.
109. A. Zainudin, L. A. C. Ahakonye, R. Akter, D.-S. Kim, and J.-M. Lee, "An efficient hybrid-dnn for ddos detection and classification in software-defined iiot networks," *IEEE Internet of Things Journal*, 2022.
110. S. Duraibi, W. Alhamdani, and F. T. Sheldon, "Replay spoof attack detection using deep neural networks for classification," in *2020 International Conference on Computational Science and Computational Intelligence (CSCI)*, pp. 170–174, IEEE, 2020.
111. W.-H. Lin, H.-C. Lin, P. Wang, B.-H. Wu, and J.-Y. Tsai, "Using convolutional neural networks to network intrusion detection for cyber threats," in *2018 IEEE International Conference on Applied System Invention (ICASI)*, pp. 1107–1110, IEEE, 2018.

112. D. Nedeljkovic and Z. Jakovljevic, "Cnn based method for the development of cyber-attacks detection algorithms in industrial control systems," *Computers & Security*, vol. 114, p. 102585, 2022.
113. X. Kan, Y. Fan, Z. Fang, L. Cao, N. N. Xiong, D. Yang, and X. Li, "A novel iot network intrusion detection approach based on adaptive particle swarm optimization convolutional neural network," *Information Sciences*, vol. 568, pp. 147–162, 2021.
114. X. Xu, H. Hu, Y. Liu, H. Zhang, and D. Chang, "An adaptive ip hopping approach for moving target defense using a light-weight cnn detector," *Security and Communication Networks*, vol. 2021, pp. 1–17, 2021.
115. M. Kravchik and A. Shabtai, "Detecting cyber attacks in industrial control systems using convolutional neural networks," in *Proceedings of the 2018 workshop on cyber-physical systems security and privacy*, pp. 72–83, 2018.
116. H. M. Song, J. Woo, and H. K. Kim, "In-vehicle network intrusion detection using deep convolutional neural network," *Vehicular Communications*, vol. 21, p. 100198, 2020.
117. F. Xu, Z. Li, Z. Nie, H. Shao, and D. Guo, "New recurrent neural network for online solution of time-dependent underdetermined linear system with bound constraint," *IEEE Transactions on Industrial Informatics*, vol. 15, no. 4, pp. 2167–2176, 2018.
118. H. Yao, X. Tang, H. Wei, G. Zheng, Y. Yu, and Z. Li, "Modeling spatial-temporal dynamics for traffic prediction," arXiv:1803.01254, vol. 1, no. 9, 2018.
119. F. Bao, Y. Wu, Z. Li, Y. Li, L. Liu, and G. Chen, "Effect improved for high-dimensional and unbalanced data anomaly detection model based on knn-smote-lstm," *Complexity*, vol. 2020, pp. 1–17, 2020.
120. K. Praanna, S. Sruthi, K. Kalyani, and A. S. Tejaswi, "A cnn-lstm model for intrusion detection system from high dimensional data," *J. Inf. Comput. Sci*, vol. 10, no. 3, pp. 1362–1370, 2020.
121. J. Goh, S. Adepu, M. Tan, and Z. S. Lee, "Anomaly detection in cyber physical systems using recurrent neural networks," in *2017 IEEE 18th International Symposium on High Assurance Systems Engineering (HASE)*, pp. 140–145, IEEE, 2017.
122. C. Feng, T. Li, and D. Chana, "Multi-level anomaly detection in industrial control systems via package signatures and lstm networks," in *2017 47th Annual IEEE/IFIP International Conference on Dependable Systems and Networks (DSN)*, pp. 261–272, IEEE, 2017.
123. M. A. Ferrag and L. Maglaras, "Deepcoin: A novel deep learning and blockchain-based energy exchange framework for smart grids," *IEEE Transactions on Engineering Management*, vol. 67, no. 4, pp. 1285–1297, 2019.
124. E. Habler and A. Shabtai, "Using lstm encoder-decoder algorithm for detecting anomalous ads-b messages," *Computers & Security*, vol. 78, pp. 155–173, 2018.
125. B. Jiang, J. Yang, G. Ding, and H. Wang, "Cyber-physical security design in multimedia data cache resource allocation for industrial networks," *IEEE Transactions on Industrial Informatics*, vol. 15, no. 12, pp. 6472–6480, 2019.
126. A. Diro and N. Chilamkurti, "Leveraging lstm networks for attack detection in fog-to-things communications," *IEEE Communications Magazine*, vol. 56, no. 9, pp. 124–130, 2018.
127. H. Yi, G. Kim, J. Lee, S. Ahn, Y. Lee, S. Yoon, and Y. Paek, "Mimicry resilient program behavior modeling with lstm based branch models," arXiv:1803.09171, 2018.
128. A. Goknil, P. Nguyen, S. Sen, D. Politaki, H. Niavis, K. J. Pedersen, A. Suyuthi, A. Anand, and A. Ziegenbein, "A Systematic Review of Data Quality in CPS and IoT for Industry 4.0," *ACM Computing Surveys*, p. 3593043, Apr. 2023.
129. R. W. Bauer, J. L. Matuzsa, R. F. Blackmer, and S. Glucksberg, "Noise localization after unilateral attenuation," *The Journal of the Acoustical Society of America*, vol. 40, no. 2, pp. 441–444, 1966.
130. M. Panoff, R. G. Dutta, Y. Hu, K. Yang, and Y. Jin, "On sensor security in the era of iot and cps," *SN Computer Science*, vol. 2, no. 1, p. 51, 2021.
131. T. Bureš, P. Hnětynka, F. Plášil, D. Škoda, J. Kofroň, R. Al Ali, and I. Gerostathopoulos, "Targeting uncertainty in smart cps by confidence-based logic," *Journal of Systems and Software*, vol. 181, p. 111065, 2021.

132. P. Evans, "Scaling and assessment of data quality," *Acta Crystallographica Section D: Biological Crystallography*, vol. 62, no. 1, pp. 72–82, 2006.
133. Z. Li, H. Liu, J. Zhao, T. Bi, and Q. Yang, "A power system disturbance classification method robust to pmu data quality issues," *IEEE Transactions on Industrial Informatics*, vol. 18, no. 1, pp. 130–142, 2021.
134. M. F. Hasan and N. S. Al-Ramadan, "Cyber-attacks and cyber security readiness: Iraqi private banks case," *Social Science and Humanities Journal*, vol. 5, no. 8, pp. 2312–2323, 2021.
135. Ö. Durmuş and A. Varol, "Analysis and modeling of cyber security precautions," in *2021 9th International Symposium on Digital Forensics and Security (ISDFS)*, pp. 1–8, IEEE, 2021.
136. K. Bhushan and B. Gupta, "Hypothesis test for low-rate ddos attack detection in cloud computing environment," *Procedia computer science*, vol. 132, pp. 947–955, 2018.
137. Y. Zhou, L. Lamont, and L. Li, "Wormhole attack detection based on distance verification and the use of hypothesis testing for wireless ad hoc networks," in *MILCOM 2009-2009 IEEE Military Communications Conference*, pp. 1–7, IEEE, 2009.
138. K. Han, Y. Duan, R. Jin, Z. Ma, H. Wang, W. Wu, B. Wang, and X. Cai, "Attack detection method based on Bayesian hypothesis testing principle in cps," *Procedia Computer Science*, vol. 187, pp. 474–480, 2021.
139. M. Haddadi and R. Beghdad, "A confidence interval based filtering against ddos attack in cloud environment: A confidence interval against ddos attack in the cloud," *International Journal of Information Security and Privacy (IJISP)*, vol. 14, no. 4, pp. 42–56, 2020.
140. L. Wang and Y. Liu, "A ddos attack detection method based on information entropy and deep learning in sdn," in *2020 IEEE 4th Information Technology, Networking, Electronic and Automation Control Conference (ITNEC)*, vol. 1, pp. 1084–1088, IEEE, 2020.
141. S. Sambangi and L. Gondi, "A machine learning approach for ddos (distributed denial of service) attack detection using multiple linear regression," in *Proceedings*, vol. 63, p. 51, MDPI, 2020.
142. A. Nagaraja, U. Boregowda, and R. Vangipuram, "Study of detection of ddos attacks in cloud environment using regression analysis," in *International Conference on Data Science, E-learning and Information Systems 2021*, pp. 166–172, 2021.
143. R. F. Fouladi, O. Ermiş, and E. Anarim, "A ddos attack detection and defense scheme using time-series analysis for sdn," *Journal of Information Security and Applications*, vol. 54, p. 102587, 2020.
144. M. Cheng, Y. Ling, and W. B. Wu, "Time series analysis for jamming attack detection in wireless networks," in *GLOBECOM 2017-2017 IEEE Global Communications Conference*, pp. 1–7, IEEE, 2017.
145. M. Gniewkowski, "An overview of dos and ddos attack detection techniques," in *International Conference on Dependability and Complex Systems*, pp. 233–241, Springer, 2020.
146. G. Münz, S. Li, and G. Carle, "Traffic anomaly detection using k-means clustering," in *Gi/itg workshop mmbnet*, vol. 7, 2007.
147. Z. Muda, W. Yassin, M. Sulaiman, and N. Udzir, "Intrusion detection based on k-means clustering and naïve bayes classification," in *2011 7th international conference on information technology in Asia*, pp. 1–6, IEEE, 2011.
148. F. Zhang, Z. Zhang, P. Zhang, and S. Wang, "Ud-hmm: An unsupervised method for shilling attack detection based on hidden markov model and hierarchical clustering," *Knowledge-Based Systems*, vol. 148, pp. 146–166, 2018.
149. J. Song, Z. Zhu, and C. Price, "Feature grouping for intrusion detection system based on hierarchical clustering," in *Availability, Reliability, and Security in Information Systems: IFIP WG 8.4, 8.9, TC 5 International Cross-Domain Conference, CD-ARES 2014 and 4th International Workshop on Security and Cognitive Informatics for Homeland Defense, SeCIHD 2014, Fribourg, Switzerland, September 8-12, 2014. Proceedings 9*, pp. 270–280, Springer, 2014.
150. B. Purnama, E. A. Winanto, D. Stiawan, D. Hanapi, M. Y. bin Idris, R. Budiarto, et al., "Features extraction on iot intrusion detection system using principal components analysis (pca)," in *2020 7th International Conference on Electrical Engineering, Computer Sciences and Informatics (EECSI)*, pp. 114–118, IEEE, 2020.

151. E. Zyad, A. Taha, and B. Mohammed, "Improve r2l attack detection using trimmed pca," in *2019 International Conference on Advanced Communication Technologies and Networking (CommNet)*, pp. 1–5, IEEE, 2019.
152. S. Rathore and J. H. Park, "Semi-supervised learning based distributed attack detection framework for iot," *Applied Soft Computing*, vol. 72, pp. 79–89, 2018.
153. M. Idhammad, K. Afdel, and M. Belouch, "Semi-supervised machine learning approach for ddos detection," *Applied Intelligence*, vol. 48, pp. 3193–3208, 2018.
154. L. Khan, M. Awad, and B. Thuraisingham, "A new intrusion detection system using support vector machines and hierarchical clustering," *The VLDB journal*, vol. 16, pp. 507–521, 2007.
155. F. T. Liu, K. M. Ting, and Z.-H. Zhou, "Isolation forest," in *2008 eighth ieee international conference on data mining*, pp. 413–422, IEEE, 2008.
156. S. Hariri, M. C. Kind, and R. J. Brunner, "Extended isolation forest," *IEEE transactions on knowledge and data engineering*, vol. 33, no. 4, pp. 1479–1489, 2019.
157. J. Lesouple, C. Baudoin, M. Spigai, and J.-Y. Tourneret, "Generalized isolation forest for anomaly detection," *Pattern Recognition Letters*, vol. 149, pp. 109–119, 2021.
158. G. Staerman, P. Mozharovskyi, S. Clémençon, and F. d'Alché Buc, "Functional isolation forest," in *Asian Conference on Machine Learning*, pp. 332–347, PMLR, 2019.
159. Z. Cheng, C. Zou, and J. Dong, "Outlier detection using isolation forest and local outlier factor," in *Proceedings of the conference on research in adaptive and convergent systems*, pp. 161–168, 2019.
160. R. Ferdousi, M. A. Hossain, and A. El Saddik, "Early-stage risk prediction of non-communicable disease using machine learning in health cps," *IEEE Access*, vol. 9, pp. 96823–96837, 2021.
161. E. M. Sledjeski, L. C. Dierker, R. Brigham, and E. Breslin, "The use of risk assessment to predict recurrent maltreatment: A classification and regression tree analysis (cart)," *Prevention science*, vol. 9, pp. 28–37, 2008.
162. Y. J. Choi, B. R. Park, J. Y. Hyun, and J. W. Moon, "Development of an adaptive artificial neural network model and optimal control algorithm for a data center cyber–physical system," *Building and Environment*, vol. 210, p. 108704, 2022.
163. A. Zeinalpour and H. A. Ahmed, "Addressing the effectiveness of ddos-attack detection methods based on the clustering method using an ensemble method," *Electronics*, vol. 11, no. 17, p. 2736, 2022.
164. M. N. Kurt, O. Ogundijo, C. Li, and X. Wang, "Online cyber-attack detection in smart grid: A reinforcement learning approach," *IEEE Transactions on Smart Grid*, vol. 10, no. 5, pp. 5174–5185, 2018.
165. T. Gu, A. Abhishek, H. Fu, H. Zhang, D. Basu, and P. Mohapatra, "Towards learning-automation iot attack detection through reinforcement learning," in *2020 IEEE 21st International Symposium on "A World of Wireless, Mobile and Multimedia Networks" (WoWMoM)*, pp. 88–97, IEEE, 2020.
166. M. Frigault and L. Wang, "Measuring network security using Bayesian network-based attack graphs," in *2008 32nd Annual IEEE International Computer Software and Applications Conference*, pp. 698–703, IEEE, 2008.
167. J. Homer, S. Zhang, X. Ou, D. Schmidt, Y. Du, S. R. Rajagopalan, and A. Singhal, "Aggregating vulnerability metrics in enterprise networks using attack graphs," *Journal of Computer Security*, vol. 21, no. 4, pp. 561–597, 2013.
168. H. Meng, X. An, and J. Xing, "A data-driven Bayesian network model integrating physical knowledge for prioritization of risk influencing factors," *Process Safety and Environmental Protection*, vol. 160, pp. 434–449, 2022.
169. M. Abolghasemi, V. Khodakarami, and H. Tehranifard, "A new approach for supply chain risk management: Mapping scor into Bayesian network," *Journal of Industrial Engineering and Management (JIEM)*, vol. 8, no. 1, pp. 280–302, 2015.
170. H. A. Kholidy, "Autonomous mitigation of cyber risks in the cyber–physical systems," *Future Generation Computer Systems*, vol. 115, pp. 171–187, 2021.

171. C. Alberts, A. Dorofee, J. Stevens, and C. Woody, "Introduction to the octave approach," *Pittsburgh, PA, Carnegie Mellon University*, pp. 72–74, 2003.
172. J. Freund and J. Jones, *Measuring and managing information risk: a FAIR approach*. Butterworth-Heinemann, 2014.
173. C. McCarthy, K. Harnett, et al., "National institute of standards and technology (nist) cybersecurity risk management framework applied to modern vehicles," tech. rep., United States. National Highway Traffic Safety Administration, 2014.

Chapter 9
Future of Data-Driven Cyber Physical Systems

Abstract This concluding chapter peers into the horizon of this transformative technological landscape, offering a glimpse of what lies ahead and how DDCPS will continue to shape our world. As we venture into the uncharted territory of tomorrow, several pivotal themes and emerging trends crystallize, providing insights into the trajectory of DDCPS and its impact on industries, society, and innovation. This chapter discusses the future of DDCPS by examining emerging trends and technologies. We explore the potential impacts of DDCPS on various industries, such as manufacturing, healthcare, transportation, and energy, and how they could benefit from real-time optimization and decision-making capabilities. The chapter then examines emerging trends, including advancements in machine learning and artificial intelligence, the use of edge computing, the rise of blockchain technology, and new cybersecurity measures. It concludes by summarizing the potential benefits of DDCPS and the importance of staying up-to-date with the latest developments in the field. Societal and ethical implications: The chapter navigates the intricate terrain of societal and ethical considerations in the DDCPS future. It touches upon the importance of responsible AI, data privacy, and cybersecurity in safeguarding the integrity of these systems. It also highlights the imperative of bridging the digital divide to ensure that the benefits of DDCPS are accessible to all segments of society.

9.1 Potential Impacts

The potential impact of DDCPS on different industries is significant. One of the most promising applications is in the manufacturing industry. By collecting data from sensors in real-time, manufacturers can optimize their production processes, reduce costs, and improve quality control. DDCPS can also enable predictive maintenance, identifying potential issues before they cause downtime or quality problems. This can lead to significant improvements in productivity and efficiency.

DDCPS represent a pivotal paradigm shift in various industries, underpinned by their ability to facilitate real-time optimization and decision-making. This transformative capability engenders a host of significant advantages, with ramifications

F. Li et al., *Data-Driven Cyber Physical Systems*,
https://doi.org/10.1007/978-981-96-8709-1_9

spanning efficiency gains, resource allocation, waste reduction, enhanced safety, and heightened productivity.

Real-time optimization is the linchpin of DDCPS, enabling industries to judiciously allocate resources with surgical precision. Whether it's manufacturing, logistics, or energy management, DDCPS empower systems to dynamically adapt and fine-tune their operations in response to changing conditions. This adaptability translates into resource efficiency, where the right resources are allocated to the right tasks at precisely the right time. For example, in manufacturing, DDCPS can optimize production lines in real-time, adjusting parameters such as machine speeds, energy consumption, and material usage to minimize waste and enhance overall efficiency. The result is a leaner, more sustainable production process.

Concomitant with real-time optimization, DDCPS also foster real-time decision-making, a capability that reverberates across numerous domains. In industrial settings, rapid decision-making can mitigate risks and enhance safety. For instance, in an autonomous vehicle, DDCPS continuously processe sensor data to make split-second decisions about braking, steering, and navigation to ensure passenger safety. In healthcare, real-time data analysis can trigger early warnings for critical patient conditions, enabling swift intervention and potentially saving lives. Moreover, the industrial landscape benefits from heightened productivity as DDCPS orchestrate workflows and resources with precision. In smart agriculture, for instance, real-time decision-making based on weather data and crop conditions can optimize irrigation and fertilization schedules, boosting agricultural yields.

The convergence of real-time optimization and decision-making within DDCPS epitomizes a new era where industries can harness data as a dynamic strategic asset. The agility and responsiveness offered by these systems empower businesses to not only thrive in a rapidly changing world but also pioneer novel paradigms of efficiency, sustainability, safety, and productivity. As industries continue to embrace and refine DDCPS applications, the prospect of a more streamlined, adaptive, and data-driven future looms ever larger on the horizon. The application of DDCPS in different industries is shown in Fig. 9.1.

Manufacturing industry: DDCPS can provide real-time monitoring and optimization of production processes, leading to improved efficiency and reduced downtime. By analyzing real-time data from sensors and other sources, CPS can identify potential equipment failures or process inefficiencies before they cause major problems. This allows for timely maintenance or adjustments to be made, reducing the risk

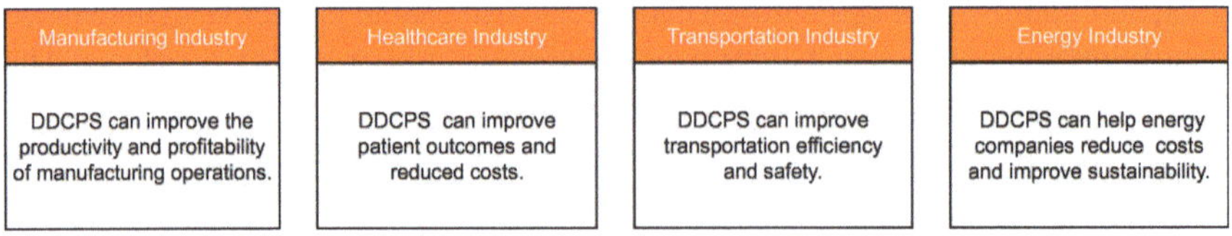

Fig. 9.1 The application of DDCPS in different industries. The advantages of DDCPS in manufacturing, healthcare, transportation and energy industries are described

of costly downtime. Additionally, by analyzing production data, CPS can identify areas for optimization and suggest changes to improve efficiency and reduce waste. Overall, DDCPS have the potential to significantly improve the productivity and profitability of manufacturing operations.

Healthcare industry: DDCPS can have a transformative impact on the healthcare industry by enabling real-time monitoring and analysis of patient data, leading to improved patient outcomes and reduced costs. For example, CPS can monitor patient vital signs and detect anomalies in real-time, allowing healthcare professionals to quickly intervene and prevent adverse events. Additionally, CPS can analyze patient data to identify patterns and risk factors, allowing for more targeted interventions and personalized treatment plans. By optimizing patient care through DDCPS, healthcare providers can reduce costs and improve patient outcomes. In the healthcare industry, DDCPS can improve patient outcomes and reduce costs. By monitoring patient health in real-time using wearable devices and sensors, healthcare providers can provide proactive care, identifying potential health issues before they become more serious. DDCPS can also help healthcare providers to improve operational efficiency, reduce waste, and better manage resources.

Transportation industry: DDCPS have the potential to improve transportation efficiency and safety by enabling real-time monitoring and analysis of vehicle data. For example, CPS can monitor vehicle performance and detect potential issues before they cause breakdowns or accidents. Additionally, CPS systems can analyze traffic patterns and optimize routing, reducing congestion and improving overall transportation efficiency. By providing real-time insights and analysis, DDCPS can help transportation companies reducing costs and improving customer satisfaction. In the transportation industry, DDCPS can be used to optimize routes and improve safety. By collecting and analyzing data from sensors on vehicles, transportation companies can optimize their fleets, reduce fuel consumption and improve overall efficiency. DDCPS systems can also enable predictive maintenance, identify potential issues before they cause breakdowns or accidents.

Energy industry: DDCPS can play a critical role in optimizing energy production and reducing waste. By analyzing real-time data from sensors and other sources, CPS can identify potential issues and optimize energy production processes in real-time. Additionally, CPS can analyze energy consumption patterns and suggest changes to improve efficiency and reduce waste. By enabling real-time optimization and decision-making, DDCPS can help energy companies reducing costs and improving sustainability. The energy industry is another area where DDCPS can have a significant impact. By optimizing energy usage, companies can reduce their carbon footprint and lower costs. DDCPS can also enable more efficient management of renewable energy sources such as solar and wind power, improving overall energy efficiency and reducing reliance on fossil fuels.

In the grand tapestry of technological evolution, DDCPS emerge as catalytic agents poised to catalyze transformation across a multitude of industries. Their core competence lies in the realm of real-time optimization and decision-making, heralding a paradigm shift that ripples through the very fabric of how these industries operate and thrive.

At the heart of this transformative potential is the ability of DDCPS to harness the real-time data torrent emanating from sensors and diverse sources. This data is the lifeblood of contemporary industries, and DDCPS act as the discerning oracle, extracting pearls of wisdom from the data deluge. It's akin to having a digital sentinel that tirelessly monitors, interprets, and acts upon the pulse of data in real-time. In manufacturing, this translates into the optimization of production processes as DDCPS continuously adjust parameters to minimize downtime and maximize resource utilization. In healthcare, it manifests as the capacity to monitor patient vitals with unparalleled precision, triggering early alerts for timely interventions.

Moreover, the insights unfurled by DDCPS extend beyond mere operational enhancements. They are beacons that guide industries toward enhanced efficiency, cost reduction, and superior outcomes. The watchful eye of DDCPS identifies inefficiencies and bottlenecks in supply chains, suggesting streamlined workflows that lead to cost savings. In energy management, it orchestrates the allocation of resources like electricity and heating in real-time, optimizing consumption patterns and reducing waste. Across domains as diverse as agriculture, transportation, and smart cities, DDCPS foster a dynamic ecosystem where real-time insights steer industries toward holistic optimization.

As DDCPS continue to mature and proliferate, it holds the promise of not merely revolutionizing industries but also rewriting the narrative of human progress. The boundless vistas of innovation it unlocks provide fertile ground for reimagining how we interact with technology and data. In this transformative journey, DDCPS emerge as an indispensable ally, helping industries navigating the complex currents of the digital age while elevating efficiency, sustainability, and the human experience to unprecedented heights.

Overall, the potential impact of DDCPS systems on different industries is significant. By using data to drive analysis and insights, companies can make real-time decisions, leading to significant improvements in productivity, efficiency, and cost savings. However, as with any new technology, there are also potential challenges and risks associated with the adoption of DDCPS systems, including data privacy and security concerns. It is essential for companies to carefully consider these issues and develop appropriate strategies for managing them.

9.2 Emerging Trends and Technologies in DDCPS

Emerging trends and technologies are transforming the field of DDCPS, creating new opportunities and challenges for researchers and practitioners. This chapter explores some of these emerging trends and technologies, highlighting their potential impact on the design, development, and deployment of DDCPS.

One of the key emerging trends is the increasing use of machine learning and artificial intelligence (AI) in DDCPS. These technologies enable CPS to process and analyze large amounts of data in real-time, making it possible to identify patterns, predict outcomes, and optimize performance. In particular, deep learning techniques, such as convolutional neural networks and recurrent neural networks, are proving to be especially effective in handling time series data, which is a key characteristic of many CPS applications.

Another emerging trend is the use of edge computing to enhance the capabilities of DDCPS. Edge computing involves processing and analyzing data at or near the edge of the network, rather than sending it to a central data center for analysis. This approach can reduce latency and bandwidth requirements, enabling DDCPS to operate more efficiently and effectively. It can also enhance security by keeping sensitive data within the network and reducing the risk of data breaches.

Emerging trends and technologies in DDCPS are continuously advancing and shaping the future of many industries. These technologies are enabling more efficient and effective decision-making, reducing costs, improving safety, and enhancing overall productivity. In this chapter, we will discuss some of the most prominent emerging trends and technologies in DDCPS. The emerging trends and technologies for DDCPS are summarized in Fig. 9.2.

One emerging trend is the integration of edge computing in CPS. Edge computing involves processing data at the edge of the network, closer to the source of data, rather than transmitting it to a centralized location for processing. This approach can reduce latency and enable real-time decision-making, critical for many CPS applications. Edge computing also allows for data reduction, filtering, and compression before transmission, reducing the bandwidth required for data transfer. With the increasing amount of data generated by CPS, edge computing can improve the scalability, performance, and security of these systems.

Another emerging trend is the use of digital twins in CPS. A digital twin is a virtual replica of a physical system or process that can be used for analysis, optimization, and monitoring. With the increasing availability of data from sensors and other sources, digital twins can be trained using machine learning algorithms to predict system behavior and optimize performance. This approach can reduce downtime, improve

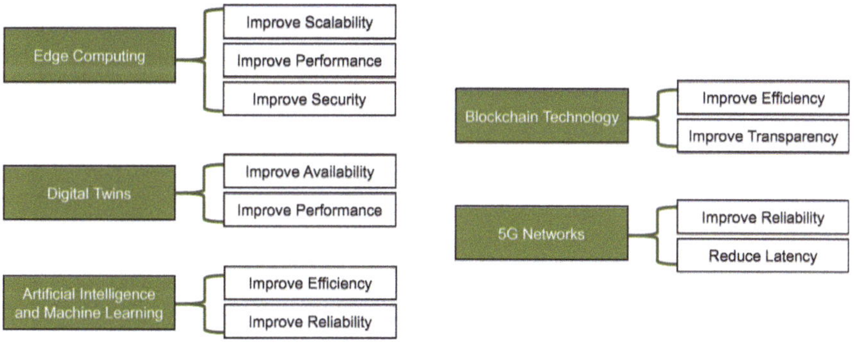

Fig. 9.2 The emerging trends and technologies for DDCPS. We describe the emerging trends and advantages of DDCPS from five aspects: edge computing, digital twins, artificial intelligence and machine learning, blockchain technology and 5G networks

maintenance, and enhance overall system performance. Digital twins can also be used for simulations and testing, enabling the development of new CPS without disrupting physical systems.

The use of AI and ML techniques is another emerging trend in DDCPS. AI and ML can be used to analyze vast amounts of data generated by CPS and extract insights to optimize system performance. These techniques can enable predictive maintenance, real-time monitoring, and anomaly detection, improving the efficiency and reliability of CPS. AI and ML can also be used to develop control systems that adapt to changing conditions and improve overall system performance.

The integration of blockchain technology is another emerging trend in DDCPS. Blockchain technology enables secure, transparent, and decentralized data storage and sharing, which can be critical for applications such as supply chain management, logistics, and smart contracts in CPS. By enabling secure and tamper-proof data sharing, blockchain technology can improve trust and reduce the risk of fraud or hacking. Blockchain technology can also improve the efficiency and transparency of transactions in CPS, enabling more efficient and secure data sharing among multiple stakeholders.

The use of 5G networks is also an emerging trend in DDCPS. 5G networks offer higher bandwidth, lower latency, and improved reliability compared to previous generations of wireless networks. This technology can enable real-time data transfer and enable applications such as autonomous vehicles, remote surgery, and smart cities. The use of 5G networks can also enable more efficient and reliable communication among multiple CPS devices, improving overall system performance.

Finally, the integration of augmented reality (AR) and virtual reality (VR) technologies is an emerging trend in DDCPS. AR and VR technologies can enable remote maintenance, training, and collaboration in CPS applications. By providing a virtual representation of physical systems and processes, AR and VR can improve decision-making, reduce downtime, and enhance safety in CPS applications. These technologies can also enable the development of new CPS applications, enabling more efficient and effective design and testing.

Emerging trends and technologies in DDCPS are continuously advancing and improving the performance, efficiency, and safety of many industries. Edge computing, digital twins, AI and ML, blockchain technology, 5G networks, and AR and VR technologies are just a few of the emerging trends shaping the future of CPS. As these technologies continue to evolve, they will enable new applications, improve system performance, and enhance overall productivity.

9.3 Societal and Ethical Implications

The chapter on the "Future of Data-driven Cyber-Physical Systems (DDCPS)" delves deep into the intricate terrain of societal and ethical considerations that loom large on the horizon of this transformative technological future. As DDCPS continue to

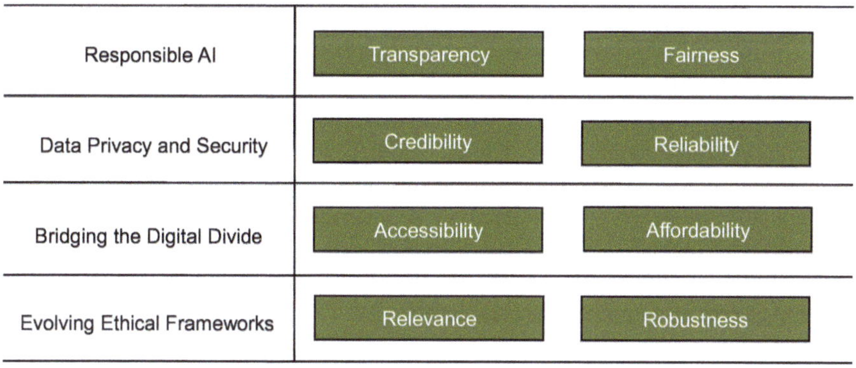

Fig. 9.3 The societal and ethical implications of DDCPS. The social and ethical implications include four components: responsible AI, data privacy and security, bridging the digital divide and evolving ethical frameworks

evolve and permeate every facet of our lives, it becomes increasingly critical to navigate the ethical and societal implications to ensure a future that is both progressive and responsible. The societal and ethical implications of DDCPS is shown in Fig. 9.3.

Responsible AI

At the heart of the societal and ethical discourse surrounding DDCPS is the concept of responsible AI. As DDCPS systems become increasingly autonomous and data-driven, questions about their decision-making processes and accountability come to the forefront. Responsible AI underscores the need for transparency, fairness, and accountability in the algorithms and models that underpin these systems. It challenges us to ensure that DDCPS decisions align with human values, avoid biases, and are explainable and interpretable. This ethical imperative safeguards against unintended consequences and ensures that AI augments human capabilities rather than replacing them.

Data privacy and security

Another pivotal consideration is data privacy and security. DDCPS thrive on data, and this data often includes sensitive information about individuals and organizations. Safeguarding this data against breaches, unauthorized access, and misuse is paramount. Ethical data handling practices, robust encryption, and stringent access controls become not just good practices but ethical imperatives. Moreover, the chapter emphasizes the need for comprehensive cybersecurity measures to protect DDCPS against cyber threats, which have the potential to disrupt critical systems and compromise safety.

Bridging the digital divide

As DDCPS proliferate, there is a pressing need to bridge the digital divide to ensure equitable access to the benefits of these technologies. While DDCPS can empower industries, enhance healthcare, and optimize transportation for those who have access, it also has the potential to exacerbate disparities for those left behind. The chapter underscores the imperative of ensuring that DDCPS applications are designed with inclusivity in mind, considering factors like accessibility and affordability. Initiatives to provide digital literacy and access to underserved communities become ethical imperatives in this context.

Evolving ethical frameworks

In a rapidly evolving technological landscape, ethical considerations must adapt and evolve. The chapter advocates for the continuous development of ethical frameworks and guidelines that keep pace with advancements in DDCPS. It calls for interdisciplinary collaboration among technologists, ethicists, policymakers, and society at large to ensure that ethical norms remain relevant and robust.

As DDCPS propels us into a future where the boundaries between the physical and digital worlds blur, ethical and societal considerations must remain at the forefront of our collective consciousness. The chapter serves as a call to action, a reminder that while the potential of DDCPS is vast, it is our responsibility to harness this potential in ways that uplift humanity, promote fairness, protect privacy, and ensure that the benefits of DDCPS are accessible to all, leaving no one behind in the journey towards progress and innovation.

9.4 Concluding Remarks

As we conclude this comprehensive exploration of DDCPS, we stand at the nexus of a profound technological evolution that holds immense promise for the future. Throughout the pages of this book, we've delved deep into the intricate tapestry of DDCPS, uncovering its principles, design methodologies, and real-world applications. As we reflect upon this journey, several salient themes and takeaways crystallize, underscoring the transformative power and boundless potential of DDCPS.

A revolution in real-time insights

At the core of DDCPS lies the ability to transmute data into actionable insights with unparalleled immediacy. Whether in manufacturing, healthcare, transportation, energy, or myriad other domains, the capacity to optimize processes and make informed decisions in real-time is the elixir that fuels innovation and efficiency. DDCPS herald a new era where data isn't just a resource but a dynamic ally in the quest for excellence.

Interdisciplinary synergy

Our exploration has illuminated the interdisciplinary nature of DDCPS. It draws upon the bedrock of cyber-physical systems, merges with the robust methodologies of data science, and infuses the alchemy of machine learning and artificial intelligence. DDCPS thrives at the intersection of these disciplines, weaving a rich tapestry that transcends traditional boundaries, fostering collaborations that catalyze breakthroughs.

Challenges as Catalysts

Throughout our journey, we've encountered challenges that are as formidable as they are instructive. Handling vast data volumes in real-time, integrating heterogeneous data sources, preserving data quality and integrity, ensuring security and privacy, and harmonizing human-machine collaboration—these challenges are not obstacles to be circumvented but crucibles for innovation. DDCPS thrive on these challenges, emerging stronger and more resilient with each trial.

The path forward

As we gaze into the future, it is abundantly clear that DDCPS are not a culmination but a waypoint in the ever-evolving landscape of technology. Emerging trends such as edge computing, quantum computing, and the relentless proliferation of the Internet of Things (IoT) will converge with DDCPS, charting a course toward uncharted frontiers. The potential for DDCPS to orchestrate smarter cities, more sustainable industries, personalized healthcare, and safer transportation systems is limitless.

In closing, the journey through the realm of DDCPS leaves us with a profound realization—a realization that the fusion of data and physical systems is not merely a technological trend but a transformative force that will reshape industries, societies, and the very essence of human experience. It is an ode to the indomitable spirit of innovation, a testament to the boundless human capacity to harness knowledge and technology to overcome challenges and forge a brighter future.

As we bid adieu to these pages, we do so with the unwavering belief that the future belongs to those who dare to dream, to those who harness the power of data-driven insights to paint a canvas of progress and prosperity. With DDCPS as our compass, we embark upon this journey into the unknown with optimism, resilience, and an unquenchable thirst for discovery. The road ahead is fraught with challenges, but it is also laden with opportunities, and it is in the crucible of these challenges that the true potential of DDCPS will be realized.

Competing Interests

1. The research is supported by the National Key Research and Development Project under Grant 2023YFB3307300, the National Natural Science Foundation of China under Grants 62373014, 62422301, 62125301, and 92467205.
2. The authors have no conflicts of interest to declare that are relevant to the content of this chapter.

MIX
Papier aus verantwortungsvollen Quellen
Paper from responsible sources
FSC® C105338

If you have any concerns about our products,
you can contact us on
ProductSafety@springernature.com

In case Publisher is established outside the EU,
the EU authorized representative is:
Springer Nature Customer Service Center GmbH
Europaplatz 3, 69115 Heidelberg, Germany

Printed by Libri Plureos GmbH
in Hamburg, Germany